ENVIRONMENTAL CONTAMINATION FROM THE FUKUSHIMA NUCLEAR DISASTER

The 2011 accident at the Fukushima Daiichi Power Station led to serious radioactive contamination of the environment. Due to transportation by seasonal wind and ocean currents, these radioactive materials have now been observed in many places in the Northern Hemisphere. This book provides a unique summary of the environmental impact of the unprecedented accident. It covers how radioactive materials were transported through the atmosphere, oceans and land. The techniques used to investigate the deposition and migration processes are also discussed including atmospheric observation, soil mapping, forest and ecosystem investigations and numerical simulations. With chapters written by international experts, this is a crucial resource for researchers working on the dispersion and impact of radionuclides in the environment. It also provides essential knowledge for nuclear engineers, social scientists and policymakers to help develop suitable mitigation measures to prepare for similar large-scale natural hazards in the future.

TERUYUKI NAKAJIMA is an Emeritus Professor at the University of Tokyo. He is currently serving as Chief Scientist of the Earth Observation Research Center (EORC) at the Japan Aerospace Exploration Agency (JAXA). At the time of the Fukushima accident he was a member of the Science Council of Japan, Section President of Atmospheric and Hydrospheric Sciences at the Japan Geoscience Union and an executive member of the Japan Meteorological Society. In these roles, he helped investigate and organise the emergency response to the disaster. He is a fellow of the American Geophysical Union, and in 2017 he was awarded the 2017 Japan Purple Ribbon medal.

TOSHIMASA OHARA is Research Director of the Fukushima Branch at the National Institute for Environmental Studies (NIES). He leads the Environmental Emergency Research Program that contributes to environmental recovery and renovation in Fukushima. After the Fukushima accident, his group worked on atmospheric simulations of radionuclides from the disaster, and published the first result of temporal and spatial variations of deposition rates on a regional scale. He is President of the Japan Society for Atmospheric Environment and serves as a member of the Science Advisory Committee of the Acid Deposition Monitoring Network in East Asia (EANET).

MITSUO UEMATSU is Emeritus Professor and former Director of the Centre for International Collaboration at the Atmosphere and Ocean Research Institute at the University of Tokyo. His major research interests include the long-range transport of natural and anthropogenic substances over the ocean and the properties of marine aerosols, including their impact on the marine environment. He has received several awards from Japanese societies and international organisations. He has served as the president of the Oceanographic Society of Japan, a member of the Scientific Committee of the International Geosphere–Biosphere Programme (IGBP SC) and chair of the Japanese National Committee for Intergovernmental Oceanographic Commission (IOC) of UNESCO.

YUICHI ONDA is Chief Administrator of the Center for Research and Environmental Dynamics, and a professor at the Graduate School of Life and Environmental Sciences, both at the University of Tsukuba. He specialises in hydrogeomorphology and geomorphic development. After the Fukushima accident, he started an interdisciplinary research project on gamma-emitting radionuclides released into the environment in order to study the behaviour of radionuclides in terrestrial and marine environments. The results of the study are expected to contribute to the reconstruction of the contaminated environment.

CAMBRIDGE ENVIRONMENTAL CHEMISTRY SERIES

The Environmental Chemistry Series

This wide-ranging series covers all areas of environmental chemistry, placing emphasis on both basic scientific and pollution-orientated aspects. It comprises a central core of textbooks, suitable for those taking courses in environmental sciences, ecology and chemistry, as well as more advanced texts (authored or edited) presenting current research topics of interest to graduate students, researchers and professional scientists. Books cover atmospheric chemistry, chemical sedimentology, freshwater chemistry, marine chemistry, and soil chemistry.

Recent books in the series

W. Davison, *Diffusive Gradients in Thin-Films for Environmental Measurements*

P. G. Coble, J. Lead, A. Baker, D. M. Reynolds and R. G. M. Spencer, *Aquatic Organic Matter Fluorescence*

S. Roy, C. A. Llewellyn, E. S. Egeland and G. Johnsen, *Phytoplankton Pigments: Characterization, Chemotaxonomy and Applications in Oceanography*

E. Tipping, *Cation Binding by Humic Substances*

D. Wright and P. Welbourn, *Environmental Toxicology*

S. J. de Mora, S. Demers and M. Vernet, *The Effects of UV Radiation in the Marine Environment*

T. D. Jickells and J. E. Rae, *Biogeochemistry of Intertidal Sediment*

ENVIRONMENTAL CONTAMINATION FROM THE FUKUSHIMA NUCLEAR DISASTER

Dispersion, Monitoring, Mitigation and Lessons Learned

Edited by

TERUYUKI NAKAJIMA
Japan Aerospace Exploration Agency, Japan

TOSHIMASA OHARA
National Institute for Environmental Studies, Japan

MITSUO UEMATSU
University of Tokyo, Japan

YUICHI ONDA
University of Tsukuba, Japan

CAMBRIDGE
UNIVERSITY PRESS

CAMBRIDGE
UNIVERSITY PRESS

University Printing House, Cambridge CB2 8BS, United Kingdom

One Liberty Plaza, 20th Floor, New York, NY 10006, USA

477 Williamstown Road, Port Melbourne, VIC 3207, Australia

314–321, 3rd Floor, Plot 3, Splendor Forum, Jasola District Centre, New Delhi – 110025, India

79 Anson Road, #06–04/06, Singapore 079906

Cambridge University Press is part of the University of Cambridge.

It furthers the University's mission by disseminating knowledge in the pursuit of education, learning, and research at the highest international levels of excellence.

www.cambridge.org
Information on this title: www.cambridge.org/9781108475808
DOI: 10.1017/9781108574273

First published 2019

Printed in the United Kingdom by TJ International Ltd. Padstow Cornwall

A catalogue record for this publication is available from the British Library.

ISBN 978-1-108-47580-8 Hardback

Contents

Contributors

Aono, Tatsuo
National Institutes for Quantum and Radiological Science and Technology, Chiba, Japan

Aoyama, Michio
Fukushima University, Fukushima, Japan

Baumann, Zofia
University of Connecticut, Groton, CT, USA

Chino, Masamichi
National Institutes for Quantum and Radiological Science and Technology, Takasaki, Japan

Didier, Damien
Institute for Radioprotection and Nuclear Safety (IRSN), Paris, France

Ebihara, Mitsuru
Waseda University, Tokyo, Japan

Elwood, James A.
Meiji University, Tokyo, Japan

Evrard, Olivier
Laboratoire des Sciences du Climat et de l'Environnement (LSCE/IPSL), Paris, France

Fisher, Nicholas S.
Stony Brook University, Stony Brook, NY, USA

Fujiwara, Mamoru
Osaka University, Osaka, Japan

Goto, Daisuke
National Institute for Environmental Studies (NIES), Tsukuba, Japan

Hamajima, Yasunori
Kanazawa University, Kanazawa, Japan

Igarashi, Yasuhito
Kyoto University, Kyoto, Japan

Imada, Masatoshi
University of Tokyo, Tokyo, Japan

Ishimaru, Takashi
Tokyo University of Marine Science and Technology, Tokyo, Japan

Kajino, Mizuo
Meteorological Research Institute, Tskuba, Japan

Kanda, Jota
Tokyo University of Marine Science and Technology, Tokyo, Japan

Kawano, Takeshi
Japan Agency for Marine-Earth Science and Technology (JAMSTEC), Yokosuka, Japan

Kita, Kazuyuki
Ibaraki University, Mito, Japan

Laceby, J. Patrick
Alberta Environment and Parks, Calgary, Alberta, Canada

Madigan, Daniel J.
Harvard University, Cambridge, MA, USA

Masumoto, Yukio
University of Tokyo, Tokyo, Japan

Mathieu, Anne
Institute for Radioprotection and Nuclear Safety (IRSN), Paris, France

Matsunaka, Tetsuya
Kanazawa University, Kanazawa, Japan

Moriguchi, Yuichi
University of Tokyo, Tokyo, Japan

Morino, Yu
National Institute for Environmental Studies (NIES), Tsukuba, Japan

Nagai, Haruyasu
Japan Atomic Energy Agency (JAEA), Tokai, Japan

Nagao, Seiya
Kanazawa University, Kanazawa, Japan

Nakajima, Teruyuki
Japan Aerospace Exploration Agency (JAXA), Tsukuba, Japan

Nakamura, Hisashi
University of Tokyo, Tokyo, Japan

Ohara, Toshimasa
National Institute for Environmental Studies (NIES), Tsukuba, Japan

Onda, Yuichi
University of Tsukuba, Tsukuba, Japan

Otsuka, Takaharu
University of Tokyo, Tokyo, Japan

Oura, Yasuji
Tokyo Metropolitan University, Tokyo, Japan

Quélo, Denis
Institute for Radioprotection and Nuclear Safety (IRSN), Paris, France

Saito, Kimiaki
Japan Atomic Energy Agency (JAEA), Kashiwa, Japan

Sakaguchi, Aya
University of Tsukuba, Tsukuba, Japan

Sasa, Kimikazu
University of Tsukuba, Tsukuba, Japan

Saunier, Olivier
Institute for Radioprotection and Nuclear Safety (IRSN), Paris, France

Shibata, Tokushi
Chiyoda Technol Corporation, Tokyo, Japan

Shimoura, Susumu
University of Tokyo, Tokyo, Japan

Shinohara, Atsushi
Osaka University, Osaka, Japan

Takahashi, Tomoyuki
Kyoto University, Kyoto, Japan

Takahashi, Yoshio
University of Tokyo, Tokyo, Japan

Takemura, Toshihiko
Kyushu University, Fukuoka, Japan

Takenaka, Chisato
Nagoya University, Nagoya, Japan

Takigawa, Masayuki
Japan Agency for Marine-Earth Science and Technology (JAMSTEC), Yokohama, Japan

Tanaka, Kazuya
Japan Atomic Energy Agency (JAEA), Tokai, Japan

Tanaka, Taichu Yasumichi
Meteorological Research Institute, Tsukuba, Japan

Tanihata, Isao
Osaka University, Osaka, Japan
Beihang University, Beijing, China

Tsuda, Atsushi
University of Tokyo, Kashiwa, Japan

Tsumune, Daisuke
Central Research Institute of Electric Power Industry, Chiba, Japan

Tsuruta, Haruo
Remote Sensing Technology Center of Japan, Tokyo, Japan

Uematsu, Mitsuo
University of Tokyo, Kashiwa, Japan

Watanabe, Akira
Fukushima University, Fukushima, Japan

Yamazawa, Hiromi
Nagoya University, Nagoya, Japan

Yokoyama, Hiromi
University of Tokyo, Kashiwa, Japan

Yoshida, Naohiro
Tokyo Institute of Technology, Yokohama, Japan

Preface

A large area that includes Fukushima Prefecture was seriously contaminated by radioactive materials emitted into the atmosphere and the ocean by the accident at the Fukushima Daiichi Nuclear Power Station (hereafter, FDNPS[1]) of the Tokyo Electric Power Company, which was caused by the Tohoku Region Pacific Coast Earthquake in 2011. The emitted radioactive materials were transported by seasonal wind and ocean currents to a wide area of the globe and have been observed in various places in the Northern Hemisphere. These materials have also been detected in soil, forests, lakes, rivers and seas due to fallout and direct discharge, and continued movement in the environment.

Radioactive materials emitted by the accident were thus transported widely and exist in our environment in different forms. Various efforts have been made to accurately understand the material transport; such efforts are indispensable for determining suitable mitigation measures. However, the devastating accident related to the radioactive materials was their first experience of such for most scientists, and researchers have since fumbled with trial-and-error activities. This confusing situation also occurred in the actions taken by the government and related organisations, as manifested in the accident investigations by private, government and Diet committees that began in 2012. After more than seven years since the accident, it is important for long-term mitigation and impact assessment to reflect on the current and past, as well as to analyse what we did, what we understood and what was not sufficient.

This motivation drove the development of this publication. The original Japanese book was published in 2014, and the English edition was started shortly thereafter. Professionals in each scientific field drafted this book to summarise the scientific knowledge accumulated in the years since the event, as well as the trajectories of the research community's activities. The main theme of the present

[1] There are several other expressions, such as FDNPP. See also the Glossary.

book, which was drafted by groups of professionals specialising in the study of the atmosphere, ocean, land and radioactive materials, is the migration of the radioactive materials emitted into the environment by the FDNPS accident. Other books by professionals in the fields of nuclear power engineering and nuclear reactor physics should be consulted to understand the events that played out inside the power station.

This book comprises three parts. Part I discusses the migration of the radioactive materials in the environment. Chapter 1 summarises the basic concepts and fundamental knowledge regarding the environmental migration related to the accident. Chapter 2 estimates the amount of radioactive material emitted by the accident. Chapter 3 shows investigation results of radioactive material pollution in the atmosphere and simulation of atmospheric transport, and deposition using atmospheric transport models, while Chapter 4 presents global model simulation results. Chapter 5 describes investigation results in the ocean and marine ecosystem, and presents simulation analyses using ocean transport models. Chapter 6 treats deposition and migration processes of radioactive materials and describes investigations of soil mapping, forest investigation and migration of radioactive materials into the forest biosphere and agricultural products.

Part II examines the current status of the infrastructure for disaster prevention and discusses the problems and issues with improving the system. Chapter 7 introduces the current status of the monitoring system for nuclear power stations and discusses how this could be improved. It was found that numerical simulations are useful for investigating the wide area of contamination. In reality, there was confusion related to the governmental use of the emergency rapid radioactivity impact prediction network system (SPEEDI). Chapter 8 presents the thoughts of scientists related to the SPEEDI problem. Chapter 9 summarises the decontamination actions that are useful for reconstructing living areas.

Part III reflects on the thoughts and actions from the perspective of scientists related to the future, based on an analysis of inadequate activities. Chapter 10 discusses the method of dissemination of scientific knowledge that has the largest societal impact. Chapter 11 provides a detailed report of urgent activities that immediately followed the accident and cross-field research activities by volunteer researchers, with messages for future generations.

There were many instances of extremely sensitive care in the investigations and actions by the government and related organisations that were reflected in the psychological reaction to the radioactivity resulting from the disaster, which was akin to that of an atomic bomb. Difficult situations were encountered when attempting to distribute scientific facts and limit harmful rumours. Throughout the development of the accident investigations, the public wanted to receive information that would help them take actions of their own instead of being restricted. The disseminated information was also useful for efficient mitigation actions. Hence, we thought we should express frankly to the public what we thought as scientists. At the same time, we made notes for future generations based on our urgent actions and our thoughts regarding the large-scale environmental pollution. It will be important to prepare for similar large-scale disasters such as earthquakes, tsunami, torrential rain, volcanic eruptions and others. From this perspective, as scientists, we wrote our message to society and policymakers.

The role of scientists is to present accurate facts and disseminate information, but scientific investigations require much time. We should be persistent in solving problems. This effort is needed for the dissemination of important scientific knowledge to the public for long-term mitigation.

Important progress has been made in the last three years, following the publication of the first edition; namely, the establishment of action guidelines for emergent large-scale disasters by the Science Council of Japan (SCJ) and the Japan Academic Network for Disaster Reduction, which includes 56 academic societies. The SCJ guideline was invoked for the first time following the Kumamoto earthquake in 2015, when the SCJ collaborated with the academic network to provide well-prepared dissemination of scientific information to the public. We have incorporated these new events and scientific studies that were published after the publication of the Japanese edition. In addition, new international co-authors have provided their knowledge and recommendations from an international viewpoint, as presented in Chapter 12.

The cover images are the composite of the geographical and temporal changes of the air dose rate map on land (see Section 1.10) and the radiocaesium concentration of the coastal surface sediments off Fukushima (see Section 5.6) in the summer of 2011, the summer of 2012 and the winter of 2014–15. The images show the radionuclide migration from land to the coastal sea over time, with radioactive decay and dilution.

Acknowledgements

We are grateful to the people, groups and institutions who provided us with their data and knowledge necessary for drafting this book. We also acknowledge Hiromi Yamazawa, Yuichi Moriguchi, Haruo Tsuruta, Masayuki Takigawa, Yasuhito Igarashi, Tatsuo Aono, the late Yasuyuki Muramatsu, who passed away on 2 July 2016, and Daisuke Goto for their support in editing the entire manuscript and preparing the Glossary. We are also grateful to James Elwood, the chief English editor, for his contribution to the English edition.

We acknowledge that this translated version was financially supported by MEXT/KAKENHI/Innovative Areas 2409 (PI: Yuichi Onda), MOEJ/ERTDF/S-12 (PI: Teruyuki Nakajima) and MOEJ/ERTDF/ 5-1501 (PI: Yuichi Moriguchi).

Part I

Transport of Radioactive Materials in the Environment

Part I presents scientific knowledge related to environmental contamination by radioactive materials emitted from the accident at the Fukushima Daiichi Nuclear Power Station of the Tokyo Electric Power Company. Chapter 1 summarises basic concepts for understanding the event. Chapter 2 estimates the amount of emitted radioactive materials. Chapters 3 through 6 describe the transport of radioactive materials in the regional atmosphere, global atmosphere, ocean and land areas.

1

Introduction

Basic Concepts Regarding the Fukushima Accident and Radiation and Radioactivity

This chapter presents an overview of the accident and basic scientific concepts that are important for understanding the migration processes of radioactive materials in the environment.

1.1 Overview of the Fukushima Accident

TERUYUKI NAKAJIMA, TOSHIMASA OHARA, MITSUO UEMATSU AND YUICHI ONDA

The accident of the Fukushima Daiichi (First) Nuclear Power Station (FDNPS) of the Tokyo Electric Power Company (hereafter, Fukushima accident) transpired after the Tohoku Region Pacific Coast Earthquake occurred in March 2011. Table 1.1 summarises the main events of the accident. After the earthquake occurred at 14:46 on 11 March 2011, tsunami waves of 13 m in height arrived at the FDNPS (TEPCO, 2011); the diesel power engine stopped at 15:41. Due to this electricity loss, the nuclear reaction became uncontrollable. The Fukushima Daini (Second) Power Station was able to make a controlled stop for cooling even after the intrusion of seawater from a tsunami wave with a height of 9 m. The estimated maximum height in the design of the Daiichi and Daini Power Stations was 5.1 m. In contrast, the estimated maximum tsunami height in the design of the Onagawa Nuclear Power Station of the Tohoku Electric Power Company, which avoided serious damage, was 14.8 m (Matsumoto, 2007).

A hydrogen explosion occurred at the first power unit of the FDNPS at 15:36 on 12 March and in the third power unit at 11:01 on 14 March. Radioactive materials were emitted in large quantities into the atmosphere due to these events. It is suggested from monitoring data that there were several other emission sources, such as water vapour release by vent operations in the reactor pressure vessel and direct discharge of leaked high-level contaminated cooling water from the reactor into the ocean.

Table 1.1 *Main events associated with the Fukushima accident in 2011, based on government reports (NERH, 2011a; 2011b; NSC, 2012; CSIC, 2012)*

11 March
- 14:46: Tohoku Region Pacific Coast Earthquake
- ~15:30: Tsunami wave arrives at the FDNPS
- 15:42: Electrical power lost at the FDNPS; special measures concerning nuclear emergency preparedness enacted; Article 10 issued by TEPCO and related organisations
- 16:45: Water injection into the reactor is not possible; Nuclear Disaster Special Measures Law Article 15 is issued by TEPCO and related organisations
- 19:03: Emergency declaration of nuclear disaster countermeasures; headquarters for the FDNPS implemented by the prime minister
- 20:50: Fukushima prefecture governor requests the evacuation of Okuma Town and Futaba Town within 2 km distance of the FDNPS
- 21:23: Evacuation of residents within 3 km of the FDNPS instructed by the Nuclear Disaster Headquarters, with a sheltering distance of 10 km

12 March
- 5:44: Instruction of evacuation to residents within 10 km of the FDNPS by the Nuclear Disaster Headquarters
- 15:36: Hydrogen explosion in the FDNPS first power unit
- 18:35: Instruction of evacuation to residents within 20 km of the FDNPS by the Nuclear Disaster Headquarters
- 20:20: Seawater injection into the FDNPS first power unit

14 March
- 11:01: Hydrogen explosion of the FDNPS third power unit

15 March
- 00:00: Dry vent operation of the FDNPS second power unit
- 11:00: Shelter instructions to residents within 20–30 km of the FDNPS by the Nuclear Disaster Headquarters; the Japan Coast Guard sets the navigation danger zone off the coast of the FDNPS; Fukushima Prefecture Fishermen's Union Federation consults with the Fukushima Prefecture Agriculture, Forestry and Fisheries Department, resulting in self-restraint operations

17 March
- The Ministry of Health, Labour and Welfare notifies local governments to carry out regulations pertaining to the Food Sanitation Law Article 6 No. 2 so as not to use food that exceeds the provisional regulation value of 'the food intake restriction index value' indicated in the 'Disaster prevention measures of nuclear facilities' of the Nuclear Safety Commission

17–19 March
- A radiation measurement survey is carried out by US military aircraft of the US Department of Energy (DOE)

21 March
- Instructions issued on shipping restrictions of certain foods by the Nuclear Emergency Response Headquarters

22 March
- Reports are made of detection of radioactive iodine and caesium exceeding regulation levels from seawater near the nuclear power plant's southern outlets outside the peripheral monitoring area defined by TEPCO

Table 1.1 (*cont.*)

23 March	
	The marine radioactivity survey based on the sea monitoring plan begins
	The results of the first SPEEDI calculations by the Nuclear Safety Commission are published
25 March	
	Promotion of the voluntary evacuation of residents within 20–30 km of the FDNPS by the Nuclear Emergency Response Headquarters
5 April	
	Radioactive iodine exceeding the provisional regulation values of radioactive substances in food from sand lance (Kounago) obtained off the coast of Ibaraki Prefecture are detected
6–29 April	
	Airborne monitoring is first carried out by the US Department of Energy
13 April	
	There is a joint planning meeting for soil investigation by the Safety Confirmation Project and Earth Science Project
14 April	
	Marine observations are carried out by the oceanographic research vessel *Mirai* on cruise MR11-03
25 April	
	Daily publication of the SPEEDI diffusion simulation results by the Nuclear Safety Commission begins
18–26 May	
	The second airborne monitoring campaign by MEXT and the US Department of Energy takes place
31 May to 2 July	
	The third airborne monitoring campaign by MEXT takes place
6–14 June	
	The first investigation campaign of the radioactive material distribution by MEXT begins, conducting soil measurements and research into the atmosphere and rivers within 2 km of FDNPS
13 June	
	Start of operation of the circulating seawater purification system for radioactive material
27 June to 8 July	
	The second investigation campaign by MEXT investigating radioactive material distribution begins
19 July	
	The Ministry of Agriculture, Forestry and Fisheries begins the national survey for the radioactive substance contamination of rice straw
10 August	
	Operation of the circulation cooling system for all fuel pools of the FDNPS begins
25 October to 5 November	
	The fourth airborne monitoring campaign after a typhoon by MEXT takes place
16 December	
	Completion of step two of 'Path to convergence of the accident' (cold shutdown state, controlled radioactive material release and others), as announced by the Nuclear Disaster Headquarters

The total amount of atmospheric emissions of radioactive materials were estimated to be 9–37 (mean and RMSD: 17 ± 8) PBq (PBq = 10^{15} Bq) for radioactive caesium ^{137}Cs (Stohl *et al.*, 2012; Terada *et al.*, 2012; SCJ, 2014; Katata *et al.*, 2015). The large uncertainty in the estimated value is related to various causes, such as data loss from monitoring posts and weather stations due to the tsunamis and power outages, and a lack of observations over the transported area of the Pacific Ocean. The amount of direct discharge into the ocean was estimated to be 3–6 PBq (Kawamura *et al.*, 2011; Estournel *et al.*, 2012; JAEA, 2012; Tsumune *et al.*, 2012, 2013; Miyazawa *et al.*, 2013; Aoyama *et al.*, 2016).

The spring wind regime at the time of the accident was that of a winter-like regime with stronger-than-normal winds that transported large amounts of the emitted radioactive materials in the atmosphere to the ocean region (Takemura *et al.*, 2011), with only 12–37% being deposited onto the Japanese land area (SCJ, 2014). These radioactive materials were transported by winds associated with various pressure systems and deposited on the land and ocean surface in a complex distribution (e.g. see Figure 3.1, which shows that an area of elevated ^{137}Cs contamination exceeding 1000 kBq/m^2 was distributed beyond 30 km around the FDNPS in the northwest direction). This distribution pattern closely overlaps the total accumulation distribution in March to April as simulated by SPEEDI system (see Chapter 8 for details). The SPEEDI simulation results were made available to the public every day after 25 April 2011, which was more than one month after the accident.

The government set the Evacuation Order Zone (20 km) and the Indoor Evacuation Zone (20–30 km) until 15 March, and defined the Planned Evacuation Zone in the northwestern area, which included Kawamata Town, Iitate Village and Katsurao Village, until 22 April. During this period, people evacuated in these areas were considered to have received radioactive exposure. There were also shipping restrictions for agricultural and livestock products, and occurrences of rice straw and gravel contamination. The number of displaced persons in the areas inside and outside Fukushima Prefecture reached 160,000 in 2012 (Fukushima Prefectural Government, 2018). Large land use changes in Fukushima areas after 2011 have been reported, detected by high-resolution satellite observations as shown in Figure 1.1 (Ishihara and Tadono, 2017); it is highly possible that this can be attributed to the disaster comprising the large earthquake, tsunami, and Fukushima accident.

Furthermore, 98% of the deposited radioactive materials on soil were adsorbed by the 5 cm surface layer (Kato *et al.*, 2012a; Takahashi *et al.*, 2015), while the portion deposited on the forest remained in the canopy for a long time. The mean lifetime of the radioactive material in the forest canopy reached 200 days following the Chernobyl event, while there were reports of the material remaining for more

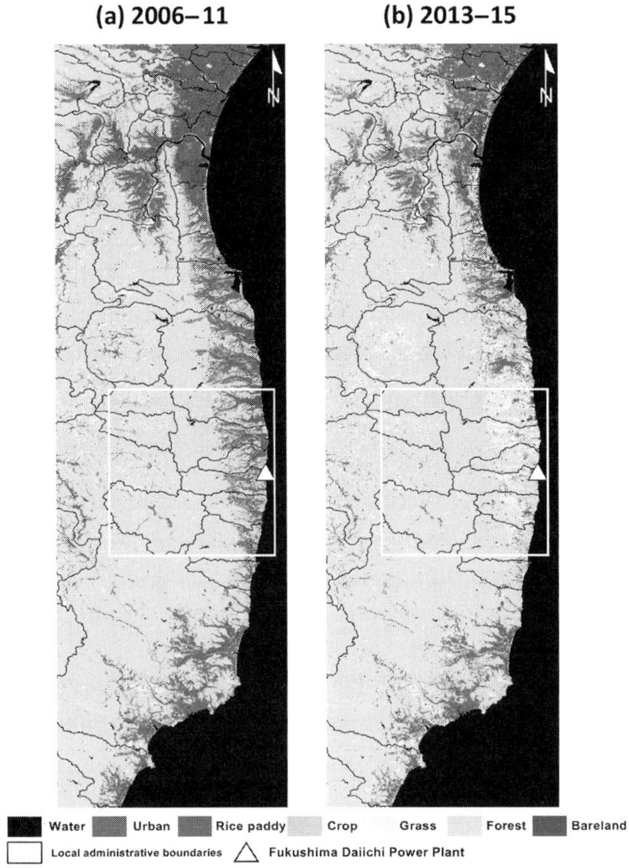

Figure 1.1 Land use changes before and after 2011 by (a) ALOS AVNIR-2 observation in the years 2006–11 and (b) Landsat-8 OLI in 2013–15 as found by Ishihara and Tadono (2017).
Provided by JAXA. A black and white version of this figure will appear in some formats. For the colour version, refer to the plate section.

than 300 days following the Fukushima event (Kato *et al.*, 2012b), with differences depending on the type of forests (Imamura *et al.*, 2017; Kato *et al.*, 2017). These radioactive materials migrate in the environment and flow into lakes and rivers and finally are carried to the sea unless they undergo decontamination. The third and fourth MEXT airborne monitoring campaigns of the radiation dose rate (May to July 2011 and October to November 2011) indicated that the concentration decreased in the 80 km inland area of the Abukuma Mountains, whereas that in rivers and coasts increased, suggesting that migration and river inflow of the radioactive materials had occurred over a long period (MEXT, 2011; NRA, 2013). A MEXT river measurement of solid suspended particles indicated a particle concentration of 55 kBq/kg ^{137}Cs in the main stream of the Abukuma

River. A riverbed ^{137}Cs concentration of up to 16 kBq/kg was also observed, with a good linear correlation with the suspended particle concentration, indicating that large amounts of radioactive materials remained in the riverbed. Moreover, a good linear correlation between suspended ^{137}Cs and the riverbed concentration was observed. Continued outflow of radioactive materials is expected from paddy fields (Yoshimura *et al.*, 2015) and organic materials from forests.

There were two main pathways to the ocean: deposition of the atmospheric radioactive materials ejected by the explosions and direct discharge of polluted water from the FDNPS. Shipboard measurements from research and volunteer vessels after early April 2011 observed radioactive caesium (^{134}Cs and ^{137}Cs) in the surface water over a wide area of the North Pacific Ocean. The ^{137}Cs concentration reached 196 Bq/m^3 in some areas, which is two orders of magnitude higher than that in the surrounding ocean areas (Aoyama *et al.*, 2012, 2016). Radioactive caesium was detected from the suspended materials and zooplankton sampled in a North Pacific area (47°N and 167°E) 2300 km from the FDNPS (Honda *et al.*, 2012). A shipping restriction on seafood was issued after a concentration exceeding the regulation level (100 Bq/kg) was detected in June and August 2012 from Pacific cod at the Hachinohe port (NERH, 2012). In August 2012, a radioactive caesium level of 25 kBq/kg was detected from greenlings caught 1 km offshore of the Ohta River mouth in Fukushima Prefecture (TEPCO, 2012).

A full analysis of the Fukushima accident requires investigation in many fields, such as reactor phenomena, health impacts and societal impacts (SCJ, 2012); here, we focus on environmental aspects during the first year following the event.

1.2 Radioactive Elements, Radioactive Nuclides and Radioactive Substances

MITSURU EBIHARA AND ATSUSHI SHINOHARA

An element is the minimum unit of tangible components comprising a substance. As of 2019, there were 118 elements listed in the Periodic Table, and all elements were given formal nomenclature by the IUPAC (International Union of Pure and Applied Chemistry). Two-thirds of the elements are stable, and the remainder are so-called radioactive elements. Some elements have several isotopes that have different mass numbers, with each isotope being called a nuclide. Over 90% of nuclides known today are radioactive. Only a few of these radioactive elements and radioactive nuclides exist in the natural environment.

One of the major elements comprising organisms such as humans is carbon. In living organisms, there are three types of carbon nuclides (^{12}C, ^{13}C and ^{14}C) that have different mass numbers, that is 12, 13 and 14. Of these, ^{12}C and ^{13}C are stable nuclides, while ^{14}C is radioactive, with a half-life of 5700 years. ^{14}C is constantly generated in the upper atmosphere by nuclear reactions between neutrons derived from cosmic radiation and ^{14}N existing abundantly in the air; it is regarded as a representative natural radioactive nuclide. Because ^{14}C inevitably exists in living organisms, archaeological age can be traced by measuring the concentration of ^{14}C remaining in relics. Radioactive nuclides produced in artificial environments, such as nuclear explosions and nuclear reactors in nuclear power plants, are known as artificial radioactive nuclides, being represented by ^{137}Cs and ^{90}Sr.

Substantial amounts and different types of artificial radioactive nuclides that accumulated in the nuclear reactors were released into the environment at the time of the FDNPS accident. Regarding the abundance of radioactive nuclides and the abundance ratios among nuclides in the environment, various organisations conducted measurements long before the accident at the FDNPS. The measuring objects were radioactive nuclides derived from atmospheric nuclear tests carried out on a worldwide scale and radioactive nuclides released into the environment regularly or accidentally from nuclear-related facilities. Some of the radioactive nuclides recently released into the environment include those from the nuclear meltdown in Chernobyl in the former Soviet Union in 1986 and the criticality incident at the Tokai-mura JCO in 1999. Such radioactive nuclides released into the environment by nuclear experiments and accidents in the past at nuclear-related facilities attenuate according to their half-lives. Some of these materials have already been substantially annihilated, while others remain in the environment without decaying. To describe the diffusion and subsequent migration of the radioactive nuclides released into the environment by the accident at the FDNPS, it will be increasingly important in the future to consider the radioactive nuclides that existed prior to the Fukushima accident.

In this book, the term radioactive nuclide is used almost synonymously with radioactive isotope, but differently from the term radioactive substance. Materials containing radioactive nuclides are radioactive substances. As referred to later, the amount of radioactive nuclides is typically expressed as the number of disintegrations per unit time – namely, in the rate of disintegration – and this amount is defined as radioactivity. Moreover, the amount of a radioactive substance is expressed as the amount of the substance per unit mass. Numerous radioactive substances were released, and radioactive nuclides were widely dispersed into the environment. In considering the impact on human health due to the radiation emitted by disintegrating radioactive nuclides, it is critically important to understand that the diffusion and subsequent migration of radioactive nuclides released into the environment relies heavily on measuring radioactive nuclides and obtaining analysis results of high reliability.

In the following, a general explanation is given in regard to radioactive nuclide measurements and how to obtain correct measurements. Further explanations are given on how radioactivity values (disintegration rates) and radioactivity concentrations (radioactivity per mass or volume) are sought from the measurement rates (number counts per unit time) of radiation, as well as on the relations between radioactivity and dose of radiation.

1.3 Measurement of Radiation

MITSURU EBIHARA AND ATSUSHI SHINOHARA

Radioactive nuclides change (disintegrate – or debacle or collapse, corresponding to the English term decay) into other nuclides over time. Nuclides that are produced after disintegration, if radioactive, repeat the disintegration process and eventually settle into stable nuclides. The speed of attenuation of radioactive nuclides is related to each nuclide and expressed in half-life or average lifetime. The half-life is literally defined as the time required for the number of radioactive nuclides to decrease by half, and the lifetime is the time required for the number of pieces to fall to 1/e (approximately 0.368). When radioactive nuclides disintegrate, radiation is discharged. Because the radiation energy is related to each radioactive nuclide, the radioactive nuclide can be identified by measuring its radiation energy.

To express a radioactive nuclide quantitatively, the disintegration rate (radioactivity) of the radioactive nuclide (becquerel, or Bq) is commonly used rather than the amount of the radioactive nuclide (i.e. mole). Therefore, as discussed later, it can be determined from the count rate of radiation. The quality and quantity of radioactive nuclides can be obtained by measuring radiation emitted from radioactive nuclides. As radiation enters a substance, it interacts with molecules and atoms comprising the substance and creates excited-state molecular species, such as ions and radicals. In measuring radiation, the interaction between radiation and the substance of interest is utilised. The radioactive nuclides released into the environment by the Fukushima accident include three types of radioactive rays, α, β and γ, depending on the type of radioactive nuclide. These radioactive rays have different patterns of interactions with substances, which can be utilised in measuring radiation and determining the quality and quantity of radioactive nuclides emitting radiation.

1.3.1 α-Ray Measurement

An α-ray is a particle emitted as an atomic nucleus goes through α-disintegration, consisting of an accelerated atomic nucleus of ^{4}He ($^{4}He^{2+}$). Most α-rays emitted from radioactive nuclides have energy falling within the range 4–7 MeV, which is narrower than the energy fluctuation ranges of β-rays and γ-rays emitted through radiation disintegration. Measurements of radiation energy to examine the intensity distribution of energy, which is defined as radiation spectrometry, are used for qualitative and quantitative analyses of radioactive rays; for example, α-rays and γ-rays, which have mono-energy depending on their radioactive nuclides. α-rays have a large mass and are thus absorbed into a substance more easily than are β-rays and γ-rays. Therefore, samples containing radioactive nuclides emitting α-rays do not produce accurate measurements if the α-rays are measured as is. In the case of solid environment samples such as soil samples, chemical separation and refinement processes are required for α-ray-emitting radioactive nuclides. Ultimately, α-ray-emitting radioactive nuclides are coated on a metal plate for sampling, and α-ray spectrometry is conducted.

Among the uranium nuclear fuels used at nuclear power stations, ^{235}U and ^{238}U generate nuclides of elements heavier than uranium through neutron capture reaction and β-disintegration in the following process:

$$
\begin{aligned}
&{}^{235}\mathrm{U} \xrightarrow{+\mathrm{n}} {}^{236}\mathrm{U}\left(2.342 \times 10^{7}\,\mathrm{y}\right) \xrightarrow{+\mathrm{n}} {}^{237}\mathrm{U}(6.75\ \mathrm{d}) \xrightarrow{\beta^-} {}^{237}\mathrm{Np}\left(2.14 \times 10^{6}\,\mathrm{y}\right)\\
&\quad \xrightarrow{+\mathrm{n}} {}^{238}\mathrm{Np}(2.12\ \mathrm{d}) \xrightarrow{\beta^-} {}^{238}\mathrm{Pu}(87.7\ \mathrm{y}) \xrightarrow{+\mathrm{n}} {}^{239}\mathrm{Pu}(24,100\ \mathrm{y})\\
&{}^{238}\mathrm{U} \xrightarrow{+\mathrm{n}} {}^{239}\mathrm{U}(23.5\,\mathrm{m}) \xrightarrow{\beta^-} {}^{239}\mathrm{Np}(2.35\ \mathrm{d}) \xrightarrow{\beta^-} {}^{239}\mathrm{Pu}(24,100\ \mathrm{y}) \xrightarrow{+\mathrm{n}} {}^{240}\mathrm{Pu}(6560\ \mathrm{y})\\
&\quad \xrightarrow{+\mathrm{n}} {}^{241}\mathrm{Pu}(14.4\ \mathrm{y}) \xrightarrow{\beta^-} {}^{241}\mathrm{Am}\ (433\ \mathrm{y}) \xrightarrow{+\mathrm{n}} {}^{242}\mathrm{Am}(16.0\ \mathrm{m}) \xrightarrow{\beta^-} {}^{242}\mathrm{Cm}(162.8\ \mathrm{d})
\end{aligned}
$$

Among the plutonium isotopes, ^{239}Pu is used as a nuclear fuel because it causes nuclear fission by neutrons. Reactors nos 1, 2 and 4 at the FDNPS used a regular uranium fuel, whereas the reactor no. 3, a so-called pluthermal type, used MOX fuel of uranium mixed with ^{239}Pu. In nuclear fuels used for prolonged hours, as shown in the above reaction, radioactive isotopes of americium and curium exist in minimal amounts, in addition to plutonium.

From the soil surrounding the FDNPS, plutonium was chemically separated and tested using α-ray spectrometry. As a result, in addition to the plutonium isotopes derived from nuclear experiments conducted in the 1950s, additional ^{238}Pu, ^{239}Pu and ^{240}Pu were confirmed. However, the quantity was estimated to be 1/10 000th of that released by the accident at the Chernobyl atomic power station, which was found to be a level with almost no biological impact. α-ray spectrometry is not capable of separately measuring ^{239}Pu and ^{240}Pu due to the close proximity in the α-ray energies. In recent years, high-resolution multi-collector inductively coupled plasma mass spectrometers (HR-MC-ICP-MSs) capable of simultaneously measuring many isotopes have been widely used for high-sensitivity and high-accuracy measurements of isotope compositions of ^{238}Pu, ^{239}Pu, ^{240}Pu and ^{242}Pu.

1.3.2 *β-Ray Measurement*

A β-ray is the radiation emitted through β-disintegration. There are three patterns in β-disintegration: β^-(electron emission), β^+(positron emission) and EC (electron capture). Ordinarily, β-disintegration means β^-disintegration. β-rays emitted through β^-disintegration represent a flow of electrons that do not have mono-energy, which differs from α-rays and γ-rays; instead, they have energy varying from zero to the maximum energy value per disintegration. The maximum energy is related to each nuclide, and the value is defined as the β-ray energy. In the case of β-rays with less energy than the maximum, the neutrino emits energy simultaneously from its nucleus to compensate for the difference. In measuring β-rays, a GM counter tube that utilises the ionisation effect of a gas by β-rays or a liquid scintillation detector (dubbed as liquid scin) that utilises the luminous phenomenon of organic matter excited by radiation is used. To identify a β-ray-emitting nuclide, the β-ray's (maximum) energy needs to be determined on an absorption curve drawn using the GM counter and a β-ray absorber plate or via pulse height analysis (for energy discrimination of luminescence intensity as converted into pulse voltages) using a liquid scin. When several radioactive nuclides coexist, it is necessary to separate them in advance by conducting radiochemical separation.

^{90}Sr is a representative β-ray-emitting radioactive nuclide found for the FDNPS accident and has 546 keV of β-ray energy. The fission yield of ^{90}Sr in the thermal neutron-induced fission reaction of ^{235}U is significant at 5.818%. Because strontium belongs to the same group as calcium (Group 2) in the Periodic Table, ^{90}Sr easily deposits in bones when orally ingested. Moreover, its half-life is 28.79 years, which means that internal exposure causes problems. ^{90}Sr transforms into ^{90}Y (half-life: 2.669 days) through β-disintegration and turns into stable ^{90}Zr through yet another β-disintegration. Through these disintegration processes, ^{90}Sr does not emit γ-rays, and ^{90}Y emits almost no γ-rays. Therefore, β-ray measurement is required to determine the amount of ^{90}Sr. Typically, in determining the quantity of ^{90}Sr, measurements are done by examining the β-rays emitted through the disintegration process of ^{90}Y into ^{90}Zr and not the β-rays emitted at the time of disintegration into ^{90}Y. This is because the ^{90}Y β-rays have approximately four times the energy of ^{90}Sr at 2.26 MeV, allowing for a higher-sensitivity determination of ^{90}Sr.

To determine the quantity of ^{90}Sr using β-ray measurements, it is necessary to remove other coexistent radioactive nuclides through a chemical separation process and refine ^{90}Sr. Therefore, coexistent metallic elements and organic substances must first be removed, and alkaline earth elements (Ca, Sr and Ba) must be separated. Then, Sr can be separated and refined using the fuming nitric acid method, oxalic acid precipitation method, ion-exchange method or others. After being left intact for approximately two weeks, radioactive equilibrium (a condition in which the speeds of radiative disintegration among disintegrating nuclides are equalised) is attained between ^{90}Sr and ^{90}Y. ^{90}Y is then separated for β-ray measurements. As seen above, determining the quantity of radioactive nuclides via β-ray measurements requires radiochemical separation processes. Thus, it takes some time before measurements can be obtained.

1.3.3 γ-Ray Measurement

A γ-ray is an electromagnetic ray of short wavelength. Through α- or β-disintegration, most radioactive nuclides are left in a highly excited state before transitioning into a lower excited or ground state. γ-rays are emitted during the energy state transition of nuclides, and the energy is mono-energy and related to each nuclide, which is analogous to that of α-rays. Because it is an electromagnetic ray of short wavelength, it has a high transmission capacity to pass through substances, allowing for as-is measurements of γ-ray-emitting nuclides contained in bulky samples. Except for ^{90}Sr, most radioactive nuclides emit γ-rays as they disintegrate, and the γ-ray measurement method is most frequently used to analyse radioactive nuclides. γ-ray spectrometry that uses a germanium semiconductor detector is widely utilised for γ-ray determination because of its high energy resolution.

1.4 Example of γ-Ray Spectrometry to Determine Accurate Radioactivity Values

MITSURU EBIHARA, ATSUSHI SHINOHARA AND YASUNORI HAMAJIMA

Residents in the vicinity of the FDNPS were most affected by the accident, and they are still looking for answers to their questions. How long do they have to wait before they can return to regular life? How much radioactive exposure will they be subject to, and for how long? Behind those who are enduring these difficult circumstances, there are more people who are concerned about radiation and radioactivity. To answer these questions and to relieve their worries, numerical values for radiation and radioactivity (or radioactive concentration) are frequently cited. At times, they are presented as reference values. In this section, we will address how we can collect accurate and reliable measurements of radiation and radioactivity by referring to γ-ray spectrometry using a germanium semiconductor detector.

Figure 1.2 A sectional view of the germanium semiconductor detector for detection of radiation (γ-ray) shown by arrows emitted from a point source (centre of arrow).

Figure 1.3 Measurement of a soil sample in a U8 container with a vertical-type germanium semiconductor detector. (a) A lead shield covers the detector mounted on top of the liquid nitrogen Dewar vessel. (b) An enlarged view of a U8 container with a metal weight.
Provided by Ebihara's laboratory, Tokyo Metropolitan University.

1.4.1 Germanium Semiconductor Detector

There are three types of germanium semiconductor detectors: coaxial, planar and well types (Figure 1.2). Of the three, the coaxial type and well type are used for measuring γ-ray-emitting nuclides contained in environmental samples. A coaxial detector has germanium crystals formed cylindrically, with the crystals wrapped in aluminium. Depending on where the detector is installed, it comes in upright or horizontal versions, with the upright type being more convenient for measuring large or liquid samples. To measure γ-ray-emitting nuclides in soil samples, cylindrical containers made of plastics, commonly called U8 containers, are universally used; the upright detector is also easier to handle in this case. Figure 1.3 shows a typical example of the detector used for the measurement of soil samples in U8 containers.

Meanwhile, the well-type detector is one that has germanium crystals formed in a well shape, with the window made of thin aluminium inside the well. Samples are inserted into the well for measurement, which allows germanium crystals to be

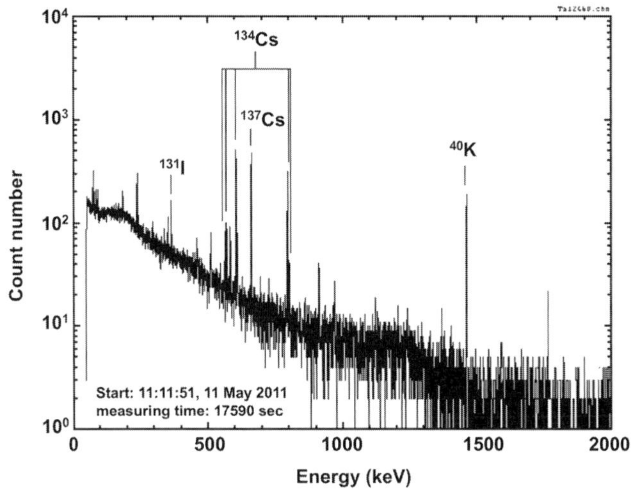

Figure 1.4 An example of a γ-ray spectrum obtained by measuring a soil sample contaminated by radionuclides released by the FDNPS with a germanium semiconductor detector.
Provided by Ebihara's laboratory, Tokyo Metropolitan University.

exposed with a higher probability to incident radioactive rays, and the γ-ray count efficiency is higher than that for the coaxial-type detector, although the sum effect is liable to occur, as discussed later. In the case of a well-type detector, sample sizes are limited. Usually, the nuclide to be measured is radiochemically separated and refined prior to measuring the γ-rays by a well-type detector. Thus, the sample size is substantially reduced, and interference by other coexisting radioactive nuclides is curtailed. Therefore, the method is often used to determine the concentration of ^{137}Cs and ^{134}Cs dissolved in seawater. The detector is integrated with a liquid nitrogen Dewar vessel and a vacuum cooler (cryostat) to cool the detector.

1.4.2 γ-Ray Spectrum

Figure 1.4 shows an example of an γ-ray spectrum obtained by a germanium semiconductor detector from measurements of samples contained in the U8 container of soil contaminated by radioactive nuclides emitted in the accident at the FDNPS. The horizontal axis represents the γ-ray energy, while the vertical axis denotes the measured counts. The spectrum indicates a slowly declining change from the low-energy side to the high-energy side, with several sharp peaks. These peaks of γ-rays with mono-energy are observed thanks to the excellent energy resolution of the germanium semiconductor detector. The gradual change supporting the peaks is called the background, where γ-rays are detected as they jump in from around the detector or as they are emitted by the peaking γ-rays in cross-interactions with the detector and surrounding substances. Radioactive nuclides existing in the sample can be identified from the peak positions, namely the γ-ray energy.

In Figure 1.4, the spectrum is presented for measurements of soil samples collected approximately two months after the accident, with the γ-ray energy at the peak points highlighting the existence of ^{131}I, ^{134}Cs and ^{137}Cs in the samples. These nuclides generally go through β-disintegration, and the γ-rays denoted in the figure were emitted as they went through β-disintegration. A large peak is also observed at 1461 keV, which represents γ-rays emitted by ^{40}K that existed in the soil sample or around the detector.

1.4.3 Radioactivity Calculation

The quantity of a radioactive nuclide is typically represented by the disintegration count per unit time (disintegration rate); as a unit of radioactivity under the SI unit system, disintegration per second (dps) is defined as a becquerel (Bq). To find the radioactivity of radioactive nuclides in the soil based on a spectrum similar to that shown in Figure 1.4, it is necessary to obtain the count rate from the net measurement count (deducting the contribution from the background) of γ-ray peaks and then convert it to the disintegration rate. The disintegration rate is determined from the soil sample, although the count rate differs based on the measurement conditions, such as the type of detector or the relative position of the samples vis-à-vis the detector. The conversion of the count rate into the disintegration rate can be computed in principle, although it is actually done by using a reference sample that is commonly called a count rate–disintegration rate calibration sample.

In the case of a soil sample (real sample) contained in a U8 container, the following procedure is used. A simulated sample (comparative standard sample) is prepared by adding to simulated soil of the same size as the real sample a radioactive nuclide of the measuring object whose disintegration rate (radioactivity) is known, which is used for γ-ray measurements under the same conditions as for the real sample. Under the same measurement conditions, the count rate and the disintegration rate are proportional. Therefore, by comparing the count rate, the radioactivity of the radioactive nuclides in the soil samples can be computed. Adjustments for attenuation by disintegration are made as needed. To obtain an accurate radioactivity value, it is essential that the configurations of the real sample and the simulated sample be equal for calibration, with no differences in the distributions of radioactive nuclides.

1.4.4 Causes of Error in Radioactivity Measurement

Some errors in seeking radioactivity values by means of γ-ray spectrometry are excludable, while the rest may not be. Disintegration of radioactive nuclides is a probability event, and the next disintegration cannot be accurately predicted. Therefore, after N disintegration events have been observed, the measured values entail $\sqrt{N}$ times indeterminacy at one standard deviation (1σ). The numerical value represented by $\sqrt{N}$ is called a counting error. This counting error is one that cannot be excluded. For example, when 100 disintegration events have been observed, the probability of the true number of disintegration events falling between 90 and 110 is only 68%. To ensure 95% probability, the fluctuation range must be broadened to between 80 and 120. The same applies to radioactivity defined by the disintegration rate, and the value always involves errors attributable to counting errors. In particular, the smaller the count rate (as in the case of a low radioactivity value), the larger the error is likely to be. Therefore, care must be taken in making notations of significant figures.

Regarding γ-ray spectrometry, it is important to be aware of errors resulting from the sum coincidence effect (sum effect). When a radioactive nuclide emits many γ-rays almost simultaneously during disintegration, several γ-rays overlap in entering the detector, and the observation is composed of a sum of the overlapped γ-ray energies and is not indicative of each γ-ray's energy. The effect caused by this phenomenon is called the sum effect. Due to the sum effect, peaks in γ-rays are less than/lower than their original strengths and, as a result, a smaller radioactivity value is often obtained.

Unless adjustments for such attenuation are made, a large error is obtained in the determined quantity. Among the radioactive nuclides released at the time of the FDNPS accident, the major ones determined by γ-ray spectrometry include ^{131}I, ^{134}Cs and ^{137}Cs, as represented in the spectrum in Figure 1.4. Among these, no sum effect is found for ^{137}Cs because it emits only one γ-ray of 661.6 keV as it disintegrates. ^{131}I releases many γ-rays, although, as shown in Figure 1.4, its peak strength is outstanding at 364.5 keV, while the next strong γ-ray peaks at 284.3 keV and 637.0 keV are an order of magnitude lower than that at 354.5 keV. Therefore, the determined quantity will have no significant error even if the sum effect is disregarded. In contrast, an adjustment should be required for ^{134}Cs for its count rate due to the sum effect. The γ-ray emission pattern is complicated for ^{134}Cs when it disintegrates into ^{134}Ba. For one disintegration, there are times when it consecutively emits γ-rays of 795.9 keV and 604.7 keV, while in some cases it emits three γ-rays of 569.3 keV, 795.9 keV and 604.7 keV, and at other times three γ-rays of 801 keV, 563.2 keV and 604.7 keV. In determining the quantity of ^{134}Cs, γ-rays of high strength at 604.7 keV and

795.9 keV are used. When measured using the coaxial-type germanium semiconductor detector (Figure 1.2(a)), if the sample and the detector are placed close together (see Figure 1.3), the γ-ray measurement rate is reduced by 20–30%, for which adjustments are needed. This is especially true when a well-type germanium detector (Figure 1.2(c)) is used, in which the probability of simultaneously measuring several γ-rays tends to be high, resulting in significant sum effects, and further care must be taken.

The following procedure is used to make adjustments for the attenuation in the count rate caused by the sum effect. First, a reference sample of known radioactivity is prepared, and γ-ray measurements are conducted as far away from the detector as possible (more than 20 cm) to negate the sum effect. Next, the reference sample is measured at regular positions. In advance, variances in the detection efficiency are determined at two different positions; these values are used in determining the attenuated count rate due to the sum effect.

There is a phenomenon called the pulse pile-up effect, which is similar to the sum effect. This effect results from the simultaneous entry into the detector of γ-rays emitted from numerous radioactive nuclides. In measuring samples with high count rates, caution is required. For the samples related to the Fukushima accident, which generally exhibited low count rates, such caution is hardly necessary. Excludable errors such as the sum effect are called systematic errors. These can be avoided without much attention.

1.5 Radioactivity and Radiation Dose

MITSURU EBIHARA AND ATSUSHI SHINOHARA

Regarding radiation exposure, measuring the dose of radiation derived from the disintegration of radioactive nuclides is much more important than that from radioactivity. To discuss radiation dose, different units are used depending on the focus: radiation or exposure to radiation. The latter becomes crucial in evaluating the influences of radiation on the environment and from the standpoint of providing protection from radiation. In the following, some terms for the radiation dose are summarised.

- Exposure radiation dose (unit: C/kg): a unit that focuses on radiation. It indicates the dose signifying the strength of photons from X-rays and γ-rays, as defined by the extent of ionisation in the atmosphere. In the SI unit system, C/kg is used. Roentgen (R), which was previously used, is used as a supplementary unit.
- Absorbed dose (unit: J/kg = Gray (Gy)): a unit that focuses on the recipient of radiation. It is a numerical value that represents the extent of absorption of

radiation by the substance exposed to radiation, signifying the absorbed energy per unit mass; it is widely used for radiation in general.

- Equivalent dose (unit: J/kg = Sievert (Sv)): the dose for assessment using a universal scale that indicates exposure to radiation. It was introduced in consideration of the impact of radiation on human tissues and organs, which varies by radiation type and energy. It is calculated as the absorption dose multiplied by the radiation weighting factor (a numerical value determined by the biological impact on different organs, which varies by radiation type and energy). The radiation weighting factor of photons such as γ-rays is set to 1 regardless of their energy. The equivalent dose is a numerical value calculated for each organ and tissue of the human body.
- Effective dose (unit: J/kg = Sievert (Sv)): the effective dose is the most frequently used concept of a dose in terms of radiation protection and is calculated by multiplying the equivalent dose of each tissue by its weighting factor and summing the values for all tissues of the body. The tissue weighting factor is a relative value determined for each tissue based on the different sensitivities to radiation, with the total value for all organs set to 1. Based on the effective dose, the influences of exposure are evaluated based on the differences in radiation effects and whether they affect the whole body homogeneously or inhomogeneously.

At the sites of nuclear power-related plants, facilities such as monitoring posts and monitoring stations are set up to measure the radiation at all times, typically using the absorbed dose rate (unit: Gy/h), which is the absorption dose per unit time. After the accident at the FDNPS, air dose rates in the surrounding areas of the nuclear reactors, as well as other locations in the Kanto and Tohoku districts, were measured for publication. In most of these cases, air dose rates were represented using units of Sv/h. This stems from the general idea that homogeneous external exposure to γ-rays over the whole body suffices to cover the effects on the human body of radioactive nuclides emitted into the environment in the nuclear power station accident. This approach was also derived from the practice of making the radiation weighting factor 1 and assuming that the absorption dose (rate) is equal to the effective dose (rate).

The radioactivity value (disintegration rate) (a numerical value in Bq or Bq/kg) and the air dose rate (a numerical value related to the effective dose rate in Sv/h) are essentially based on separate phenomena, and no direct conversions mutually occur, similar to the conversion from the radiation count rate to radioactivity. However, because the dose rate measured for the FDNPS accident represents the dose of radiation released through the radiation disintegration of the radioactive nuclides discharged into the environment by the accident, it is possible to link the disintegration rate (radioactivity) with the dose rate using several assumptions or simplifications. Publication of the converted numerical values should be done with caution. Currently, with equipment and instruments for measuring radioactivity values and air dose rates being plentifully available, direct measurements of respective values individually may provide more reliable numerical values.

1.6 Effects of Radioactive Substances on Humans

YASUHITO IGARASHI AND TATSUO AONO

Radiation exposure refers to a living human body being subjected to ionising radiation. The types of radiation exposure are roughly classified into internal and external ones. Internal exposure is related to radioactive substances being taken into the body via food and drink (hereafter, ingestion) and breathing (hereafter, inhalation). For example, knowing that radioactive iodine, typically ^{131}I, accumulates in thyroids and caused cancer after the Chernobyl nuclear accident is important. As a protective measure, saturating one's thyroid with non-radioactive iodine by taking an iodine pill as soon as possible is needed to limit the effects of radioactive iodine, which is anticipated to be massively emitted. However, an overdose of iodine may cause side-effects and can be harmful to human bodies. Therefore, it is very important that the dosage should be carefully monitored by doctors or other specialists. External exposure, on the other hand, comes from outside the body, such as from cosmic rays and the radiation emitted from naturally occurring radionuclides contained in soil, rocks and structures in the natural environment. For the Fukushima accident we need to consider radiation exposure from the radioactive plume as well as radioactive contamination on and in the ground. The first and latter terms are often referred to as sky shine and ground shine, respectively.

For both external and internal exposure, the same measure of ionising radiation is required to assess the human exposure dose. Thus, the International Commission on Radiological Protection (ICRP) has established a unified series of the radiation dose system – i.e. Sievert (Sv) – which is the unit used for the effective dose, committed dose and organ-equivalent dose. Humans are always exposed to ionising radiation from nature, even when there is no pollution caused by a nuclear accident or test. The global average of this background radiation dose for one person has been previously evaluated: for external exposure, 0.4 mSv from cosmic rays and 0.5 mSv from the ground are common, and for internal exposure, 1.2 mSv from radon and its decay products in air by inhalation and 0.3 mSv from food and water ingestion (^{40}K, ^{210}Pb, ^{210}Po, etc.), totalling 2.4 mSv per year (UNSCEAR, 2000), are common. Because radon is a radioactive noble gas (^{220}Rn and ^{222}Rn are naturally occurring isotopes) and its decay products are also radioactive and mostly α-emitters, the internal exposure dose from inhalation is relatively high. These background doses depend on regions of the world and living environments, food and water consumption, etc.; the same level is not always recorded around the world. The Japanese average annual exposure is estimated to be 3.8 mSv as an effective dose, with under half from the natural radiation mentioned above (1.5 mSv) and more than half from medical exposure (2.3 mSv).

Measuring internal exposure as an effective dose is quite difficult and is often done by calculating the amount of radioactive substances taken into the human body (it is also difficult to measure directly the internally deposited radionuclides, except for specific ones). The simple method for conducting these estimations relies on the dose conversion coefficients; the ICRP recommends and tabulates such values of the dose coefficients in Sv/Bq. These conversion coefficients are obtained by model calculations assuming a reference human body with specific dimensions, metabolism, functions, etc. The coefficient means that the effective dose is recommended as the committed dose based on 1 Bq of activity via oral or inhalation pathways for individual radionuclide intake for children and the general public, including adults. For example, if a person drinks two litres of water with a ^{131}I concentration of 300 Bq/L per day for one month, this would result in 0.4 mSv by multiplying the dose coefficient (2.2×10^{-8} Sv/Bq) by the total intake (600 Bq/d × 30 d). Similarly, in the case of 200 ml of water with a ^{131}I concentration of 300 Bq/L per day for one month, this would result in 0.04 mSv, and the ingestion of 50 grams of spinach with a ^{131}I concentration of 2000 Bq/kg per day for one month would result in 0.07 mSv. Here, 0.07 mSv is almost identical to the external exposure dose of a person who remained in a place with an air dose rate of 0.1 μSv/h for one month.

When a person is exposed to ionising radiation, high-energy molecules are generated in the body, such as hydrated electrons and radicals, which are caused by ionisation and excitation. Radiation effects on the human body begin with DNA damage due to the excited molecules as well as direct ionisation damage of DNA by radiation. However, there are repair and recovery functions in each molecular and cellular structure as well as in organs; not all damage leads to physical impairment. If repaired and restored, no damage remains; however, if this process is not successful, it will cause mutation and cell death. If the human body only loses a few cells, there would be no effect.

There are two different types of effects from ionising radiation on the human body: physical and genetic effects. The acute damage from physical effects appears within a few days to a few months after exposure. Lethal effects appear if exposed to 4–8 Sv over the entire body. In the event that the entire body or a large portion of it is exposed to a very high dose level of radiation (several Gy (Gray)), leukopenia

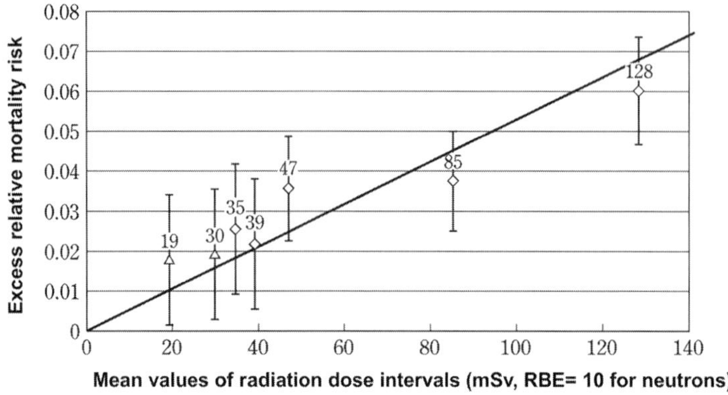

Figure 1.5 Excess relative mortality risk of solid cancer for atomic bomb survivors in low radiation dose areas: a survey covering 1950–97. Numbers in the figures are average doses (mSv).◊: statistically significant ($p < 0.05$) and Δ: not significant ($p > 0.05$). The diagonal line is with a risk coefficient of 0.53/Sv.
Cited from Encyclopaedia of Nuclear Energy ATMOICA. Original data source is Preston *et al.* (2003).

and alopecia will appear as the typical symptoms. The late effects are the physical ones that appear after recovery from acute radiation damage or appear with long-term incubation after exposure to relatively low doses of radiation – cataracts and cancers are typical effects. Genetic effects influence descendants via chromosome abnormality or genetic mutation in the case of reproductive cells exposed to ionising radiation. Acute damage and late effects, such as cataracts, are deterministic ones, which are certain to occur when the exposure exceeds certain thresholds (the smallest effective value). However, cancers and the genetic effect are considered to be stochastic because they increase in probability in proportion to exposure dose. Deterministic effects are related to dysfunction in living tissue caused by extensive cell damage. While stochastic effects are related to damage in which only a few cells are exposed, the remaining effect will be amplified later on.

The radiation protection exposure dose limit is set based on the stochastic effect, which is based on the curve of the dose and excessive risk obtained from available data. The curve is based on the excess death rate of exposed people (atomic bomb survivors) in the Hiroshima and Nagasaki cases (Figure 1.5). However, even if a person is exposed to the same dose of ionising radiation, comparing the acute and long-term (chronic) effects, smaller effects typically arise in the latter case. This is called the dose rate effectiveness, and it is thought that chronic exposure results in more recovery from the radiation damage. The factor indicating the degree of dose rate effectiveness is called the dose rate effectiveness factor (also called the reduction factor); the ICRP uses a value of 2. This means that the slope of the excessive risk curve obtained from atomic bomb survivors is reduced by half.

Moreover, from a radiation protection perspective, the ICRP recommends that radiation protection based on the LNT model (linear non-threshold model) linearly increases with dose, although there is no threshold for the stochastic effect, even for low-level doses and excessive cancer risk. Thus, the ICRP estimates that the excessive risk is 5% for 1 Sv exposure, which is 0.5% at 100 mSv.

Although there is no threshold for the stochastic effect, it does not mean that every person exposed to the pollution of the Fukushima accident will be affected in a similar manner. Instead, the risk should be considered to be relative. For example, one of the most common causes of death in Japan is cancer, accounting for approximately 30% of the current death toll. The principal causes of cancer are considered to be bad lifestyle, smoking, viruses and bacteria. For a person exposed to a low dose of 100 mSv with a low dose rate throughout his or her life, the likelihood of death owing to cancer increases from 30% to 30.5% (assuming that the risk is additive). However, the exposure data indicate that younger people are at higher risk (see figure 2 in RERF, 2013); thus, children's exposure should not be regarded in the same way as that of adults. Furthermore, we must be aware that various uncertainties are involved.

To evaluate the low-dose ionising radiation effect in a statistical and significant manner, as many samples as possible are required, which makes this problem harder to solve. Consequently, various discussions have occurred. The low-dose radiation impacts on the human body are not simple enough to clarify within a few pages here. There is a great interest in low-dose ionising radiation effects, which are related to the increased risk of cancer. Immediately following the accident, a few specialists claimed that less than 100 mSv would be harmless, and objections were raised by the general public. Moreover, there are still discussions and divided opinions about the appropriateness of the LNT model, despite the ICRP's conclusion. It is difficult to evaluate the risk in a precise manner for low doses, as mentioned above. In addition, at present, with clear evidence lacking, no conclusion is possible about the positive effects of low-dose radiation, such as that indicated by the hormesis hypothesis. There is an increasing need for research about long-term ionising radiation exposure with low-dose rates. The UNSCEAR (2013) reported on the radiation exposure effect caused by the Fukushima accident, which could serve as a good reference for the reader.

1.7 Environmental Transfer of Radioactive Substances

YASUHITO IGARASHI

Because medium- and long-lived anthropogenic radionuclides remain in the environment for a long time, they can migrate from the atmospheric environment to the

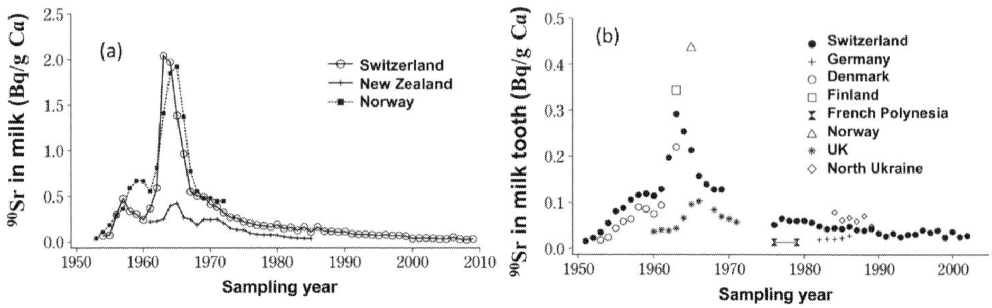

Figure 1.6 (a) Variation of the ^{90}Sr concentration in milk, and (b) in milk tooth. Cited from Froidevaux *et al.* (2012).

hydrosphere, the terrestrial environment and, finally, to humans through the food and water chains – gradually but continuously. The atmosphere is important as a gateway to every environment or sphere (in many cases, emissions begin in the atmosphere). It is desirable to manage continuous observations of atmospheric concentrations and/or radioactive fallout, which are input data for the radionuclide environmental dynamics model. The Environmental Measurements Laboratory in the USA and the Atomic Energy Research Establishment in the UK have conducted continuous observations of the radioactive fallout from the early atomic bomb (a-bomb) and hydrogen bomb (H-bomb) experiments in the atmosphere. Unfortunately, those institutes withdrew from these observations when the global fallout level declined significantly during the 1990s. However, observation of either or both the atmospheric concentration and fallout have continued in Nordic countries (Kulan, 2006; Barescut *et al.*, 2009; Paatero *et al.*, 2010; Outola and Saxén, 2012), France (Masson, 2005) and Japan (Igarashi *et al.*, 2015). These data indicate the environmental dynamics of radioactive substances, such as ^{90}Sr concentrations in milk and human baby teeth (Froidevaux *et al.*, 2012), as shown in Figure 1.6. ^{90}Sr that originated from nuclear tests (1945 to the 1970s) has been continuously detected even in the 2000s, although at very low concentrations. The most important research target of environmental impacts is exposure of the general public. It is crucial that observations of the temporal variations in concentrations and other related factors are numerous in the context of the environmental chain. Thus, Japan has implemented and kept its environmental radioactivity surveillance system for many years since the 1950s. Even after the Fukushima accident, researchers in the environmental radioactivity field, as well as earth scientists, were tasked with challenging tasks, namely describing and understanding in a scientific manner the temporal change of radioactive substances added to the atmosphere, hydrosphere and terrestrial environments.

1.7.1 Basic Concept of Environmental Contamination by Radioactive Substances

It is not easy to find a good description of environmental contamination by radioactive substances. However, the common concepts of atmospheric and ocean circulations can be applied for transport mechanisms of radioactive substances. Therefore, in this section the degree of contamination and diffusion and the condition or process that determines the contamination are discussed.

As illustrated in Figure 1.7, the spread of contamination is determined by the emission amount, the volume of air involved (degree of transport or diffusion) and wet deposition with rainfall and snow (removal from the air). The scale of the contamination depends on the weather conditions, such as strong or weak winds, and depends on the geographical conditions, such as a basin surrounded by mountains. If the wind is strong, the radioactive plume can be blown off (diluted). However, if the wind is weak, the radioactive plume remains stagnant within a basin and maintains a high concentration. Dry deposition plays a minor role in the removal of the pollution from the atmosphere, although it can be important for surface contamination in the area near the accident. Transport and diffusion change as time elapses; thus, further examination of these factors is needed. In environmental contamination research concerning nuclear disasters, there are two approaches for clarifying each factor: observations and model simulations. The latter should be based on the observed data. In this portion of the book, we describe observations and model calculations mainly from this perspective. The emission inventory, which is essential input for model calculation, can be obtained using inversion analysis, which uses the relationship expressed by the determinant as described below (e.g. Stohl *et al.*, 2012; Maki *et al.*, 2013; Saunier *et al.*, 2013; Yumimoto *et al.*, 2016).

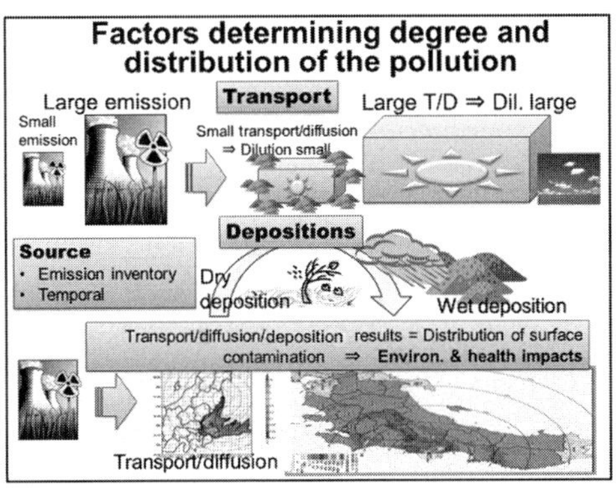

Figure 1.7 Conceptual diagram on radioactive contamination in the environment.

With the determinant expression (matrix), observational results are regarded as the final expression of occurrence in which the temporal variations in the emission inventory are influenced by transport, diffusion and deposition (removal from the air) processes, which can be described as follows:

$$\begin{aligned} &\begin{pmatrix} \text{Observation determinant} \vdots & \ddots & \vdots\, [\text{functions of observation points and temporal variations}] \end{pmatrix} \\ &\quad = \begin{pmatrix} \text{Transport and diffusion determinant for an individual point and time} \vdots & \ddots & \vdots \end{pmatrix} \\ &\quad\quad \times \begin{pmatrix} \text{Temporal variation of the emission} \vdots \end{pmatrix}. \end{aligned}$$

Building on these relationships, the atmospheric transport and diffusion (T/D) matrix can be obtained using a model simulation. Here, the T/D matrix is an operator for the emissions. Thus, the observations are a result of this matrix product. The relationship is applicable if the inversion matrix can be multiplied by the T/D matrix, which cancels the T/D matrix; finally, we obtain the emission matrix. Although the real computations are more complicated, only the principle is given here. Thus, based on the coupling of computational models and observations, radioactive contamination caused by the Fukushima accident is revealed via quantitative analysis.

1.8 Temporal Trends of Radioactive Substances after and before the Fukushima Daiichi Nuclear Power Plant Accident : Quantitative Comparison

YASUHITO IGARASHI, MICHIO AOYAMA AND MASAYUKI TAKIGAWA

To fully understand the situation of the Fukushima accident and the type and amounts of radionuclides emitted into the air and oceans, we present in Table 1.2 the data released by the former Nuclear and Industrial Safety Agency (NISA) on 20 October 2011. These are calculated data that may be revised in the future (see Chapter 2). Although the estimate includes errors and uncertainty, further discussion is warranted.

To help comprehend the scale of the accident, Table 1.3 illustrates the emission inventory of the Fukushima case compared with past cases using ^{131}I and ^{137}Cs as representative examples. Among the significant past nuclear power plant accidents, Three Mile Island (TMI) in Pennsylvania, USA, in 1979 and Chernobyl in the former Soviet Union (Ukraine at present) in 1986 are worth mentioning. For reference, the estimated emissions from the atomic bombing at Hiroshima in 1945 are also shown.

From this table, we can see that the Fukushima accident was second only to the Chernobyl accident and is expected to have longer-term environmental impacts on

Table 1.2 *Estimates (Bq) of atmospheric release of radionuclides in the Fukushima accident (NISA, 2011; half-life values from Knolls Atomic Power Laboratory, 2010).*

Nuclides	Half-life	(Period)	No.1	No.2	No.3	Total
^{133}Xe	5.243	d	3.4×10^{18}	3.5×10^{18}	4.4×10^{18}	1.1×10^{19}
^{134}Cs	2.065	y	7.1×10^{14}	1.6×10^{16}	8.2×10^{14}	1.8×10^{16}
^{137}Cs	30.07	y	5.9×10^{14}	1.4×10^{16}	7.1×10^{14}	1.5×10^{16}
^{89}Sr	50.61	d	8.2×10^{13}	6.8×10^{14}	1.2×10^{15}	2.0×10^{15}
^{90}Sr	28.8	y	6.1×10^{12}	4.8×10^{13}	8.5×10^{13}	1.4×10^{14}
^{140}Ba	12.75	d	1.3×10^{14}	1.1×10^{15}	1.9×10^{15}	3.2×10^{15}
^{127m}Te	106	d	2.5×10^{14}	7.7×10^{14}	6.9×10^{13}	1.1×10^{15}
^{129m}Te	33.6	d	7.2×10^{14}	2.4×10^{15}	2.1×10^{14}	3.3×10^{15}
^{131m}Te	1.36	d	2.2×10^{15}	2.3×10^{15}	4.5×10^{14}	5.0×10^{15}
^{132}Te	3.20	d	2.5×10^{16}	5.7×10^{16}	6.4×10^{15}	8.8×10^{16}
^{103}Ru	39.27	d	2.5×10^{9}	1.8×10^{9}	3.2×10^{9}	7.5×10^{9}
^{106}Ru	1.017	y	7.4×10^{8}	5.1×10^{8}	8.9×10^{8}	2.1×10^{9}
^{95}Zr	64.02	d	4.6×10^{11}	1.6×10^{13}	2.2×10^{11}	1.7×10^{13}
^{141}Ce	32.50	d	4.6×10^{11}	1.7×10^{13}	2.2×10^{11}	1.8×10^{13}
^{144}Ce	284.6	d	3.1×10^{11}	1.1×10^{13}	1.4×10^{11}	1.1×10^{13}
^{239}Np	2.356	d	3.7×10^{12}	7.1×10^{13}	1.4×10^{12}	7.6×10^{13}
^{238}Pu	87.7	y	5.8×10^{8}	1.8×10^{10}	2.5×10^{8}	1.9×10^{10}
^{239}Pu	24100	y	8.6×10^{7}	3.1×10^{9}	4.0×10^{7}	3.2×10^{9}
^{240}Pu	6560	y	8.8×10^{7}	3.0×10^{9}	4.0×10^{7}	3.2×10^{9}
^{241}Pu	14.29	y	3.5×10^{10}	1.2×10^{12}	1.6×10^{10}	1.2×10^{12}
^{91}Y	58.5	d	3.1×10^{11}	2.7×10^{12}	4.4×10^{11}	3.4×10^{12}
^{143}Pr	13.57	d	3.6×10^{11}	3.2×10^{12}	5.2×10^{11}	4.1×10^{12}
^{147}Nd	10.98	d	1.5×10^{11}	1.3×10^{12}	2.2×10^{11}	1.6×10^{12}
^{242}Cm	162.8	d	1.1×10^{10}	7.7×10^{10}	1.4×10^{10}	1.0×10^{11}
^{131}I	8.023	d	1.2×10^{16}	1.4×10^{17}	7.0×10^{15}	1.6×10^{17}
^{132}I	2.283	h	1.3×10^{13}	6.7×10^{6}	3.7×10^{10}	1.3×10^{13}
^{133}I	20.8	h	1.2×10^{16}	2.6×10^{16}	4.2×10^{15}	4.2×10^{16}
^{135}I	6.57	h	2.0×10^{15}	7.4×10^{13}	1.9×10^{14}	2.3×10^{15}
^{127}Sb	3.84	d	1.7×10^{15}	4.2×10^{15}	4.5×10^{14}	6.4×10^{15}
^{129}Sb	4.40	h	1.4×10^{14}	5.6×10^{10}	2.3×10^{12}	1.4×10^{14}
^{99}Mo	2.7476	d	2.6×10^{9}	1.2×10^{9}	2.9×10^{9}	6.7×10^{9}

Table 1.3 *Amounts of radionuclides released into the atmosphere (PBq).*

	Fukushima	Hiroshima atomic bomb	Chernobyl	Three Mile Island	1970 nuclear tests
^{131}I	160	63	1760	0.0056	
^{137}Cs	15	0.089	85	Not significant	
Total release	11,300		13,200		>795

the environment compared to the TMI accident and the Hiroshima atomic bomb cases. Compared with nuclear tests, environmental contamination due to nuclear power plant accidents has relatively longer and more serious effects because more radionuclides with mid- to long half-lives exist in nuclear reactor cores, where nuclear fission continues. The total amount of emitted radionuclides from the Fukushima accident was only 20% less than that from the Chernobyl accident, although most were noble gases, such as xenon-133 (^{133}Xe), having almost no impact on the environment and human body except over very short periods of time and under highly concentrated radioactive air parcel (plume) conditions. The Chernobyl accident, the TMI accident and the Hiroshima atomic bomb exhibit similar characteristics, whereas the Fukushima accident differs regarding the radionuclide abundance for each emission event or time period because the accident was composed of multiple reactor failures, explosions and continuous emissions.

The total radionuclide abundance accumulated in the reactor (which is called the core inventory) immediately following the accident was analysed by the Tokyo Electric Power Company and other institutes. In reactor units 1 to 3, the FDNPS was estimated to contain 6100 petabecquerel (1 PBq = 10^{15} Bq), nearly twice as much (i.e. 3200 PBq) as at Chernobyl reactor no. 4. However, the emission inventory of the radionuclides to the environment was relatively small because the reactor damage was not as serious as in the Chernobyl accident.

1.8.1 What Indicates the Long-Term Variation in Radioactive Substances in the Atmosphere?

In this section, we focus on the temporal trends in atmospheric radioactivity data, emphasising the importance of continuous observations and describing long-term variations in radioactive fallout and how the FDNPS accident affected the situation.

The Meteorological Research Institute (MRI) of Japan started observing total β (i.e. every β radioactive source) for atmospheric fallout in April 1954. Radionuclide analysis (^{90}Sr, ^{137}Cs, etc.) replaced the role of total β observation in 1957 and continued even after the institute relocated to Tsukuba; the institute has continued trying to quantify atmospheric radionuclides instead of describing them as 'lower than detection limit' or 'not detected; ND', despite the concentration level dropping nearly to the 'ND' level. These time series data truly reflect the environmental dynamics of anthropogenic pollutants released into the global environment. They can be likened to the global-scale experiment of anthropogenic radionuclides being considered as tracers, which trace material flow or transport over the globe. Fallout observations aim to show the dynamics via fallout using the addition of a tracer.

In this sense, the global fallout originating from atmospheric nuclear tests was the first global environmental issue confronting humans.

1.8.2 Change in Radionuclide Concentrations in the Atmosphere before the Fukushima Accident

Organised research regarding environmental radioactivity in Japan began in March 1954, when a large-scale atmospheric test was conducted in the Bikini Islands in the northern equatorial Pacific. According to Miyake (1954a; 1954b), rainfall contained anthropogenic radionuclides, which were detected over Japan after 14 May 1954. Many other districts and national research institutions reported similar findings. The maximum value recorded was 0.5×10^{-6} Ci (Curie)/L, which corresponds to 18,500 Bq/L.

As a result of the Cold War, large-scale atmospheric nuclear tests were repeatedly conducted by the USA and the former Soviet Union, and fallout reached a peak in June 1963. The monthly fallout of ^{90}Sr reached approximately 170 Bq/m^2, whereas that of ^{137}Cs was approximately 550 Bq/m^2 in Tokyo, Japan (Katsuragi, 1983), as depicted in Figure 1.8. China (in Taklimakan) and France (mainly in the Southern Hemisphere) continued nuclear tests even after the USA and the former Soviet Union stopped. Japan has been repeatedly affected by nuclear tests in China since the mid-1960s. Increased fallout was observed from the sixth (June 1967), eighth (December 1968) and tenth (September 1969) nuclear tests by China, as recorded in the MRI fallout time series (Katsuragi, 1983).

Due to the high temperature of the fireball generated by a nuclear explosion, most of the radioactive substances are launched high into the atmosphere, that is,

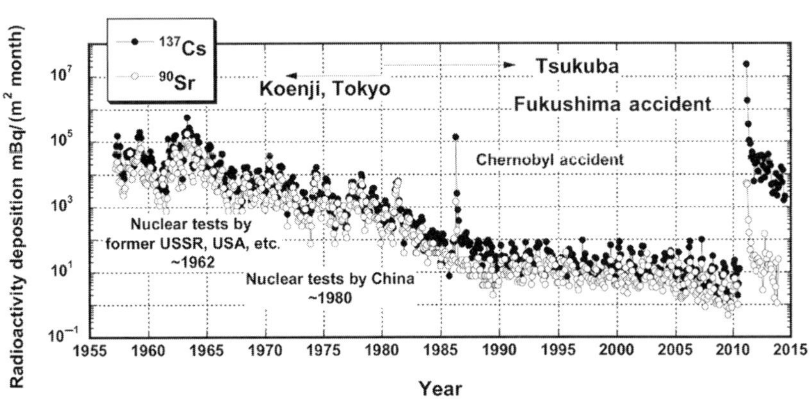

Figure 1.8 Monthly deposition of ^{90}Sr and ^{137}Cs observed at the MRI, Japan since 1954, showing the temporal change of the anthropogenic radionuclides levels in air over Japan.

into the stratosphere. The air in the stratosphere and that in the troposphere do not mix easily for mechanical (thermodynamical) reasons. Therefore, inflow of the stratospheric air to the troposphere (which is accompanied by cyclone development) became a rate-controlling factor for the surface atmospheric radioactivity concentration and the amount of fallout during the 1960s and 1970s. In this situation, the monthly fallout amount peaked in spring, when active exchange occurred between the stratosphere and the troposphere. Since the twenty-sixth Chinese nuclear test in October 1980 (the final test), all further nuclear tests have been conducted underground, meaning the amount of radioactive substances emitted into the air has reduced tremendously, and atmospheric fallout also decreased exponentially after the peak in the spring of 1981, with an approximately one-year half-reduction time. An interesting result of these tests was the growth in scientific knowledge about stratosphere–troposphere interchange processes (Igarashi, 2009).

On 26 April 1986, an accident occurred at the nuclear power plant at Chernobyl. After 3 May 1986, the MRI in Tsukuba detected ^{131}I and ^{132}Te–^{132}I in the atmosphere and in the ocean. In May 1986 the monthly fallout of ^{131}I was approximately 5900 Bq/m^2, while that of ^{137}Cs was approximately 130 Bq/m^2 (Aoyama *et al.*, 1986). As illustrated in Table 1.3, the total emission inventory of ^{137}Cs to the environment from the accident at the Chernobyl nuclear power plant was 85 PBq; for comparison, the nuclear tests in the Northern Hemisphere generated 765 $\pm$ 30 PBq in January 1970 (Aoyama *et al.*, 2006). Reflecting on the difference, the influence of the Chernobyl accident is regarded as only a single peak, as shown in Figure 1.8. The long-term impact of the Chernobyl accident was not the same as that of the atmospheric nuclear tests because not many radioactive substances were transported into the stratosphere following the Chernobyl accident.

Since the 1990s, monthly fallout has ranged between several mBq/m^2 to several tens of mBq/m^2 in Japan, which had no longer been considered to be ‘radioactive’ until the Fukushima accident (Igarashi, 2009). Thus, apart from the observations conducted by the MRI over 4 m^2 of a large basin for sampling, the fallout was often acknowledged to be less than the detection limit (‘ND’).

1.8.3 What Indicates the Time Series Data during the 1990s?

Although several per cent of the radioactive substances originating from the Chernobyl accident were transported into the lower stratosphere, the annual fallout largely has exceeded the estimate from the residence time in the stratosphere since 1994. This was mainly due to the influence of the resuspension process (the phenomenon in which a substance, once deposited on the surface of the ground, becomes suspended again along with surface soil particles in the atmosphere). The main source and host particles of resuspension (carrier of the radionuclides) was thought to be surface soil dust particles from neighbouring farmland. However, the radioactivity ratio $^{137}Cs/^{90}Sr$ of the fallout sampled in Tsukuba did not match that of the surface soil sampled in Tsukuba or any other surface soil from Japan during the late 1980s to 1990s. Therefore, the results led to the conclusion that resuspension may have origins other than neighbouring sampling locations (e.g. Igarashi, 2004; Igarashi *et al.*, 2005; 2011). We gradually learned that, over long distances, wind-blown dust such as Kosa–Asian dust (meaning large-scale wind-blown surface soil particles, which is referred to as Aeolian dust; one example is Sahara dust) brought slight but detectable amounts of radioactive substances from the atmospheric nuclear tests (Igarashi, 2004; Igarashi *et al.*, 2005; 2011). Because there was a major outbreak of Asian dust over the continent during the early 2000s, ^{137}Cs was detected all over Japan, resulting in more studies related to the hypothesis of wind-blown dust because it became a hot topic, which led to improvements in atmospheric transport/diffusion models.

Although the contamination was not as serious as that of the Fukushima accident, in the late 1990s there were several accidents in Japanese nuclear facilities (Aoyama *et al.*, 1999; Igarashi *et al.*, 1999; Komura *et al.*, 2000); in addition, there were other events, such as underground tests conducted by North Korea in the mid-2000s. However, the emission inventory of radioactive substances from these events was so small that there was no impact on atmospheric radioactive fallout in Japan, and there is no trace or evidence in the time series depicted in Figure 1.8.

1.9 Characteristics of Anthropogenic Radionuclides in the Atmosphere after the Fukushima Daiichi Nuclear Power Plant Accident

YASUHITO IGARASHI

1.9.1 Observation of Radionuclides Activity Concentrations in Tsukuba, Ibaraki Prefecture

At the MRI in Tsukuba, Ibaraki, sampling of atmospheric aerosols and the analysis of radioactivity were continued before and after the Fukushima accident, and the results were summarised and published by Igarashi *et al.* (2015). The variety of detected γ-ray-emitting radionuclides included ^{99}Mo–^{99m}Tc, ^{129m}Te–^{129}Te, ^{132}Te–^{132}I, ^{131}I, ^{133}I, ^{134}Cs, ^{136}Cs and ^{137}Cs. The time series data from Tsukuba indicated concentrations of these radionuclides being elevated by a factor of two during March 2011. These concentration peaks corresponded to a remarkable process of advection of the radioactive plume (see Chapter 3) over the Kanto Plain from the Fukushima accident (Igarashi *et al.*, 2015). Based on each emission event during the accident, which discharged different substances, the data indicated different radionuclide compositions during each increase in concentration. This difference is thought to have resulted in the regional difference in radioactive fallout compositions between the Kanto and Tohoku districts (Kinoshita *et al.*, 2011).

Monthly fallout of ^{137}Cs at the MRI in Tsukuba reached 23 $\pm$ 0.9 kBq/m^2 in March 2011. This is six or seven digits larger than the level before the Fukushima accident and approximately 50 times larger than the maximum value of the monthly fallout that originated from the atmospheric nuclear tests. In May 2011, the level matched that of June 1963, when the maximum global fallout was recorded following the atmospheric nuclear tests. Because the source was nearby in the case of the Fukushima accident, the degree of fallout increase in Tsukuba was far greater than that from the past nuclear tests. However, the spatial representativeness was small (regionally imbalanced). Thereafter, the fallout followed the same footprint as the Chernobyl accident: it rapidly decreased due to the residence time in the troposphere (Figure 1.8). The ^{137}Cs fallout (considering no decay) from 1957 to 2010 was approximately 7 kBq/m^2. Taking the radioactive decay of the individual monthly ^{137}Cs deposition into account, one can show that the total contribution would be 2.3 kBq/m^2. Thus, the Fukushima accident brought at least 10 times more ^{137}Cs deposition over the surface than that due to the global fallout. Furthermore, similar ^{134}Cs fallout was also observed from the Fukushima accident. Therefore, the contamination of the ground from both radionuclides was approximately 50 kBq/m^2, which nearly matches the value of the aerial radiation mapping by aircraft conducted by the Ministry of Education, Culture, Sports, Science and Technology (Torii *et al.*, 2013; Sanada *et al.*, 2014).

Compared to ^{137}Cs, the ^{90}Sr monthly fallout in Tsukuba was 4.4 $\pm$ 0.1 Bq/m^2 in March 2011, corresponding to 0.02% of the ^{137}Cs fallout in the same month (Igarashi *et al.*, 2015). However, the available data for ^{90}Sr remain limited. This fallout level was some thousands to tens of thousand times higher than the level before the Fukushima accident, although its impact on the environment was not as large as that of the Cs isotopes. The emission inventory of ^{90}Sr was approximately one-hundredth of that estimated for ^{137}Cs by the Nuclear and Industrial Safety Agency (NISA, 2011), which could be partly attributed to the fact that Sr has a higher melting point than Cs. The radioactive Sr particle size was larger than that of the Cs in the Chernobyl accident (Hirose *et al.*, 1993), which is probably related to the abovementioned volatility. The fallout impact was much smaller for radioactive strontium over the Kanto area, which may have been due to the fractionation processes of Cs during emission and transport. The same discussion is applicable to the atmospheric concentration. Therefore, it can be reasonably assumed that the inhalation exposure should be relatively small for ^{90}Sr in comparison with radioactive caesium and iodine. At the end of 2011, the fallout level decreased by one-thousandth to ten-thousandths, although it was still around the level of the 1970s to early 1980s (Igarashi *et al.*, 2015), when atmospheric nuclear tests were actively conducted by China.

Because ^{90}Sr and ^{137}Cs are long-lived radionuclides, they must be monitored in the environment for the purpose of protection from radiation. Continuous monitoring is very important. The variation in the ecosystem is described using a linear vertical scale, as shown in Figure 1.6. However, the variations in the atmosphere are described using a logarithmic scale, as depicted in Figure 1.8. The use of a logarithmic scale has the benefit of displaying details. If the plot were depicted using a linear scale, it would include only a few spikes of global fallout in 1963, the Chernobyl accident in 1986 and the Fukushima accident in 2011, and there would be a lengthy line along the bottom of the figure. Even if the activity concentration of radionuclides decreases exponentially in the atmosphere, its decrease would be slowed in the ecosystem due to a buffer effect, and the pollution in the ecosystem would occur over a longer period of time. Additionally, ^{90}Sr and ^{137}Cs in the environment can be used as indicators of atmospheric circulation or environmental change, as described above (see also, e.g. Aoyama *et al.*, 2012). It is very important to continue observations of ^{90}Sr and ^{137}Cs not only for the purpose of radiation protection but also for monitoring the global environment.

Furthermore, because resuspension (e.g. Igarashi, 2009; Garger *et al.*, 2012) continues for a long time, we need to be aware of future changes. Apart from continental weather conditions, the Japanese environment is characterised by relatively humid air and weak winds, and it remains unclear whether surface soil is the only major source of resuspension. Radioactive caesium is possibly volatilised by burning fields and garbage incineration (e.g. Bourcier *et al.*, 2010) and can become resuspended from the terrestrial ecosystem (Kajino *et al.*, 2016). We need to investigate which process is the main source of resuspension through observations and both data and modelling analyses (see Chapter 3).

1.9.2 Findings on Physical and Chemical Form of Radionuclides: Case for Radiocaesium

One of the notable findings achieved by Japanese scientists is that there is an insoluble type of radiocaesium-bearing aerosols (Adachi *et al.*, 2013), on the contrary to the conventional understanding (e.g. Kaneyasu *et al.*, 2012). It has been known that autoradiography of various environmental samples with radioactive contamination by the imaging plate (IP) technique gives dotty images of radionuclides (e.g. Itoh *et al.*, 2014; Niimura *et al.*, 2015). These dotty IP images indicate that radiocaesium is not only carried by submicron particles (in the case the contamination 'painted unevenly') but is also of highly concentrated form. However, what gives such activity spots was not revealed so easily. The first isolation and determination were accomplished from high-volume (HV) filter samples obtained during the early stage of the Fukushima accident at about 170 km south–south west from the accident site (Tsukuba, Ibaraki Prefecture). Adachi *et al.* (2013) showed, using SEM-EDX (scanning electron microscopy with energy dispersive X-ray spectroscopy), that such water-insoluble radiocaesium-bearing particles (a few Bq) were spherical and of 2–3 μm in diameter, composed of Fe, Zn, O and highly concentrated radiocaesium (up to 10^{11} Bq/g) (Figure 1.9). Abe *et al.* (2014) indicated that these particles are amorphous, relatively oxidised and contain heavy elements including fission products and reactor materials – even uranium – by microbeam X-ray analysis at the synchrotron radiation facility.

Water solubility of the radioactive materials is an important controlling factor for the dynamics in the environment, as well as in the human body, resulting in different internal exposure levels. It was not presumed before the accident that such radioactive insoluble aerosols were released, and it should be noted that the property was quite different from the so-called Chernobyl hot particle. The coarser particles (up to some hundreds of microns) having higher Cs activity (sometimes more than a few kBq) and an irregular shape have been found in surface soil and dust samples taken in the vicinity of the accident site within a few kilometres; a major component is Si (Satou *et al.*, 2016, 2018; Yamaguchi *et al.*, 2016). Based on the detailed analysis of $^{134/137}$Cs activity ratio, fine and coarse particles (often referred to as Type A and B (S), respectively) were assigned as being from reactor nos 2 or 3 and reactor no. 1, respectively (e.g. Satou *et al.*, 2018). These particles maintained an unchanged shape for more than a few years, so they are relatively stable in the environment, as anticipated (Satou *et al.*, 2016).

The insoluble radiocaesium bearing solid microparticles also provides valuable information on the accident scenario, and so has attracted the interest of various people. It is convincing that such insoluble microparticles were formed through liquefying and evaporation processes of the reactor materials, which mixed with radiocaesium. Research on the insoluble solid micro-radiocaesium-bearing particles is still ongoing, using state-of-the-art methodologies (e.g. Kogure *et al.*, 2016; Yamaguchi *et al.*, 2016; Furuki *et al.*, 2017), as researchers seek to learn more about the source, distribution, fate, etc.

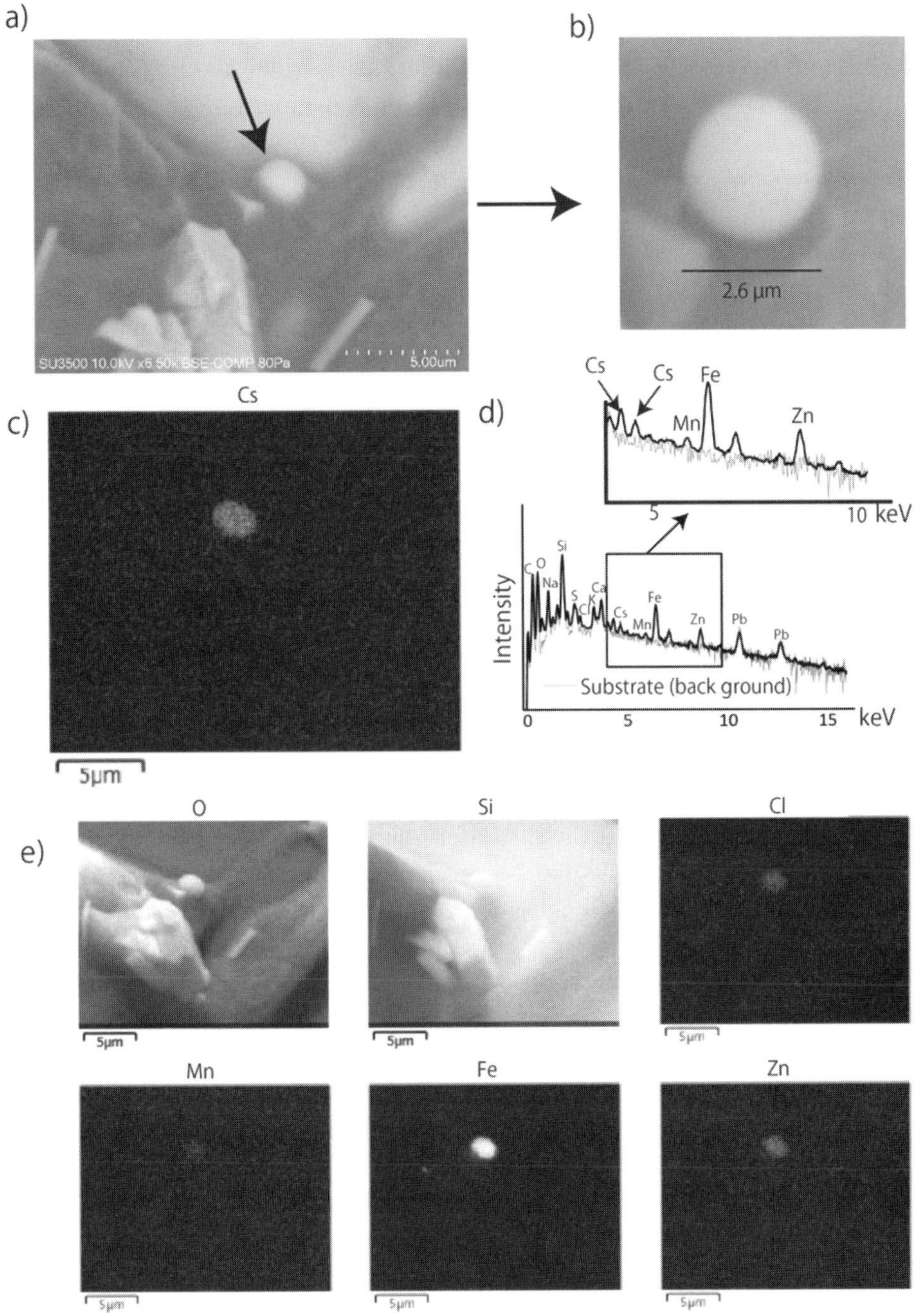

Figure 1.9 SEM and EDX mapping images of a radioactive Cs-bearing particle from the sample collected between 21:10 on 14 March and 09:10 on 15 March 2011. (a) A Cs-bearing particle partially embedded within a carbon paste. (b) The same Cs-bearing particle as (a) but measured the next day; the particle has a spherical shape. (c) An elemental mapping (Cs) of the particle in (a). (d) The EDX spectrum of the particle in (a) (black line). The light grey line shows the spectrum from the glass substrate. The Cs in the particle shows multiple peaks. (e) An elemental mapping of the other elements within the area; O, Si, Cl, Mn, Fe and Zn are possibly coexistent with Cs within the particle.

1.10 Time-Dependent Change of Radiation Levels in the 80 km Zone for Five Years after the Fukushima Accident

KIMIAKI SAITO

1.10.1 Large-Scale Environmental Monitoring in Fukushima

Following the Fukushima accident, large-scale environmental monitoring projects were implemented by the Japan Atomic Energy Agency (JAEA) in collaboration with many organisations commissioned by the Japanese government (Saito and Onda, 2015). Repeated environmental measurement of air dose rates and radionuclide deposition in the project has clarified the features of radiological conditions around the Fukushima site.

Air dose rates have been measured using four different methods: (1) measurements in undisturbed fields using a survey meter; (2) car-based surveys; (3) walking surveys; and (4) unmanned helicopter surveys. Measurements in undisturbed fields are carried out at about 6500 locations in the 80 km zone (Mikami *et al.*, 2015a). One appropriate location in every area of a 1 km^2 grid was selected to give a standard radiation level for the area. Car-based surveys have targeted wide regions in eastern Japan, reaching tens of thousands of kilometres in each measurement campaign (Andoh *et al.*, 2015). Air dose rates are measured every three seconds and averaged for an area of 100×100 m^2 to eliminate statistical fluctuations. A massive amount of data obtained in the car-based surveys has been effectively utilised for statistical analysis of contamination conditions. Walking surveys have been carried out in about 600 areas in the 80 km zone to obtain data in environments related to various human lives. The data obtained are averaged over an area of 20×20 m^2, considering the walking speed. Unmanned helicopter surveys are intended to cover the 3 km zone from the Fukushima site, where survey crews encounter difficulty entering by land (Sanada and Torii, 2015).

Two kinds of information have been obtained concerning radionuclide deposition: (1) radionuclide deposition densities, and (2) depth profiles. In the first campaign of the project, five soil samples per location were collected at about 2200 locations around the Fukushima site and analysed to determine radionuclide deposition densities (Saito *et al.*, 2015). Since the second campaign, *in-situ* measurements using a portable Ge detector have been employed; these can determine the average deposition density of a location by detecting γ-rays coming from a wide area in the location (Mikami *et al.*, 2015b). The measurements were carried out at about 600 locations in the 80 km zone. Depth profiles of radiocaesium have also been investigated by collecting soil samples at different depths using a scraper plate at more than 80 locations covering the whole 80 km zone (Matsuda *et al.*, 2015).

1.10.2 Decreasing Tendency of Air Dose Rates

Figure 1.10 shows air dose rate distribution maps within the 80 km zone from the Fukushima site for eight different occasions through 2015. These air dose rate maps were constructed by integrating results from undisturbed flat fields with those obtained by aerial monitoring. In areas where ground-measured data existed, those data were used, whereas in areas where they did not exist, aerial monitoring data were used after some adjustments. The aerial monitoring data tended to be systematically higher than those measured on the ground, especially in various environments related to human living.

Air dose rates in environments related to human living have decreased much faster than expected from radioactive decay of radiocaesium, which is summarised in Figure 1.11. The average air dose rate in undisturbed fields in the 80 km zone decreased to about one-quarter of the air dose rate of June 2011 in the four years from 2011 to 2015, while the air dose rate was expected to decline to a level just above 40% of the initial value due to radioactive decay. The average air dose rate above roads measured during car-based surveys became one-fifth of that in June 2011. These measured air dose rates all include contributions of background γ-rays from natural radionuclides, and if the contributions are excluded the reduction in the average air dose rate is even

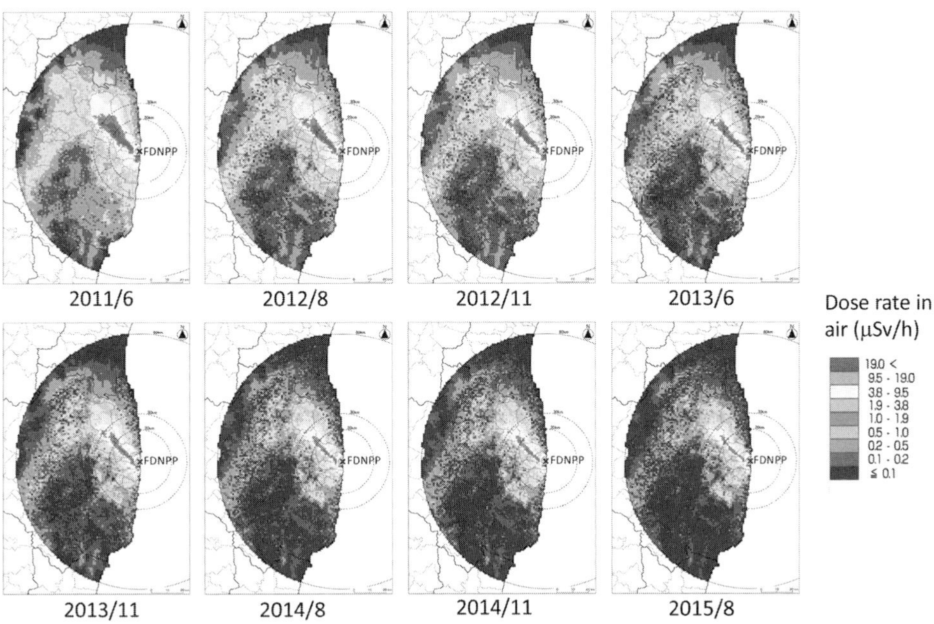

Figure 1.10 Air dose rate maps constructed by integrating data from undisturbed field measurements and aerial monitoring. A black and white version of this figure will appear in some formats. For the colour version, refer to the plate section.

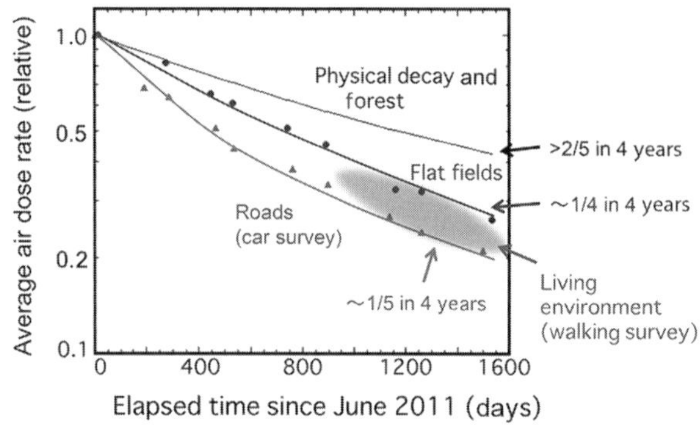

Figure 1.11 Temporal change of the average air dose rates within the 80 km zone for different conditions.

more rapid. The same kind of analysis cannot be done for walking survey results because the surveys started in 2013; the results of walking surveys were found to lie between the results from undisturbed field measurements and those from car-based surveys. Thus, air dose rates in various human-related environments were confirmed to decline very quickly. On the other hand, the air dose rates in pure forest areas were found to decrease almost in accordance with the rate of radioactive decay, even though small differences depending on the species of trees may be observed.

On the basis of statistical analyses for car-borne survey data, the dose rate reduction tendency differs according to land use surrounding the survey road. The reduction was most rapid in urban areas, followed by farmland areas and then forested areas. Furthermore, the reduction tendencies were obviously different inside and outside the evacuation area, indicating that human activities, including decontamination, tend to accelerate air dose rate reduction. However, quantitative analysis of human activity effects on the reduction of air dose rates is not easy and remains an important future challenge.

1.10.3 Change in Radionuclide Deposition Density and Depth Profile

In the first campaign carried out in June 2011, the following radionuclides were detected at many locations and were indicated on a map: ^{134}Cs, ^{137}Cs, ^{131}I, ^{129m}Te, ^{110m}Ag, ^{89}Sr, ^{90}Sr, $^{238+239}$Pu and ^{240}Pu. Due to space limitations the whole map is not shown here, but Figure 1.12 shows the distribution of ^{137}Cs as of June 2011 (Saito *et al.*, 2015). This distribution is similar to that of air dose rates on the same occasion, since radiocaesium was already the main contributor to the air dose rate. A rough estimation of effective doses for the 50 years following June 2011 due to the observed radionuclides using the maximum deposition density indicates that radiocaesium is far more important than the other observed radionuclides from the viewpoint of long-term exposure. The total deposition levels of ^{137}Cs in the 80 km

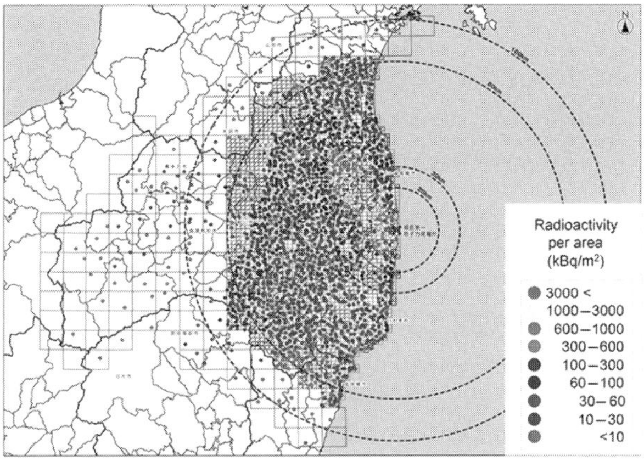

Figure 1.12 ^{137}Cs deposition density map as of June 2011, constructed using measurements of soil samples collected at more than 2200 locations.

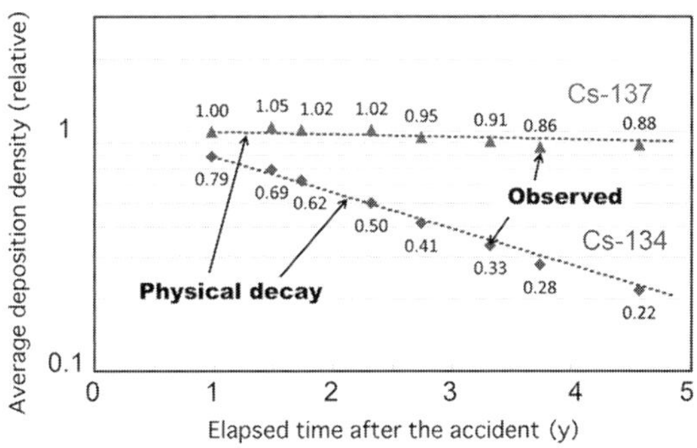

Figure 1.13 Temporal change of the average deposition density within the 80 km zone for ^{137}Cs and ^{134}Cs, determined by *in-situ* measurements using a portable Ge detector. The deposition density was normalised to that of ^{137}Cs in March 2012, when the first *in-situ* measurement data were compiled.

zone and in eastern Japan are estimated to be 1.6 and 2.0 PBq, respectively. In the 80 km zone, about 70% of radiocaesium is estimated to have been deposited in forested regions, 20% in agricultural regions and 5% in urban regions, which is nearly proportional to the distribution of these area types.

Figure 1.13 indicates the time-dependent trend of the average radiocaesium deposition density within the 80 km zone. The data for locations where decontamination took place or the surface conditions were artificially changed were excluded to elucidate natural weathering effects. The average deposition densities decreased in accordance with radioactive decay, suggesting that the migration of radiocaesium in horizontal directions is small in undisturbed fields. This finding agrees with results from migration studies, which found that the discharge rate of deposited radiocaesium from undisturbed fields is on the order of 0.1% per year

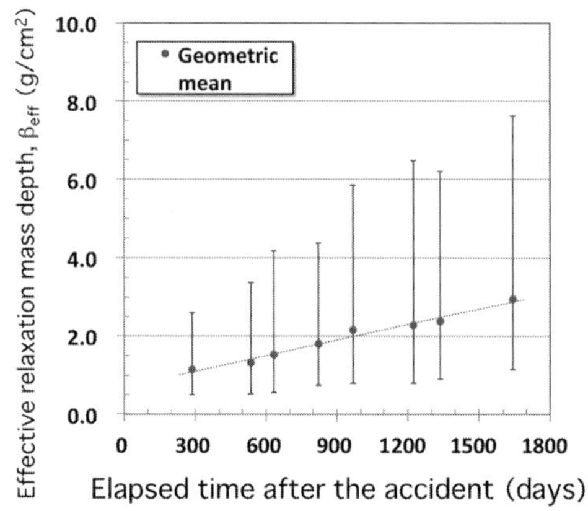

Figure 1.14 Increase in the effective relaxation mass depth with elapsed time for some 80 locations in the 80 km zone.

(Yoshimura *et al.*, 2015). In forests, radiocaesium only very slowly migrates through the system; moreover, the discharge rates from the forest system have also been found to be small. On the contrary, radiocaesium deposited on artificial structures such as roads and houses generally tends to migrate at high rates over a short time period.

Radiocaesium deposited on the surface has penetrated into the ground with time. There exist two kinds of typical radiocaesium depth profiles in the ground: a profile whose radiocaesium concentration decreases exponentially with depth; and a profile having a concentration peak at a certain depth (Matsuda *et al.*, 2015). The latter type of profiles was found to be well approximated by a hyperbolic secant function which asymptotically approaches an exponential function with increasing depth. The proportion of the latter type profiles increased within a few years after the accident, but it seems to have stabilised in recent years, perhaps indicating saturation. The effective relaxation depth, which is an indicator of radiocaesium penetration into the ground, has increased with elapsed time, as shown in Figure 1.14. However, the average value of 90% depth defined as the depth up to which 90% of deposited radiocaesium is contained was still less than 5 cm in 2016.

1.10.4 Prediction of Air Dose Rate Distribution

A numerical model to simulate the trend of air dose rate distribution has been developed by employing an empirical formula with two exponential functions representing fast and slow dose reduction components (Kinase *et al.*, 2014). The empirical formula was fitted to a massive amount of car-based survey data chronologically obtained in the 80 km zone around the Fukushima site, and the

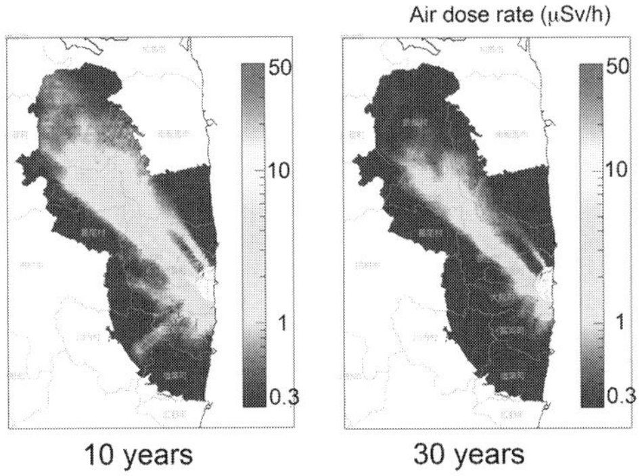

Figure 1.15 Predicted air dose rate distributions in the evacuated zone 10 years and 30 years after the accident.

parameters determined thereby were statistically analysed considering the conditions of land use and evacuation situations. As already discussed, the decreasing tendency of air dose rates depends on the land use, and differs by area – namely, outside the evacuation area and in the three kinds of evacuation areas classified by radiation level. The frequency distribution of the parameters is summarised according to the classification of these conditions. The median value of the parameters in each classified condition is usually used for predicting air dose rate distributions up to 30 years after the accident. This statistical analysis enables us to predict air dose rate distribution of 100×100 m^2 grids with confidence intervals. Figure 1.15 shows predicted air dose rate distribution in the 80 km zone. In 30 years, high dose rate regions of more than 10 μSv/h are predicted to have almost disappeared.

1.10.5 Data Provision

The data obtained through large-scale environmental measurements have been provided to the public in multiple ways. Maps showing the distribution of air dose rates or radionuclide deposition densities with diverse scales can be accessed on the website of the Nuclear Radiation Authority (NRA, 2017). The JAEA (2017) developed and has run a database for providing environmental monitoring data concerning the Fukushima accident online in diverse forms such as numerical data, maps and figures. For this, reliable environmental monitoring data have been collected from different ministries, agencies and municipalities. The system is flexibly designed; for example: data are recorded in common formats to compare different datasets readily; graphs to indicate data trends are created as a user requests; tools for simple data analysis are utilised; and meteorological and geographical data are prepared for data analysis. The stored data are frequently updated following the release of new monitoring data.

References

Abe, Y., Y. Iizawa, Y. Terada, *et al.* (2014). Detection of uranium and chemical state analysis of individual radioactive microparticles emitted from the Fukushima nuclear accident using multiple synchrotron radiation X-ray analyses. *Anal. Chem.*, **86**(17), 8521–5, doi:10.1021/ac501998d.

Adachi, K., M. Kajino, Y. Zaizen and Y. Igarashi (2013). Emission of spherical cesium-bearing particles from an early stage of the Fukushima nuclear accident. *Sci. Rep.*, **3**, 2554, doi:10.1038/srep02554.

Andoh, M, Y. Nakahara, S. Tsuda, *et al.* (2015). Measurement of air dose rates over a wide area around the Fukushima Dai-ichi Nuclear Power Plant through a series of car-borne surveys. *J. Radiat. Res.*, **139**, 266–80.

Aoyama, M., K. Hirose, Y. Suzuki, H. Inoue and Y. Sugimura (1986). High-level radioactive nuclides in Japan in May. *Nature*, **321**, 819–20.

Aoyama, M., T. Ohara and K. Komura (1999). Spread to the Kanto plains of the radio-cesium by the Power Reactor and Nuclear Fuel Development Corporation Tokai accident. *Kagaku*, **69**(1), 16–21 (in Japanese).

Aoyama, M., K. Hirose and Y. Igarashi (2006). Re-construction and updating our understanding on the global weapons tests ^{137}Cs fallout. *J. Environ. Monit.*, **8**, 431–8.

Aoyama, M., D. Tsumune and Y. Hamajima (2012). Distribution of ^{137}Cs and ^{134}Cs in the North Pacific Ocean: Impacts of the TEPCO Fukushima-Daiichi NPP accident. *J. Radioanal. Nucl. Chem.*, **296**, 535–9, doi:10.1007/s10967-012-2033-2.

Aoyama, M., M. Kajino, T. Tanaka, *et al.* (2016). ^{134}Cs and ^{137}Cs in the North Pacific Ocean derived from the March 2011 TEPCO Fukushima Dai-ichi Nuclear Power Plant accident, Japan: Part two. Estimation of ^{134}Cs and ^{137}Cs inventories in the North Pacific Ocean. *J. Oceanogr.*, **72**, 67–76, doi:10.1007/s10872.

Barescut, J., T. K. Ikäheimonen, R. Mustonen and R. Saxén (2009). Half a century of radioecological research and surveillance at STUK. *Radioprotection*, **44**, 607–12.

Bourcier, L., K. Sellegri, O. Masson, *et al.* (2010). Experimental evidence of biomass burning as a source of atmospheric ^{137}Cs, puy de Dome (1465 m a.s.l.), France. *Atmos. Environ.*, **44**, 2280–6.

CSIC (2012). *Final Report, Investigation Committee on the Accident at the Fukushima Nuclear Power Stations of Tokyo Electric Power Company.* Cabinet Secretariat, 23 July. www.kantei.go.jp/jp/noda/actions/201207/23kenshou.html (accessed 19 September 2018).

Estournel, C., E. Bosc, M. Bocquet, *et al.* (2012). Assessment of the amount of cesium-137 released into the Pacific Ocean after the Fukushima accident and analysis of its dispersion in Japanese coastal waters. *J. Geophys. Res. Oceans*, **117**, C11014, doi:10.1029/2012JC007933.

Froidevaux, P., M. Haldimann and F. Bochud (2012). Long-term effects of exposure to low-levels of radioactivity: a retrospective study of ^{239}Pu and ^{90}Sr from nuclear bomb tests on the Swiss population. In P. Tsvetkov (ed.) *Nuclear Power: Operation, Safety and Environment.* London: InTech Open, chapter 14, doi:10.5772/19058.

Fukushima Prefectural Government (2018). Steps on revitalization in Fukushima. www.pref.fukushima.lg.jp/uploaded/attachment/264747.pdf (accessed 19 September 2018).

Furuki, G., J. Imoto, A. Ochiai, *et al.* (2017). Caesium-rich micro-particles: a window into the meltdown events at the Fukushima Daiichi Nuclear Power Plant. *Sci. Rep.*, **7**, 42731, doi:10.1038/srep42731.

Garger, E. K., Y. I. Kuzmenko, S. Sickinger and J. Tschiersch (2012). Prediction of the ^{137}Cs activity concentration in the atmospheric surface layer of the Chernobyl exclusion zone. *J. Environ. Radioact.*, **110**, 53–8.

Hirose, K., S. Takatani and M. Aoyama (1993). Wet deposition of radionuclides derived from the Chernobyl accident. *J. Atmos. Chem.*, **17**(1), 61–71.

Honda, M. C., T. Aono, M. Aoyama, *et al.* (2012). Dispersion of artificial cesium-134 and -137 in the western North Pacific one month after the Fukushima accident. *Geochem. J.*, **46**, e1–e9.

Igarashi, Y. (2004). Atmospheric cycles of materials and radioactive fallout. *Isotope News*, May, 2–8 (in Japanese).

Igarashi, Y. (2009). Anthropogenic radioactivity in aerosol: a review focusing on studies during the 2000s. *Japanese Journal of Health Physics (Hoken Butsuri)*, **44**(3), 315–25. www.jstage.jst.go.jp/article/jhps/44/3/44_3_313/_pdf (accessed 19 September 2018).

Igarashi, Y., M. Aoyama, T. Miyao, *et al.* (1999). Air concentration of radiocaesium in Tsukuba, Japan following the release from the Tokai waste treatment plant: comparisons of observations with predictions. *Appl. Radiat. Isotopes*, **50**, 1063–73.

Igarashi, Y., M. Aoyama, K. Hirose, P. Povinec and S. Yabuki (2005). What anthropogenic radionuclides (^{90}Sr and ^{137}Cs) in atmospheric deposition, surface soils and Aeolian dusts suggest for dust transport over Japan. *Water Air Soil Poll. Focus*, **5**, 5–69.

Igarashi, Y., H. Fujiwara and D. Jugder (2011). Change of the Asian dust source region deduced from the composition of anthropogenic radionuclides in surface soil in Mongolia. *Atmos. Chem. Phys.*, **11**, 7069–80.

Igarashi, Y., M. Kajino, K. Zaizen, K. Adachi and M. Mikami (2015). Atmospheric radioactivity over Tsukuba, Japan: a summary of three years of observations after the FDNPP accident. *Prog. Earth Planet. Sci.*, **2**, 44, doi:10.1186/s40645-015-0066-1.

Imamura, N., M. Komatsu, S. Ohashi, *et al.* (2017). Temporal changes in the radiocesium distribution in forests over the five years after the Fukushima Daiichi Nuclear Power Plant accident. *Sci. Rep.*, **7**, 8179.

Ishihara, M. and T. Tadono (2017). Land cover changes induced by the great east Japan earthquake in 2011. *Sci. Rep.*, **7**, 45769, doi:10.1038/srep45769.

Itoh, S., T. Eguchi, N. Kato and S. Takahashi (2014). Radioactive particles in soil, plant, and dust samples after the Fukushima nuclear accident. *Soil Sci. Plant Nutr.*, doi:10.1080/00380768.2014.907735.

JAEA (Japan Atomic Energy Agency) (2012). JAEA Open Workshop 'Reconstruction of the Emission and Diffusion Processes of Materials Released by the Fukushima Daiichi Atomic Power Plant Accident', 6 March 2012, Tokyo. http://bit.ly/2Vo00Lf (accessed 19 September 2018).

JAEA (Japan Atomic Energy Agency) (2017). Database for Radioactive Substance Monitoring Data. http://emdb.jaea.go.jp/emdb/en/ (accessed 19 September 2018).

Kajino, M., M. Ishizuka, Y. Igarashi, *et al.* (2016). Long-term assessment of airborne radiocesium after the Fukushima nuclear accident: re-suspension from bare soil and forest ecosystems. *Atmos. Chem. Phys.* **16**, 13149–72, doi:10.5194/acp-2016-270.

Kaneyasu, N., H. Ohashi, F. Suzuki, T. Okuda and F. Ikemori (2012). Sulfate aerosol as a potential transport medium of radiocesium from the Fukushima nuclear accident. *Environ. Sci. Technol.*, **46**, 5720–6.

Katata, G, M. Chino, T. Kobayashi, *et al.* (2015). Detailed source term estimation of the atmospheric release for the Fukushima Daiichi Nuclear Power Station accident by coupling simulations of an atmospheric dispersion model with an improved deposition scheme and oceanic dispersion model. *Atmos. Chem. Phys.*, **15**, 1029–70.

Kato, H., Y. Onda and T. Tesfaye (2012a). Depth distribution of ^{137}Cs, ^{134}Cs, and ^{131}I in soil profile after Fukushima Dai-ichi Nuclear Power Plant Accident. *J. Environ. Radioact.*, **111**, 59–64.

Kato, H., Y. Onda and T. Gomi (2012b). Interception of the Fukushima reactor accident-derived ^{137}Cs, ^{134}Cs and ^{131}I by coniferous forest canopies. *Geophys. Res. Lett.*, **39**, L20403, doi:10.1029/2012GL052928.

Kato, H., Y. Onda, K. Hisadome, N. Loffredo and A. Kawamori (2017). Temporal changes in radiocesium deposition in various forest stands following the Fukushima Dai-ichi Nuclear Power Plant accident. *J. Environ. Radioact.*, **166**, 449–57.

Katsuragi, Y. (1983). A study of ^{90}Sr fallout in Japan. *Pap. Met. Geophys.*, **33**, 277–91.

Kawamura, H., T. Kobayashi, A. Furuno, *et al.* (2011). Preliminary numerical experiments on oceanic dispersion of ^{131}I and ^{137}Cs discharged into the ocean because of the Fukushima Daiichi Nuclear Power Plant disaster. *J. Nucl. Sci. Technol.*, **48**, 1349–56, doi:80/18811248.2011.9711826.

Kinase, S., T. Takahashi, S. Sato, R. Sakamoto and K. Saito (2014). Development of prediction models for radioactive caesium distribution within the 80 km radius of the Fukushima Daiichi Nuclear Power Plant. *Radiat. Prot. Dosim.*, **160**(4), 318–21.

Kinoshita, N., K. Sueki, K. Sasa, *et al.* (2011). Assessment of individual radionuclide distributions from the Fukushima nuclear accident covering central-east Japan. *Proc. Natl. Acad. Sci. USA*, **108**(49),19526–9, doi:10.1073/pnas.1111724108.

Knolls Atomic Power Laboratory (2010). *Nuclides and Isotopes: Chart of Nuclides*, 17th edition. Niskayuna, NY: Knolls Atomic Power Laboratory.

Kogure, T., N. Yamaguchi, H. Segawa, *et al.* (2016). Constituent elements and their distribution in the radioactive Cs-bearing silicate glass microparticles released from Fukushima nuclear plant. *Microscopy (Oxf)*, **65**(5), 451–9, doi:10.1093/jmicro/dfw030.

Komura, K., M. Yamamoto, T. Muroyama, *et al.* (2000). The JCO criticality accident at Tokai-mura, Japan: an overview of the sampling campaign and preliminary results. *J. Environ. Radioact.*, **50**, 3–14.

Kulan, A. (2006). Seasonal ^{7}Be and ^{137}Cs activities in surface air before and after the Chernobyl event. *J. Environ. Radioact.*, **90**(2), 140–50, doi:10.1016/j.jenvrad.2006.06.010.

Maki, T., T. Y. Tanaka, T. T. Sekiyama, *et al.* (2013). Radioactive Nuclei Emission Analysis from Fukushima Dai-ichi Nuclear Power Plant by Inverse Model. 93rd American Meteorological Society Annual Meeting, Austin, Texas, USA, 6 January 2013, http://bit.ly/2Vp2Tvl (accessed 19 September 2019).

Masson, O. (2005). Radioecological Impact of Saharan Dusts Fallout: Case Study of a Major Event on the 21 of February 2004 in South Part of France (INIS-FR–08-1275). http://bit.ly/2Vpozrl (accessed 19 September 2019).

Matsuda, N, S. Mikami, S. Shimoura, *et al.* (2015). Depth profiles of radioactive cesium in soil using a scraper plate over a wide area surrounding the Fukushima Dai-ichi Nuclear Power Plant, Japan. *J. Environ. Radioact.*, **139**, 427–34.

Matsumoto, Y. (2007). Safety assessment and disaster prevention measures against the tsunami at Onagawa Nuclear Power Station. IAEA/JNES/NIED Seminar on Nuclear Disaster Prevention and General Disaster Prevention Against Earthquakes and Tsunamis, Tokyo, 4 December 2007. http://bit.ly/2VpoX9h (accessed 19 September 2019).

MEXT (2011). Measurement results of aircraft monitoring in Iwate, Shizuoka, Nagano, Yamanashi, Gifu and Toyama prefectures, and revision of the monitoring results considering the influence of natural nuclides. Ministry of Education, Culture, Sports, Science and Technology, http://bit.ly/2VtToeC (accessed 19 September 2019).

Mikami, S., T. Maeyama, Y. Hoshide, *et al.* (2015a). The air dose rate around the Fukushima Dai-ichi Nuclear Power Plant: its spatial characteristics and temporal changes until December 2012. *J. Environ. Radioact.*, **139**, 250–9.

Mikami, S., T. Maeyama, Y. Hoshide, *et al.* (2015b). Spatial distributions of radionuclides deposited onto ground soil around the Fukushima Dai-ichi Nuclear Power Plant and their temporal change until December 2012. *J. Environ. Radioact.*, **139**, 320–43.

Miyake, Y. (1954a). Anthropogenic radioactive rain fell in Japan (during May to July in 1954). *Tenmon to Kishou*, **20**, 1–8 (in Japanese).

Miyake, Y. (1954b). The artificial radioactivity in rain water observed in Japan from May to August, 1954. *Pap. Met. Geophys.*, **5**, 173–7.

Miyazawa, Y., Y. Masumoto, S. M. Varlamov, *et al.* (2013). Inverse estimation of source parameters of oceanic radioactivity dispersion models associated with the Fukushima accident. *Biogeosciences*, **10**, 2349–63.

NERH (2011a). Completion report of the roadmap step 2 towards convergence of accident, Tokyo Electric Power Company Fukushima Daiichi Nuclear Power Station. Nuclear Emergency Response Headquarters, 19 July 2011.

NERH (2011b). Completion report of the roadmap step 2 towards convergence of accident, Tokyo Electric Power Company Fukushima Daiichi Nuclear Power Station. Nuclear Emergency Response Headquarters, 16 December 2011.

NERH (2012). Report of the government of Japan to the IAEA Ministerial Conference on Nuclear Safety: on the accident of TEPCO Fukushima Nuclear Power Station. Nuclear Emergency Response Headquarters, www.kantei.go.jp/jp/topics/2011/iaea_houkokusho.html (accessed 19 September 2019).

Niimura, N., K. Kikuchi, N. D. Tuyen, M. Komatsuzaki and Y. Motohashi (2015). Physical properties, structure, and shape of radioactive Cs from the Fukushima Daiichi Nuclear Power Plant accident derived from soil, bamboo and shiitake mushroom measurements. *J. Environ. Radioact.*, **139**, 234–9.

NISA (2011). About some errors of radioactive material emission inventory data, http://bit.ly/2VppnMT (accessed 19 September 2018) (in Japanese).

NRA (Nuclear Regulation Authority) (2013). On implementation of aircraft monitoring in FY2013. Nuclear Regulatory Commission, http://bit.ly/2VkAiHw (accessed 19 September 2018).

NRA (Nuclear Regulation Authority) (2017). Extension site of distribution map of radiation dose, etc. http://ramap.jmc.or.jp/map/eng/ (accessed 19 September 2018).

NSC (2012). A series of Nuclear Safety Commission documents. Nuclear Safety Commission, http://bit.ly/2VkAvui (accessed 19 September 2018) (in Japanese).

Outola, I. and R. Saxén (2012). *Radionuclide Deposition in Finland 1961–2006*. Helsinki: Radiation and Nuclear Safety Authority.

Paatero, J., K. Hämeri, T. Jaakkola, *et al.* (2010). Airborne and deposited radioactivity from the Chernobyl accident: a review of investigations in Finland. *Boreal Env. Res.* **15**, 19–33.

Preston, D. L., Y. Shimizu, D. A. Pierce, A. Suyama and K. Mabuchi (2003). Studies of mortality of atomic bomb survivors: Report 13. Solid cancer and noncancer disease mortality: 1950–1997. *Radiat. Res.*, **160**, 381–407.

RERF (Radiation Effects Research Foundation) (2013). Solid cancer risks among atomic-bomb survivors, http://bit.ly/2VpqnR9 (accessed 19 September 2018).

Saito, K. and Y. Onda (2015). Outline of the national mapping projects implemented after the Fukushima accident. *J. Environ. Radioact.*, **139**, 240–9.

Saito, K., I. Tanihata, M. Fujiwara, *et al.* (2015). Detailed deposition density maps constructed by large-scale soil sampling for gamma-ray emitting radioactive nuclides from the Fukushima Dai-ichi Nuclear Power Plant accident. *J. Environ. Radioact.*, **139**, 308–19.

Sanada, Y. and T. Torii (2015). Aerial radiation monitoring around the Fukushima Dai-ichi nuclear power plant using an unmanned helicopter. *J. Environ. Radioact.*, **139**, 294–9.

Sanada, Y., T. Sugita, Y. Nishizawa, A. Kondo and T. Torii (2014). The aerial radiation monitoring in Japan after the Fukushima Daiichi nuclear power plant accident. *Progress in Nuclear Science and Technology*, **4**, 76–80, doi:10.15669/pnst.4.76.

Satou, Y., K. Sueki, K. Sasa, K. Adachi and Y. Igarashi (2016). First successful isolation of radioactive particles from soil near the Fukushima Daiichi Nuclear Power Plant. *Anthropocene*, **14**, 71–6, doi:10.1016/j.ancene.2016.05.001.

Satou, Y., K. Sueki, K. Sasa, *et al.* (2018). Two different types of radioactive cesium particles emitted at the early stage of the Fukushima Dai-ichi Nuclear Power Station accident. *Geochem. J.*, **52**, doi:10.2343/geochemj.2.0514.

Saunier, O., A. Mathieu, D. Didier, *et al.* (2013). An inverse modeling method to assess the source term of the Fukushima Nuclear Power Plant accident using gamma dose rate observations. *Atmos. Chem. Phys.*, **13**, 11403–21, doi:10.5194/acp-13-11403-2013.

SCJ (Japan Science Council) (2012). Recommendations 'To take a new step of radiation countermeasure: actions based on the scientific search of facts', 9 April 2012.

SCJ (2014). A review of the model comparison of transportation and deposition of radioactive materials released to the environment as a result of the Tokyo Electric Power Company's Fukushima Daiichi Nuclear Power Plant accident. Report of Committee on Comprehensive Synthetic Engineering, Science Council of Japan, www.scj.go.jp/ja/info/kohyo/pdf/kohyo-22-h140902-e1.pdf to www.scj.go.jp/ja/info/kohyo/pdf/kohyo-22-h140902-e7.pdf (accessed 19 September 2018).

Stallen, P. and Coppock, R. (1987). About risk communication and risky communication. *Risk Analysis* **7**, 413–414.

Stohl, A., P. Seibert, G. Wotawa, *et al.* (2012). Xenon-133 and caesium-137 releases into the atmosphere from the Fukushima Dai-ichi nuclear power plant: determination of the source term, atmospheric dispersion, and deposition. *Atmos. Chem. Phys.*, **12**, 2313–43, doi:10.5194/acp-12-2313-2012.

Takahashi, J., K. Tamura, T. Suda, R. Matsumura and Y. Onda (2015). Vertical distribution and temporal changes of ^{137}Cs in soil profiles under various land uses after the Fukushima Dai-ichi Nuclear Power Plant Accident. *J. Environ. Radioact.*, **139**, 351–61.

Takemura, T., H. Nakamura, M. Takigawa, *et al.* (2011). A numerical simulation of global transport of atmospheric particles emitted from the Fukushima Daiichi Nuclear Power Plant. *SOLA*, **7**, 101–4, doi:10.2151/sola/2011-026.

TEPCO (2011). On the submission of a report on the results of the tsunami survey at the Fukushima Daiichi Nuclear Power Station and Fukushima Daini Nuclear Power Station to the Nuclear and Industrial Safety Agency, Ministry of Economy, Trade and Industry. Press release, 8 July, www.tepco.co.jp/cc/press/11070802-j.html (accessed 19 September 2018).

TEPCO (2012). Results of nuclide analysis of fishes and shellfishes, inland waters within Fukushima Daiichi Nuclear Power Station. Supplement document, 21 August, http://bit.ly/2EABrVO (accessed 19 September 2018).

Terada, H., G. Katata, M. Chino and H. Nagai (2012). Atmospheric discharge and dispersion of radionuclides during the Fukushima Dai-ichi Nuclear Power Plant accident: Part II. Verification of the source term and analysis of regional-scale atmospheric dispersion. *J. Environ. Radioact.*, **112**, 141–54, doi:10.1016/j. jenvrad. 2012.05.023.

Torii, T., T. Sugita, C. E. Okada, M. S. Reed and D. J. Blumenthal (2013). Enhanced analysis methods to derive the spatial distribution of ^{131}I deposition on the ground by air-borne surveys at an early stage after the Fukushima Daiichi Nuclear Power Plant accident. *Health Phys.*, **105**(2), 192–200.

Tsumune, D., T. Tsubono, M. Aoyama and K. Hirose (2012). Distribution of oceanic ^{137}Cs from the Fukushima Daiichi Nuclear Power Plant simulated numerically by a regional ocean model. *J. Environ. Radioact.*, **111**, 100–8, doi:10.1016/j. jenvrad. 2011.10.007.

Tsumune, D., T. Tsubono, M. Aoyama, *et al.* (2013). One-year, regional-scale simulation of ^{137}Cs radioactivity in the ocean following the Fukushima Daiichi Nuclear Power Plant accident. *Biogeosciences*, **10**, 5601–17.

UNSCEAR (United Nations Scientific Committee on the Effects of Atomic Radiation) (2000). UNSCEAR 2000 report to the General Assembly, with scientific annexes, www.unscear.org/unscear/en/publications/2000_1.html (accessed 19 September 2018).

UNSCEAR (United Nations Scientific Committee on the Effects of Atomic Radiation) (2013). UNSCEAR 2013 report to the General Assembly, with scientific annexes, www.unscear.org/unscear/en/publications/2013_1.html (accessed 19 September 2018).

Yamaguchi, N., M. Mitome, K. Akiyama-Hasegawa, *et al.* (2016). Internal structure of cesium-bearing radioactive microparticles released from Fukushima nuclear power plant. *Sci. Rep.*, **6**, 20548, doi:10.1038/srep20548.

Yoshimura, K., Y. Onda and H. Kato (2015). Evaluation of radiocaesium wash-off by soil erosion from various land uses using USLE plots. *J. Environ. Radioact.*, **139**, 362–9.

Yoshimura, K., Y. Onda and T. Wakahara (2016). Time dependence of the ^{137}Cs concentration in particles discharged from rice paddies to freshwater bodies after the Fukushima Daiichi NPP accident. *Environ. Sci. Technol.* **50**, 4186–93, doi:10.1021/acs.est.5b05513.

Yumimoto, K., Y. Morino, T. Ohara, *et al.* (2016). Inverse modeling of the ^{137}Cs source term of the Fukushima Dai-ichi Nuclear Power Plant accident constrained by a deposition map monitored by aircraft. *J. Environ. Radioact.*, **164**, 1–12, doi:10.1016/ j.jenvrad.2016.06.018.

2

Estimation of Environmental Releases of Radioactive Materials

2.1 Release of Radioactive Materials into the Atmosphere

MASAMICHI CHINO AND HARUYASU NAGAI

The Fukushima Daiichi Nuclear Power Station (FDNPS) accident in Japan on 11 March 2011, which was triggered by a magnitude 9.0 earthquake that resulted in a tsunami, caused a month-long discharge of radioactive materials into the atmosphere. However, in the first stage of the accident, only monitoring cars near the FDNPS could collect monitoring data because of damage to the monitoring posts and stack monitor. The limited survey data from the monitoring cars from 12–13 March 2011 (NISA, 2011) showed that radioactive caesium and iodine were already detected at Okuma-machi and Namie-machi, close to the FDNPS around 8:00 JST on 12 March due to leakage from the containment vessel. In addition, increased air dose rates due to the deposition of radionuclides discharged by a hydrogen explosion at unit 1 were observed north of the FDNPS on 13 March.

Although it was an urgent task to assess the radiological doses to the public using both environmental monitoring data and computer simulations based on atmospheric dispersion modelling of radioactive materials, environmental monitoring data were limited, and systematic emergency monitoring only began on 15 March. Furthermore, the source terms essential for computer simulations (e.g. nuclides, release rates and duration) were also unavailable, although these data were expected to be provided from stack monitors and/or the Emergency Response Support System (ERSS) of the Japanese Nuclear and Industrial Safety Agency (NISA), which analyses reactor behaviour in severe accidents.

Thus, in cooperation with the Nuclear Safety Commission of Japan (NSC), a group of Japanese scientists at the Japan Atomic Energy Agency (JAEA; hereafter the JAEA group) attempted to estimate the source term. The method applied in this estimation was a reverse estimation by coupling environmental monitoring data with atmospheric dispersion simulations under the assumption of a unit release rate (1 Bq/h). The first result was obtained by the NSC on 23 March 2011. The NISA also attempted to estimate the source term using a severe accident analysis code. Both results were made available on 12 April by the NSC and NISA, and based on these results the FDNPS accident was identified as the highest level (level 7) on the International Nuclear Event Scale (INES) of the IAEA, which was the same classification given to the Chernobyl accident. After these activities, the JAEA group successively revised the source term based on new monitoring data (Chino *et al.*, 2011; 2016; Katata *et al.*, 2012; 2015; Terada *et al.*, 2012; Kobayashi *et al.*, 2013).

During the first year, some source term estimations were published, and they were compared in the workshop 'Reconstruction of Atmospheric Release and Dispersion Processes', which was held by the JAEA in March 2012 (JAEA, 2012). Furthermore, the United Nations Scientific Committee on the Effects of Atomic Radiation (UNSCEAR) published the *UNSCEAR 2013 Report* (UNSCEAR, 2014), in which 16 source term results were compared.

2.2 Reverse Estimation Method for the Source Term

MASAMICHI CHINO AND HARUYASU NAGAI

The reverse estimation described here follows the method of the JAEA group, that is coupling environmental monitoring data with atmospheric dispersion simulations, assuming a unit release rate (1 Bq/h). Release rates are obtained as the ratio of the measured-to-calculated air concentrations of nuclide i at the sampling points:

$$Q_{i,t} = M_{i,t}/C_{i,t}, \tag{1}$$

where $Q_{i,t}$ is the release rate (Bq/h) of nuclide i when discharged into the atmosphere during time period t with a constant release rate; $M_{i,t}$ is the measured air concentration (Bq/m^3) of nuclide i in the plume released during the time period t; and $C_{i,t}$ is the dilution factor (h/m^3) of nuclide i, which is equal to its air concentration in the plume released during the time period t at the measurement point calculated under the assumption of a unit release rate.

This method of using air concentration data is more reliable than methods that use air dose rates because it does not require additional assumptions regarding the composition of the radionuclides contributing to the dose. However, when air concentration data are not available, the release rates are estimated by comparing the observed spatial patterns and/or temporal changes of air dose rates from radionuclides on the ground surface (i.e. ground shine) with the calculated values. This method assumes the fractional composition of major radionuclides contributing to ground shine, for example ^{131}I, ^{132}Te (^{132}I), ^{133}I, ^{134}Cs and ^{137}Cs. To minimise the errors from the assumption, the radioactive composition is determined based on various datasets of dust samples over eastern Japan.

For atmospheric dispersion simulations, the JAEA group used their numerical simulation system, the worldwide version of the SPEEDI (WSPEEDI-II) (Terada and Chino, 2008).

2.3 Release Rates of Radionuclides from the FDNPS

MASAMICHI CHINO AND HARUYASU NAGAI

The time trend of the release rates of ^{131}I and ^{137}Cs discharged into the atmosphere from 12 March to 1 April 2011, estimated by the JAEA group, is depicted in Figure 2.1 (Katata *et al.*, 2015), which was partially revised by Chino *et al.* (2016). Data from 24 March to 1 April are by Terada *et al.* (2012). The estimated values are the sum of the releases from units 1 to 3. The release rates shown in Figure 2.1 are not normalised to the shutdown time; they are applicable at the release time. Note that the time zone used in the following sections is Japan Standard Time (JST = UTC + 9 h).

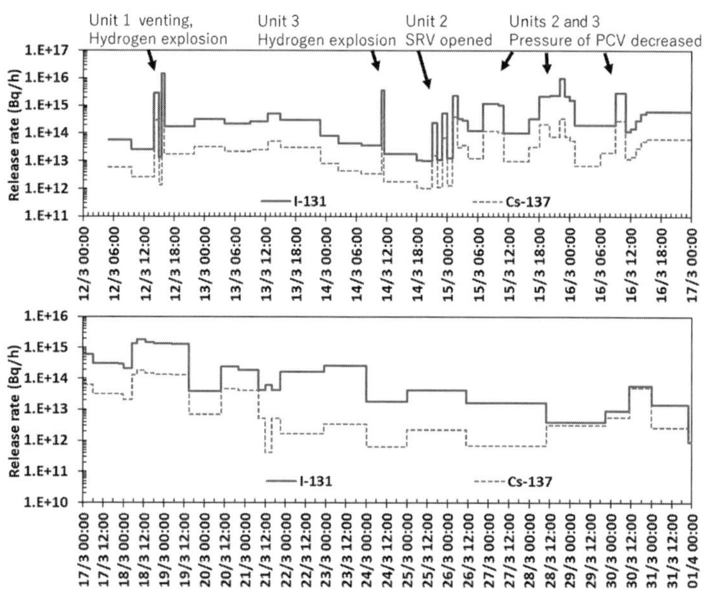

Figure 2.1 The time trend of the release rates of ^{131}I and ^{137}Cs discharged into the atmosphere from 12 March to 1 April 2011, estimated by the JAEA group (Katata *et al.*, 2015 which is partially revised by Chino *et al.*, 2016).

Data from 24 March to 1 April are by Terada *et al.* (2012). A black and white version of this figure will appear in some formats. For the colour version, refer to the plate section.

2.3.1 12 March to the Evening of 14 March

On the afternoon of 12 March, the wet venting started at 14:00, and an extreme decrease in the pressure in the primary containment vessel of unit 1 (PCV-U1) from 14:00 to 15:00 indicated an atmospheric discharge of radionuclides. The hydrogen explosion of unit 1 at 15:36 also discharged a huge amount of radionuclides into the atmosphere. According to the WSPEEDI-II simulation, the radioactive plume flowed north–northwest and increased the air dose rates downwind of the FDNPS.

On 13 March, venting operations were conducted to decrease the pressure of PCV-U3 at 9:24 and 12:30. Despite several venting operations, a hydrogen explosion also occurred in unit 3 at 11:01 on 14 March.

2.3.2 Night of 14 March to the Night of 15 March

The safety relief valve (SRV) of unit 2 was opened at 21:00 and 23:00 on 14 March and again at 1:00 on 15 March to decrease the pressure of the reactor pressure vessel; the temporal variation of observed downwind air dose rates showed three peaks due to these operations.

During the morning of 15 March, the pressure of PCV-U2 decreased between 7:00 and 12:00. This decrease corresponded with the extreme increase in the air dose rate observed at the main gate from 7:00 to 10:00, clearly indicating a huge release into the atmosphere. The second huge increase in the release rate on 15 March was in the evening. Wet venting was conducted at unit 3, corresponding to the decline in drywell pressure at unit 3 from 16:05, which finally stopped around 6:00 on 16 March. At the same time, the drywell pressure dropped rapidly at unit 2 from 18:00 on 15 March to 2:00 on 16 March. These facts suggest that the large release rate estimated during the evening originated from units 2 and 3.

2.3.3 16–31 March

The release rates remained high until 24 March. However, after 25 March the release rates gradually decreased until the beginning of April.

2.4 Evaluation of the Release Rates

MASAMICHI CHINO AND HARUYASU NAGAI

The UNSCEAR report summarised 16 estimates published on the release of radioactive material from the FDNPS. For each case, the method used and the date the estimate was made and/or the date of its publication were indicated.

According to the report, the earliest estimates were made in late March 2011, even while the accident was continuing. Others followed in subsequent months, with many being further refined over time as more information became available (relating to both the development of the accident and measurements in the environment) and as methods improved. The various estimates of the releases of the two principal radionuclides spanned relatively small ranges. For ^{131}I, the estimates ranged from approximately 100 to 500 PBq; for ^{137}Cs, the range was generally from 6 to 20 PBq.

The report also compared three temporal patterns of release based on published estimates (Stohl *et al.*, 2012; Terada *et al.*, 2012; Saunier *et al.*, 2013). While all three show peaks in the estimated release rate of ^{137}Cs that corresponded to the main events at the three reactors, these peaks were found to have occurred at different times, to have been of different durations and to have differed in magnitude at particular times by more than a factor of 10. Although the report chose to use the source term estimated by Terada *et al.* (2012), which was based on reverse or inverse modelling, the temporal variation in the source term still exhibited greater variations among the estimates, and further improvements were expected. For example, the release of ^{131}I and ^{137}Cs estimated by Terada *et al.* (2012) might have been underestimated because they used only measurement data on land, and the source term over the Pacific Ocean was interpolated from estimated values. Thus, after Terada *et al.* (2012), the JAEA group successively refined the method and used measurements of radioactive material over the Pacific Ocean and close to the FDNPS in addition to those over the Japanese mainland. Deposition modelling that considered various types of wet and dry deposition was also used (Kobayashi *et al.*, 2013; Katata *et al.*, 2015). The most recent total amounts estimated by Katata *et al.* (2015) are clearly larger (approximately 151 and 14.5 PBq for ^{131}I and ^{137}Cs, respectively) than those of Terada *et al.* (2012), although they fit within the range of the source term shown by the UNSCEAR report. For ^{137}Cs, Aoyama *et al.* (2015) also estimated the total amount of ^{137}Cs released into the atmosphere to be 15.2–20.4 PBq by coupling atmospheric and oceanic dispersion models and measurement data; ultimately, published source terms by many scientists almost all fall within this range.

For future work, we must develop a systematic source estimation system that includes environmental monitoring and atmospheric dispersion simulations and that can be used in near real time for future nuclear emergencies. For this purpose, the utilisation of aerial survey systems and online data communications is essential to quickly provide the system with measurement data. The environmental monitoring data obtained by many governmental institutions and universities are also necessary to validate the accuracy of the estimated source term. However, because these data were separately recorded in different formats during the FDNPS accident, it took much effort to find and convert them to a unified format in one database. Thus, information technology to construct a unified online database that contains various environmental monitoring data observed by many contributors is sorely needed.

2.5 Estimation of the Direct Release into the Ocean

DAISUKE TSUMUNE AND YUKIO MASUMOTO

A series of accidents at the FDNPS following the earthquake and tsunami of 11 March 2011 resulted in the release of radioactive materials into the ocean. The major radioactive materials were ^{131}I, ^{134}Cs and ^{137}Cs, which required emergency monitoring in the ocean. ^{137}Cs is one of the main radioactive materials used to estimate the released amount because ^{131}I has a short half-life (eight days), whereas that of ^{134}Cs is equal to that of ^{137}Cs.

Radioactive materials were emitted into the atmosphere and transferred to the land and ocean through wet and dry deposition. In addition, highly contaminated water was directly released into the ocean. Radioactive materials were released into the ocean via two major pathways: direct release from the site of the FDNPS accident and atmospheric deposition. It is important to distinguish the contributions of atmospheric deposition and direct release for the estimation of direct release. Direct release was visually observed on 2 April 2011. TEPCO estimated, based on visual observations, that radioactive materials were directly released during the five-day period from noon on 1 April 2011 to noon on 6 April 2011 (Japanese Government, 2011). Estimations of the beginning of the direct release have large uncertainties because of a lack of direct observational results. On 21 March 2011, TEPCO began measuring the activities of radioisotopes in the seawater adjacent to the discharge canal for reactors 5 and 6 (5–6 discharge canal) on the north side of the FDNPS site and the discharge canal for reactors 1–4 (south discharge canal) on the south side of the FDNPS site. Tsumune *et al.* (2012) analysed the ^{131}I/^{137}Cs activity ratio to differentiate the release pathways of radioactive materials for ^{137}Cs data because the ^{131}I/^{137}Cs activity ratio should not change during oceanic transport due to the weak interaction of both ^{131}I and ^{137}Cs with biogenic particles. However, the ratio might change due to the difference in the depositional behaviours of ^{131}I and ^{137}Cs during atmospheric transport. ^{131}I and ^{137}Cs were released into the atmosphere from the FDNPS reactors at very high temperatures in both gaseous and particulate forms. During transport in the atmosphere at normal temperatures, the released ^{131}I existed as a gas and as small particles less than 1 μm in size, whereas the released ^{137}Cs existed as larger particles with sizes of 1–2 μm. Because the wet deposition rate depends on the size of the particles, the ^{131}I/^{137}Cs activity ratio increases or decreases during atmospheric transport. The results of the analysis of the ^{137}Cs concentrations and $^{131}I/^{137}Cs$ activity ratios suggested that the direct release of ^{137}Cs from the FDNPS reactors occurred from 26 March to 6 April 2011. However, the direct release was apparently observed to occur only after 1 April because the radioactive dose near the screen of reactor no. 2 increased from 1.5 mSv/h on 1 April to over 1000 mSv/h on 2 April. These results suggest that there was another, non-visible pathway for the direct release from the FDNPS reactor. Moreover, we inferred that the ^{137}Cs directly released from the FDNPS reactor on 26 March was transported to the coast near the FDNPS site by 27 March and to 30 km offshore by 9 April (Tsumune *et al.*, 2013).

It was difficult to estimate the total amount of direct release using only measured data because the measured data were not sufficient during the period immediately following the accident. The total amount of direct release was estimated by several groups using different methods. Some of them used numerical simulations and compared their results with the measured data. The results with large uncertainties are summarised in Table 2.1.

TEPCO estimated that the directly released ^{137}Cs activity amounted to 0.94 PBq during the five-day period from noon on 1 April 2011 to noon on 6 April 2011 (Japanese Government, 2011). TEPCO calculated the flow rate (4.3 m^3/h) using visual information on distance and the height and diameter of the flow and then estimated the release rate of ^{137}Cs activity by multiplying the flow rate by the ^{137}Cs activity of the contaminated water (1.8×10^{12} Bq/m^3), resulting in 1.9×10^{14} Bq/day. They had visual information for five days and stopped visible leakage by injecting waterglass (a sodium silicate aqueous solution) into a pit near reactor no. 2 on 6 April 2011. TEPCO adopted the method of Tsumune *et al.* (2012) to estimate the amount of ^{137}Cs directly released from a port to the outside from 26 March to the end of September 2011 and then reported 3.6 PBq to the Japanese Government (TEPCO, 2012).

The JAEA estimated the direct release scenario of ^{137}Cs based on TEPCO's visual estimation (Kawamura *et al.*, 2011). They concluded that the release of ^{137}Cs activity amounted to 0.94 PBq from noon on 1 April to noon on 6 April 2011, according to TEPCO, and extended that estimate before and after the period in proportion to the measured activity adjacent to the FDNPS. Their estimate of the total release was 4 PBq from 21 March to 30 April 2011. They did not distinguish between direct release and atmospheric deposition in their estimate of the total amount of ^{137}Cs released. The JAEA validated their simulated results from their estimated scenario in comparison with measured ^{137}Cs activities at the Fukushima Daini nuclear power plant (FDNPP) and along the Iwasawa coast.

The Central Research Institute of Electric Power Industry (CRIEPI) applied a simple release scenario to estimate the direct release rate over the entire 12 days. In

Table 2.1 *Estimated total amount of ^{137}Cs activity directly released into the ocean by release scenario.*

Organisation	Period	Total amount of directly released ^{137}Cs (PBq)	Method	Reference
TEPCO	Noon 1 April 2011 to noon 6 April 2011	0.94	Flow rate estimated by visual observation × measured activity of contaminated water	Japanese Government, 2011
TEPCO	26 March 2011 to 30 September 2011	3.6 (11 for ^{131}I, 3.5 for ^{134}Cs)	Based on the method by Tsumune *et al.* (2012)	TEPCO, 2012
JAEA	21 March 2011 to 30 April 2011	4 (11 for ^{131}I)	Based on the estimation by TEPCO (1–6 April) and the expansion period (21 March to 30 April) in proportion to measured activity adjacent to the FDNPS*	Kawamura *et al.*, 2011
CRIEPI	26 March 2011 to 31 May 2011	3.5 ± 0.7	Inverse method based on averaged measured activity from 26 March to 6 April adjacent to the FDNPS*	Tsumune *et al.*, 2012
CRIEPI	26 March 2011 to 29 February 2012	3.6 ± 0.7 (11.1 ± 2.2 for ^{131}I; 3.5 ± 0.7 for ^{134}Cs)	Based on the method by Tsumune *et al.* (2012)	Tsumune *et al.*, 2013
Miyazawa *et al.*, 2013	21 March 2011 to 30 April 2011	5.5–5.9	Inversion method based on measured activity mainly adjacent to the FDNPS* and others	Miyazawa *et al.*, 2013
IRSN	25 March 2011 to 18 July 2011	27 (12-41)	Estimation of inventory in the ocean by observations (11 April to 30 June) and the expansion period (25 March to 18 July) in proportion to measured activity adjacent to the FDNPS*	Bailly du Bois *et al.*, 2012
Sirocco group	20 March 2011 to 30 June 2011	5.1–5.5	Inversion method based on measured activity adjacent to the FDNPS*	Estournel *et al.*, 2012
Rypina *et al.*, 2013	21 March 2011 to 30 June 2011	16.2 ± 1.6	Minimising the model–data mismatch based on the KOK cruise data (4–18 June) and the expansion period (25 March to 18 July) in proportion to measured activity adjacent to the FDNPS*	Rypina *et al.*, 2013

* Adjacent to the FDNPS; at the 5–6 and south discharge canals.

this scenario, the release of ^{137}Cs into an area covered by a mesh in front of the FDNPS site throughout all vertical layers was simulated at a constant rate of 1 Bq/s during the 12 days from 26 March to 6 April. They then calculated a scaling factor between the observed and simulated concentrations, with a constant release rate of 1 Bq/s to determine the simulated average ^{137}Cs concentrations in the mesh from 26 March to 6 April, which were equal to the ^{137}Cs concentrations observed at the 5–6 and south discharge canals near the FDNPS site. The resulting scaling factor, 2.55×10^9, was used to estimate the direct release rate of ^{137}Cs from 26 March to 6 April, which was found to be 220 TBq/day (2.2×10^{14} Bq/day). The observed ^{137}Cs concentrations decreased exponentially after 6 April. The CRIEPI converted the exponential fitted ^{137}Cs activity to the direct release rate. Their estimate of the total release of ^{137}Cs was 3.5 ± 0.7 PBq by the end of May 2011 (Tsumune *et al.*, 2012) and 3.6 ± 0.7 PBq by the end of February 2012 (Tsumune *et al.*, 2013).

The Institut de Radioprotection et de Sûreté Nucléaire (IRSN) in France estimated the direct release scenario based on measured ^{137}Cs activities from 11 April to 4 July 2011 (Bailly du Bois *et al.*, 2012). Their estimate of the total release was 27 PBq (12–41) from 25 March to 18 July 2011. They estimated the total inventory to be 11.6 PBq on 14 April 2011, by interpolating from the measured data. They then extrapolated the inventory of 11.6 PBq on 14 April to 22 PBq on 8 April 2011. They extended the estimate from 25 March to 18 July 2011, in proportion to the measured ^{137}Cs activity adjacent to the FDNPS site. The estimated total amount of directly released ^{137}Cs was 27 PBq from 25 March to 18 July 2011. Estournel *et al.* (2012) noted that the inventory of 11.6 PBq on 14 April is an overestimate because of a lack of measured activities in the northern part of the FDNPS site. Estournel *et al.* (2012) also indicated that temporal extrapolation is unreasonable because their own simulations indicated that the total inventory did not decrease between 8 and 14 April 2011. The IRSN did not show a comparison between measured ^{137}Cs activities and the ones simulated based on their estimated scenario.

The Sirocco group at Toulouse University in France estimated that the directly released ^{137}Cs amounted to 5.1–5.5 PBq (Estournel *et al.*, 2012). They used an inversion method based on measurements adjacent to the FDNPS site to estimate that the directly released ^{137}Cs amounted to 4.1–4.5 PBq. They added 1 PBq to their estimation because the simulated results were underestimated offshore compared to the measured ^{137}Cs activities.

Miyazawa *et al.* (2013) estimated that the directly released ^{137}Cs amounted to 5.5–5.9 PBq using an inversion method based on measurements not only adjacent to the FDNPS but also at other measurement sites, mainly at the FDNPP.

Rypina *et al.* (2013) estimated that the directly released ^{137}Cs amounted to 16.2 ± 1.6 PBq. Their estimate was based on minimising the model–data mismatch. The data were collected by a cruise of the research vessel *Ka'imikai-o-Kanaloa* (KOK), which covered a wider area of the open ocean during 4–18 June 2011 (Buesseler *et al.*, 2011). They estimated the release scenario for the period in proportion to the measured ^{137}Cs activity adjacent to the FDNPS.

In addition, a few studies have shown estimations of the total amount of direct release without evoking a release scenario. Dietze and Kriest (2012) used numerical simulations to estimate that the total amount of directly released ^{137}Cs was 0.94–3.5 PBq. They noted that their simulated results were not consistent with the IRSN estimate of 27 PBq for the total amount of ^{137}Cs released. Kanda (2013) estimated the direct release rate to the main harbour at the FDNPS site using the measured data in the main harbour from 3 April to 30 September 2011, and the exchange rate of harbour water with the surrounding seawater. Kanda (2013) concluded that the approach of his study resulted in an estimate fairly consistent with the CRIEPI and JAEA estimations. Kanda (2013) pointed out that direct release may continue for more than two years. Therefore, estimation of the direct release rate and the total amount of released ^{137}Cs is important for longer time scales than those discussed in previous studies.

These discrepancies reflect differences in the durations of the measured data for estimating the direct release scenario and in the analytical methods. The two larger estimations by the IRSN and Rypina *et al.* (2013) considered the measured ^{137}Cs activities for estimation from 11 April to 30 June 2011, and from 4 to 18 June 2011, respectively. They performed backward-in-time extrapolation because they did not include the major direct release period from 25 March to 6 April 2011, which might have increased the uncertainties in their estimations. In contrast, other studies considered the measured ^{137}Cs activities during the major direct release period and found results of 3–6 PBq.

References

Aoyama, M., M. Kajino, T. Tanaka, *et al.* (2015). ^{134}Cs and ^{137}Cs in the North Pacific Ocean derived from the March 2011 TEPCO Fukushima Dai-ichi Nuclear Power Plant accident, Japan. Part two: estimation of ^{134}Cs and ^{137}Cs inventories in the North Pacific Ocean. *J. Oceanogr.*, **72**, 53–65, doi:10.1007/s10872-015-0332-2.

Bailly du Bois, P., P. Laguionie, D. Boust, *et al.* (2012). Estimation of marine source-term following Fukushima Dai-ichi accident. *J. Environ. Radioact.*, **114**, 2–9, doi:10.1016/j.jenvrad.2011.11.015.

Buesseler, K., M. Aoyama and M. Fukasawa (2011). Impacts of the Fukushima nuclear power plants on marine radioactivity. *Environ. Sci. Technol.*, **45**, 9931–5, doi:10.1021/es202816c.

Chino, M., H. Nakayama, H. Nagai, *et al.* (2011). Preliminary estimation of release amounts of ^{131}I and ^{137}Cs accidentally discharged from the Fukushima Daiichi nuclear power plant into atmosphere. *J. Nucl. Sci. Technol.*, **48**(7), 1129–34.

Chino, M., H. Terada, H. Nagai, *et al.* (2016). Utilization of $^{134}Cs/^{137}Cs$ in the environment to identify the reactor units that caused atmospheric releases during the Fukushima Daiichi accident. *Sci. Rep.*, **6**, 31376, doi:10.1038/srep31376.
Dietze, H. and I. Kriest (2012). ^{137}Cs off Fukushima Dai-ichi, Japan: model based estimates of dilution and fate. *Ocean Sci.*, **8**, 319–32, doi:10.5194/os-8-319-2012.
Estournel, C., E. Bosc, M. Bocquet, *et al.* (2012). Assessment of the amount of Cesium-137 released into the Pacific Ocean after the Fukushima accident and analysis of its dispersion in Japanese coastal waters. *J. Geophys. Res. Oceans*, **117**, C11014, doi:10.1029/2012JC007933.
JAEA (2012). 'Reconstruction of Atmospheric Release and Dispersion Processes', JAEA workshop, Tokyo (6 March 2012). http://bit.ly/2Vo00Lf (accessed 19 September 2018).
Japanese Government (2011). Report of Japanese Government to the IAEA Ministerial Conference on Nuclear Safety: the accident at TEPCO's Fukushima nuclear power stations. http://bit.ly/2VkB4nU (accessed 19 September 2018).
Kanda, J. (2013). Continuing ^{137}Cs release to the sea from the Fukushima Dai-ichi Nuclear Power Plant through 2012. *Biogeosciences*, **10**(9), 6107–13, doi:10.5194/bg-10-6107-2013.
Katata, G., M. Ota, H. Terada, M. Chino and H. Nagai (2012). Atmospheric discharge and dispersion of radionuclides during the Fukushima Dai-ichi Nuclear Power Plant accident, Part I: source term estimation and local-scale atmospheric dispersion in early phase of the accident. *J. Environ. Radioact.*, **109**, 103–13.
Katata, G., M. Chino, T. Kobayashi, *et al.* (2015). Detailed source term estimation of the atmospheric release for the Fukushima Daiichi Nuclear Power Station accident by coupling simulations of an atmospheric dispersion model with an improved deposition scheme and oceanic dispersion model. *Atmos. Chem. Phys.*, **15**, 1029–70.
Kawamura, H., T. Kobayashi, A. Furuno, *et al.* (2011). Preliminary numerical experiments on oceanic dispersion of ^{131}I and ^{137}Cs discharged into the ocean because of the Fukushima Daiichi nuclear power plant disaster. *J. Nucl. Sci. Technol.*, **48**(11), 1349–56, doi:80/18811248.2011.9711826.
Kobayashi, T., H. Nagai, M. Chino and H. Kawamura (2013). Source term estimation of atmospheric release due to the Fukushima Dai-ichi Nuclear Power Plant accident by atmospheric and oceanic dispersion simulations. *J. Nucl. Sci. Technol.*, **50**, 255–64.
Miyazawa, Y., Y. Masumoto, S. M. Varlamov, *et al.* (2013). Inverse estimation of source parameters of oceanic radioactivity dispersion models associated with the Fukushima accident. *Biogeosciences*, **10**, 2349–63, doi:10.5194/bg-10-2349-2013.
NISA (2011). Results of the emergency environmental monitoring around TEPCO's Fukushima Daiichi Nuclear Power Station and Fukushima Daini Nuclear Power Station. http://bit.ly/2VpUyHJ (accessed 19 September 2018).
Rypina, I. I., S. R. Jayne, S. Yoshida, *et al.* (2013). Short-term dispersal of Fukushima-derived radionuclides off Japan: modeling efforts and model–data intercomparison. *Biogeosciences*, **10**(7), 4973–90, doi:10.5194/bg-10-4973-2013.
Saunier, O., A. Mathieu, D. Didier, *et al.* (2013). An inverse modeling method to assess the source term of the Fukushima Nuclear Power Plant accident using gamma dose rate observations. *Atmos. Chem. Phys.*, **13**, 11403–21.
Stohl, A., P. Seibert, G. Wotawa, *et al.* (2012). Xenon-133 and caesium-137 releases into the atmosphere from the Fukushima Dai-ichi nuclear power plant: determination of the source term, atmospheric dispersion, and deposition. *Atmos. Chem. Phys.*, **12**, 2313–43.

TEPCO (2012). *The Fukushima Nuclear Accidents Investigation Report.* http://bit.ly/2VxgfWH (accessed 19 September 2018).

Terada, H. and M. Chino (2008). Development of an atmospheric dispersion model for accidental discharge of radionuclides with the function of simultaneous prediction for multiple domains and its evaluation by application to the Chernobyl nuclear accident. *J. Nucl. Sci. Technol.*, **45**, 920–31.

Terada, H., G. Katata, M. Chino and H. Nagai (2012). Atmospheric discharge and dispersion of radionuclides during the Fukushima Dai-ichi Nuclear Power Plant accident, Part II: verification of the source term and analysis of regional-scale atmospheric dispersion. *J. Environ. Radioactiv.*, **112**, 141–54.

Tsumune, D., T. Tsubono, M. Aoyama and K. Hirose (2012). Distribution of oceanic ^{137}Cs from the Fukushima Dai-ichi Nuclear Power Plant simulated numerically by a regional ocean model. *J. Environ. Radioact.*, **111**, 100–8, doi:10.1016/j.jenvrad.2011.10.007.

Tsumune, D., T. Tsubono, M. Aoyama, *et al.* (2013). One-year, regional-scale simulation of ^{137}Cs radioactivity in the ocean following the Fukushima Dai-ichi Nuclear Power Plant accident. *Biogeosciences*, **10**(8), 5601–17, doi:10.5194/bg-10-5601-2013.

UNSCEAR (2014). Levels and effects of radiation exposure due to the nuclear accident after the 2011 great east-Japan earthquake and tsunami. In *Effects and Risks of Ionizing Radiation. UNSCEAR 2013 Report*, Vol. 1. New York: United Nations.

3

Diffusion in the Atmosphere

3.1 The Atmospheric Transport Process for Radioactive Substances and the Effects of Meteorological Conditions

HISASHI NAKAMURA, YU MORINO AND MASAYUKI TAKIGAWA

In nuclear power plant accidents, enormous amounts of radioactive substances are released over a relatively short period of time (several hours to days). The direction and range over which the substances are dispersed and the amount deposited on the ground surface are influenced not only by the amount that is released but also by meteorological conditions at the time of and immediately following an accident. The most important meteorological factors include wind direction, wind speed and precipitation. Whereas wind direction and speed directly affect the atmospheric transport of radioactive substances, precipitation is the predominant factor that controls the removal of these substances from the atmosphere by wet deposition.

Radioactive substances released from a point source diffuse and disperse three-dimensionally via countless microscale eddies generated throughout the atmosphere, resulting in a decrease in the concentrations over time. At higher altitudes, radioactive substances are also subject to mixing by larger-scale eddies associated with cyclones and anticyclones and by the strong wind shear on the flanks of the westerly jet streams. Even under such mixing and diffusion, the substances released into the atmosphere are not necessarily transported isotropically from the point source. Because the substances released are mostly advected by local winds, their concentrations are particularly elevated downstream of the source. Thus, atmospheric transport of radioactive substances involves wind advection, mixing and diffusion, as well as removal from the atmosphere by wet and dry deposition.

Except for the extreme case of the Chernobyl accident, a nuclear power plant accident such as the Fukushima Daiichi Nuclear Power Station (FDNPS) accident often leads to the release of radioactive substances that primarily remain within the planetary boundary layer (PBL), the lowermost layer extending to approximately 1 km above the Earth's surface. The near-surface wind distribution around the power plant is critical for the transport of these substances. Because winds near the surface are relatively weak, long-distance transport of the emitted substances is unlikely. High-concentration areas with the potential to cause severe environmental impacts are likely to be limited to the vicinity of the source, although the specific distribution can be substantially influenced by the local wind system, including land/sea breezes, and by local terrain. In addition, the boundary layer tends to deepen with increasing wind speed and decreasing static stability (stratification based on the vertical temperature distribution). For example, when the area around a power plant is covered by a high-pressure system in the cold season, winds near the ground surface weaken at night or in the early morning, and the ground is cooled strongly by radiative cooling under the clear sky. Under such circumstances, increased static stability suppresses the vertical mixing of radioactive substances in the PBL, while the dry and calm conditions prevent the substances being removed through wet deposition and advected horizontally, respectively. Therefore, the likelihood that the released substances will remain in the near-surface atmosphere within the vicinity of the power plant increases. Such atmospheric conditions are known to favour photochemical smog formation in urban areas and are also applicable to nuclear power plant accidents. In contrast, radioactive substances will be well mixed vertically within the PBL under strong winds and/or surface warming due to insolation.

Furthermore, a portion of the radioactive substances near the top of the PBL may be left as a polluted air mass in the free atmosphere as the PBL height decreases at night. Alternatively, another portion may be carried upward into the free atmosphere by ascending motions associated with a cyclone. Once reaching the free atmosphere, these substances can be transported over longer distances by strong winds. However, if a large amount of radioactive material is explosively released and reaches the free atmosphere *en masse*, as was the case in the Chernobyl accident, there is a risk of serious contamination over extensive areas.

3.2 Atmospheric Transportation and Deposition of the Radioactive Materials

YU MORINO, MASAYUKI TAKIGAWA AND HISASHI NAKAMURA

In terms of the environmental (in this case, human) impacts of radioactive substances released into the atmosphere, the most important are short-term internal exposure due to inhalation of the suspended materials (especially ^{131}I) through respiratory organs and long-term external exposure to or ingestion of materials deposited from the atmosphere onto the ground or ocean surface (especially ^{134}Cs and ^{137}Cs). To understand these long-term effects on land, it is necessary to study the movement of these materials from the soil surface to deeper soil layers; deposition from the atmosphere to forests, pastures and agricultural fields; and migration to the ocean via groundwater and rivers over long time periods and a wide geographical area. In addition, it is necessary to accurately assess the distribution of atmospheric deposition immediately following the release of the materials to establish the initial conditions that will enable subsequent estimation of radioactive material concentrations in the various pools.

Among the radioactive substances released from the FDNPS, it is believed that radiocaesium will have the most long-lasting impact. Fallout monitoring by prefectures (Nuclear Regulation Authority, 2011) and airborne monitoring conducted by the Ministry of Education, Culture, Sports, Science and Technology (MEXT, 2011) have provided invaluable information regarding changes at specific locations over time and spatial distributions at given time points.

Data from fallout monitoring by prefectures revealed that elevated ^{131}I and ^{137}Cs deposition was observed at all locations during the period 19–24 March 2011 (maximum deposition of 100 kBq ^{131}I per m^2/day and 10 kBq ^{137}Cs per m^2/day), and while elevated deposition was also observed on 30 March, 8–10 April and 18–20 April (maximum deposition of 1 kBq ^{131}I per m^2/day and 0.1–1.0 kBq ^{137}Cs per m^2/day), the measured levels were more than an order of magnitude lower than those observed during 19–24 March. Meanwhile, data from airborne monitoring revealed that large amounts of ^{134}Cs and ^{137}Cs were deposited in various locations, including the area between Iitate Village and Fukushima City, northwest of the accident site, central Fukushima Prefecture, the northern portion of Gunma and Tochigi Prefectures, the northern and southern portions of Ibaraki Prefecture and the northern portion of Miyagi Prefecture, whereas the deposition of ^{137}Cs in the remainder of the country was 10 kBq/m^2 or less, except for the Tohoku and Kanto regions (Figure 3.1(a)).

The utility of these empirical data is limited in terms of analysing deposition mechanisms. For example, no periodic fallout monitoring data exist for the period prior to 17 March. Furthermore, because the airborne monitoring data represent accumulated deposition, they do not provide any information regarding the factors leading to the formation of localised hotspots or when such hotspots were formed. To overcome these limitations, combined analysis using these data and numerical simulations based on an atmospheric diffusion model yields useful information. In the simulations, a numerical calculation programme (Figure 3.2) is used to simulate wind, precipitation, transport and diffusion of suspended radioactive substances, and deposition to the Earth's surface in a virtual space.

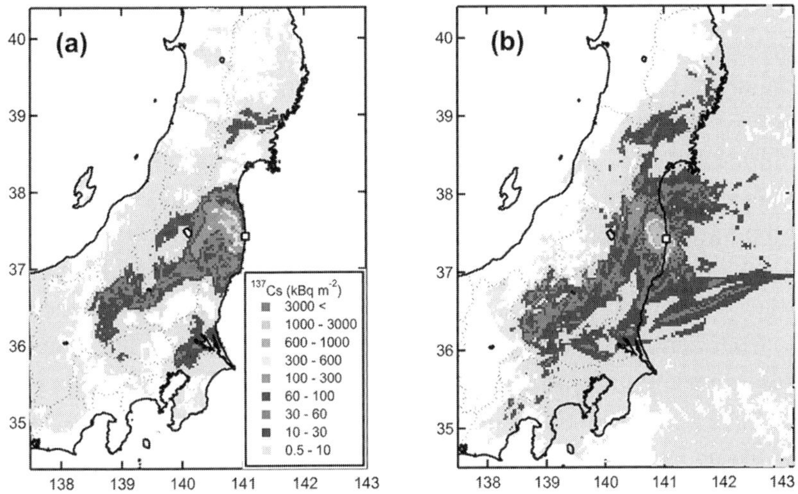

Figure 3.1 Estimated cumulative ^{137}Cs deposition based on (a) airborne monitoring survey and (b) atmospheric simulation (Morino *et al.*, 2013). Simulated period was from 11 March to 20 April. Reprinted with permission from Morino *et al.* (2013). Episode analysis of deposition of radiocaesium from the Fukushima Daiichi nuclear power plant accident. Copyright 2013 American Chemical Society.

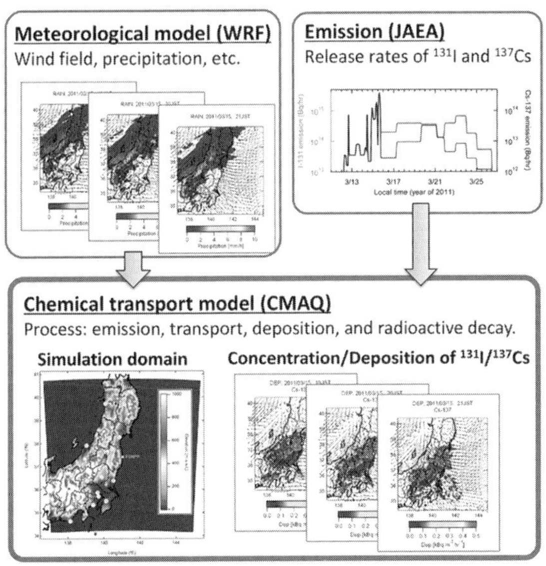

Figure 3.2 Schematic diagram of the atmospheric transport simulation.

Numerical simulations of atmospheric diffusion (Figure 3.1(b); adapted from Morino *et al.*, 2013) and airborne monitoring (Figure 3.1(a)) identified essentially the same areas as high-deposition areas: the area to the northwest of the FDNPS, central Fukushima, the three northernmost prefectures in the Kanto region, southern Ibaraki Prefecture to northwestern Chiba Prefecture, southern and northern Miyagi Prefecture, Saitama Prefecture and western Tokyo. The atmospheric diffusion model simulated ^{137}Cs deposition in high-deposition areas (receiving 10 kBq/m^2 or greater) to within one order of magnitude. Furthermore, the atmospheric diffusion model did a fair job of estimating daily deposition rates relative to measurements obtained from the prefectural fallout monitoring. The atmospheric diffusion model did a particularly good job of qualitatively simulating the increase in ^{137}Cs deposition in the Tohoku and Kanto regions between 20 and 23 March, as well as the fluctuation and spatial distribution of deposition during subsequent 10-day periods. Considering that the atmospheric diffusion simulations are based on estimates for the amount of material released over time calculated by Chino *et al.* (see Chapter 2), the simulation results were reasonable in terms of the released amount, at least for the periods during which the plume was moving over land. Estimates for ^{137}Cs deposition generated by the numerical simulations were consistent with observed values to within a factor of 10. Considering that errors on the order of a factor of 2–5 are typical for atmospheric diffusion simulations, the release amount estimates by Chino *et al.* (2011), while quantitatively including uncertainty, were qualitatively reasonable, at least for the period starting on 20 March.

The results of the numerical simulation of atmospheric diffusion indicated that ^{137}Cs deposition on land in Japan was concentrated during two periods from 15–16 March and 20–23 March and occurred mostly as wet deposition. These periods are consistent with the periods of the atmospheric ^{137}Cs plumes (Figure 3.3; Nakajima *et al.*, 2017), identified from the observation of hourly ^{137}Cs concentration at 99 stations (see Section 3.7). In the following we discuss the characteristics of the events in which atmospheric ^{137}Cs plumes were transported to and ^{137}Cs deposited on land in Japan.

3.2.1 Meteorological Conditions and Transport of Radioactive Substances, 11–16 March 2011

On 12 March, the first ^{137}Cs plume was detected in northwestern Fukushima. During this event, migrating high-pressure systems were passing the Japanese islands, and southeasterly to southerly winds transported ^{137}Cs plumes from the FDNPS to northeastern Fukushima Prefecture (Kitayama *et al.*, 2018). The ^{137}Cs concentration at the Haramachi site (25 km north of the FDNPS) reached a maximum concentration of 575 Bq/m^3 at 22:00–23:00 local time (LT). During this time frame, there was no rainfall, and thus the amount of ^{137}Cs deposition was small.

The greatest amount of radioactive material is believed to have been released from the FDNPS into the atmosphere between the evening of 14 March and the morning of 16 March (see Figure 2.1). The large amount of radioactive substances released on the morning of 15 March was carried inland from the Fukushima

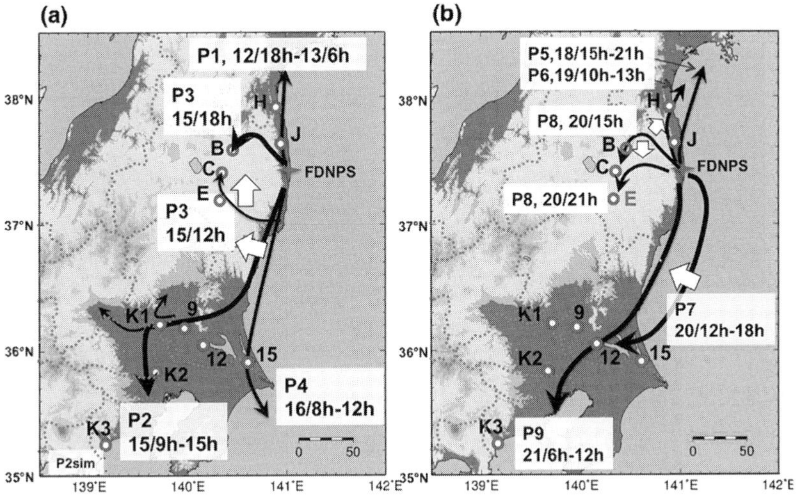

Figure 3.3 Schematic diagrams of transport routes of ^{137}Cs plumes during (a) 12–16 March and (b) 18–21 March. Thick arrows indicate the general trend of movement of the plumes (Nakajima *et al.*, 2017).

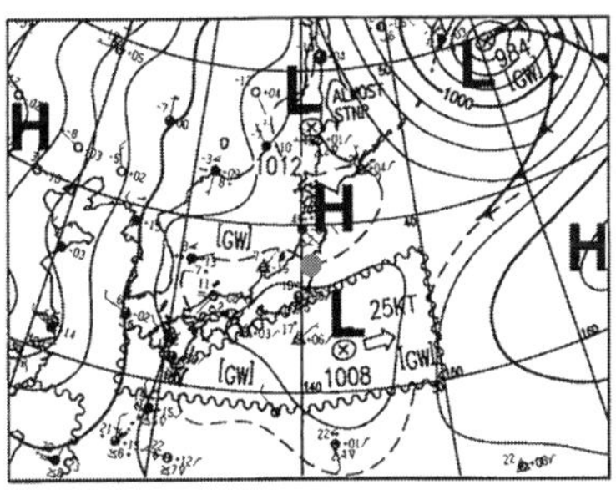

Figure 3.4 Surface weather map for 9:00 JST on 15 March 2011, based on analysis by the Japan Meteorological Agency (Takemura *et al.*, 2011). Easterly surface winds are prevailing in the region near the FDNPS (grey circle), which is located between the low pressure system (L) in the south and the weak high pressure system (H) in the north.

coastline to the northwest. In addition, a portion of this material underwent long-distance transport to North America and Europe (Takemura *et al.*, 2011). This movement can be largely attributed to the passing of a low-pressure system along the southern coast of Japan. According to the surface weather chart (Figure 3.4), a large low-pressure zone associated with a low-pressure trough spread to the northwest of this low-pressure system.

The maximum height attained by the radioactive substances released from sources including the unit 1 and unit 2 reactors on 15 March was relatively low and was estimated by the Tokyo Electric Power Company (TEPCO) to be on the

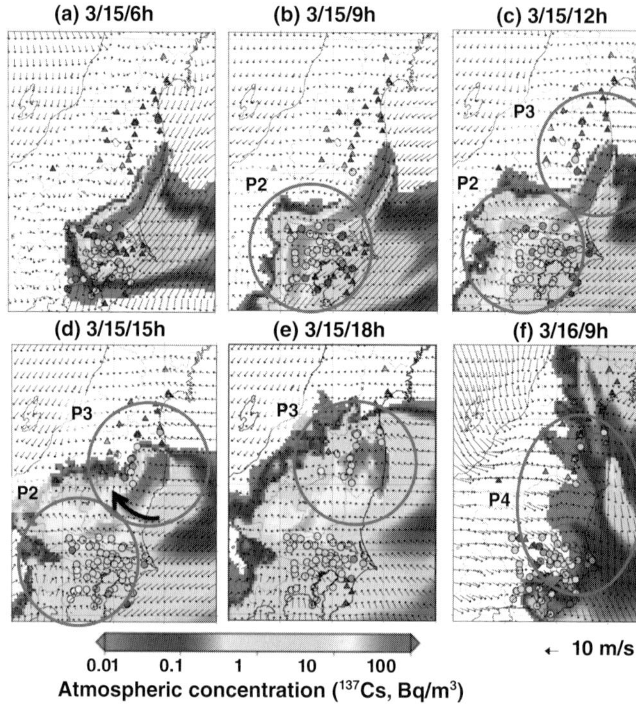

Figure 3.5 Maps of observations (coloured circles) and model ensemble means of atmospheric ^{137}Cs concentration for 15–16 March (Nakajima *et al.*, 2017). Areas identified as plumes with concentration larger than 10 Bq/m^3 are encircled and labelled with the name of the plume. JMA-MANAL wind vectors at 1000 hPa are superimposed. A black and white version of this figure will appear in some formats. For the colour version, refer to the plate section.

order of 30 m. Therefore, most of the released material was retained in the PBL and carried inland by low-level winds. Radioactive substances released up to the early morning on 15 March were carried to Ibaraki and Tochigi Prefectures via the Fukushima coast by northerly to northeasterly winds accompanying the low-pressure system that was passing along Japan's southern coast. During this time frame, the lowland areas of the Kanto region did not receive any rainfall, and although a transient increase in the ^{137}Cs concentration was observed in various locations (Figure 3.5(a)), it is believed that relatively little radioactive caesium was deposited on the ground surface. In fact, although the atmospheric concentration of ^{131}I measured at Tokai Village in Ibaraki Prefecture by the Japan Atomic Energy Agency (JAEA) reached 1600 Bq/m^3 between 6:00 and 9:00 on 15 March, the concentration fell rapidly to 39 Bq/m^3 between 9:00 and 15:00. After that, radioactive plumes were transported to the Chiba and Saitama Prefectures (Figure 3.5(b,c)). Based on these observations, we believe

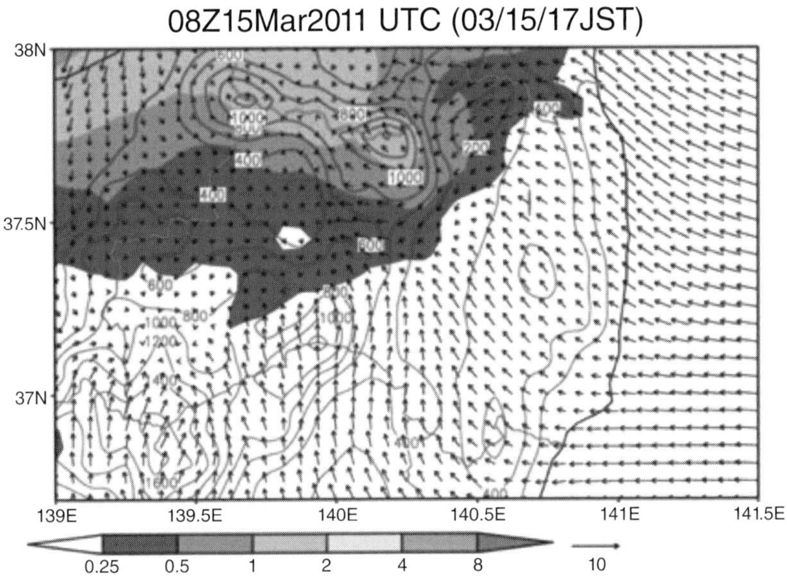

Figure 3.6 Surface wind speed (arrows, m/s) and hourly precipitation rate (shaded areas, mm) as of 17:00 JST on 15 March 2011; analysis by the Japan Meteorological Agency. Contour lines indicate surface elevation (m).

that, although a portion of the radioactive substances carried in the plume that passed the area during the morning of 15 March was deposited on the ground due to the effect of near-surface eddies and gravitational settling, because there was no rainfall the majority of the suspended material was not removed from the atmosphere and was transported inland to the northern section of the Kanto Plain.

On the afternoon of 15 March, when the low-pressure area to the northwest approached land, winds in the northern Kanto region changed to southerly winds, carrying the radioactive material from northern Kanto to central Fukushima Prefecture (Figure 3.5(d)). Simultaneously, the winds in the vicinity of the FDNPS rotated clockwise from northerly to southeasterly winds. Thus, the majority of radioactive material released in the morning and just past noon was carried northwest of the power plant, in the direction of Iitate Village. A large portion of the radioactive material that was carried inland within the PBL is believed to have been effectively removed by the precipitation (snowfall) observed widely in Fukushima Prefecture and the mountainous regions of northern Kanto between the afternoon of 15 March and the early morning of the following day, resulting in the creation of hotspots via wet deposition (Figure 3.6). According to simulations based on the atmospheric diffusion model, the deposition of ^{137}Cs peaked in various regions around 19:00 on 15 March.

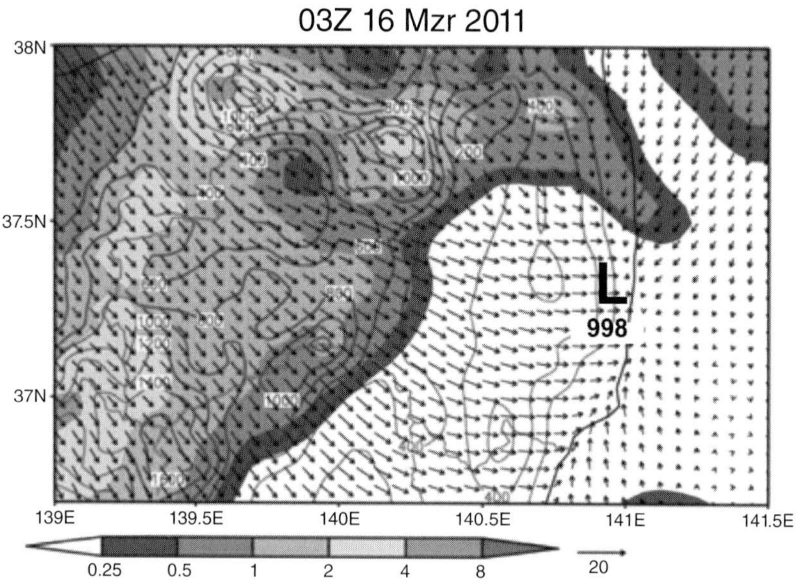

Figure 3.7 Surface wind speed (arrows, m/s) and hourly precipitation rate (shaded areas, mm) as of 12:00 JST on 16 March 2011, according to analysis by the JMA. The L indicates the meso cyclone (998 hPa) that formed around noon, and the contour lines indicate surface elevation (m).

On the morning of 16 March, the low-pressure system merged with the low-pressure area lying to its northwest and developed rapidly over the ocean to the east. Underlying this activity was an unseasonally strong northwesterly wind that carried the majority of the radioactive material released on 16 March to the Pacific Ocean. However, according to a meso-scale model (MSM) analysis by the Japanese Meteorological Agency (JMA) of the area surrounding Japan, a secondary depression (low-pressure area in the centre of a cold air mass) formed off the southern coast of Fukushima Prefecture around noon on 16 March due to a low-pressure trough aloft accompanied by a cold air mass (T. Miyasaka, personal communication, 2017). Therefore, easterly and southerly winds blew for a short period of time in a limited area from off the southern coast of Fukushima Prefecture to the vicinity of Kitaibaraki City (Figure 3.7). It is believed that a portion of the radioactive substances that had been carried offshore prior to the morning of 16 March was carried back onshore by these winds. Furthermore, the high-altitude, low-pressure trough, which was responsible for the rapid development of the low-pressure system over land lying to its southeast, also caused ascending air currents to occur over a wide area comprising eastern Japan and the area off the eastern coast. A portion of the radioactive substances that had been retained in the PBL was lifted by these air currents to the middle of the troposphere, where they were

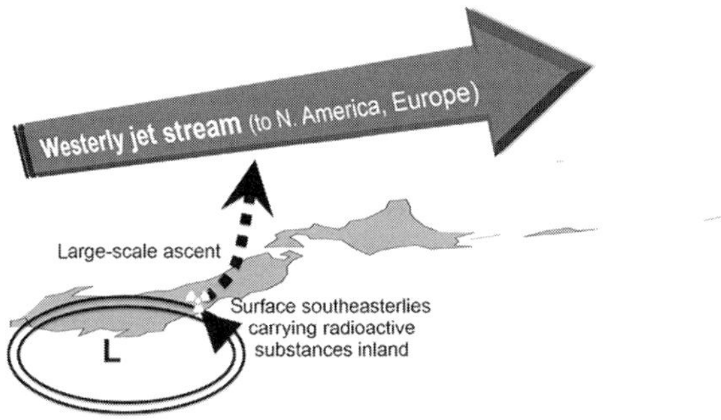

Figure 3.8 Schematic diagram of the mechanisms for long-range transport of radioactive substances released from the FDNPS under the meteorological conditions on 15 March 2011.

carried to North America and as far as Europe by stronger-than-usual westerly jet streams. Jet streams in the mid-troposphere (at a height of approximately 5.5 km) over the northern Pacific Ocean reached speeds exceeding 30 m/s. At such speeds, it is believed that the radioactive substances traversed the Pacific Ocean and reached the western coast of North America in 3–4 days (Figure 3.8; for more detail, see Chapter 4).

3.2.2 Meteorological Conditions and Transport of Radioactive Substances, 20–23 March 2011

Northeasterly winds that blew 21–23 March carried and spread the radioactive substances released from the FDNPS over a wide area of the Kanto Plain. On the morning of 15 March, when the greatest amount of radioactive material was released from the power plant, there were also northerly and northeasterly winds that carried the material from Fukushima Prefecture to the Kanto region. However, by noon, the winds changed to easterly and southerly. At that time, precipitation was limited to mountainous regions in the north; thus, significant deposition was limited to those regions. Although large-scale releases due to venting or hydrogen explosions ceased after 17 March, ^{137}Cs, which continued to be released from the FDNPS, was carried to northern Miyagi and southern Iwate Prefectures by the southerly winds that started to blow around noon on 20 March; wet deposition also occurred due to the precipitation that began in the evening of the same day. This weather front moved southward to eastern Japan as the large migratory anticyclone causing these southerly winds retreated eastward over the ocean (Figures 3.9 and 3.10).

As a result, starting around midnight on 20 March, winds off the coast of Fukushima Prefecture changed from southerly to northeasterly, and the radioactive substances began to be transported in the direction of the Kanto region. Between

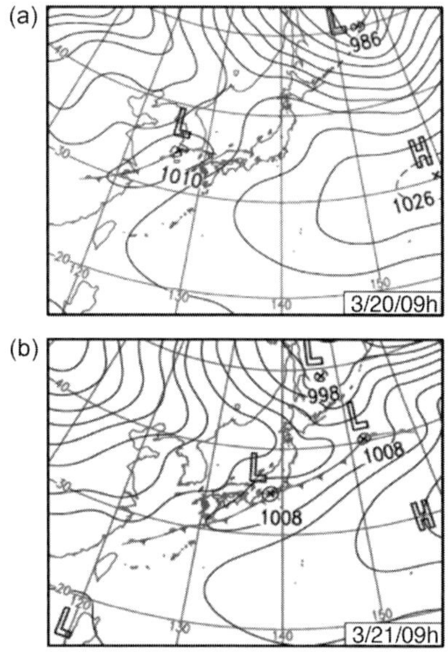

Figure 3.9 Surface weather maps for (a) 9:00 JST on 20 March 2011 and (b) 9:00 JST on 21 March reported by the JMA.

21 and 23 March, the weather front stagnated off the southern coast, causing an extended period of rainfall, which is typical for spring. On the Kanto Plain, the weather front caused wet northerly winds to blow, occasionally accompanied by rainfall that presumably resulted in intermittent wet deposition of radioactive substances over a wide area. The hotspots scattered across the Kanto Plain, including the area surrounding Kashiwa City in Chiba Prefecture, are believed to have been formed in this manner.[1] For example, on the morning of 21 March, the ^{137}Cs concentration at the Kashiwa site drastically increased from 2.2 Bq/m^3 at 6:00–7:00 LT to 161 (105) Bq/m^3 at 8:00–9:00 LT (9:00–10:00 LT), then decreased to 1.8 Bq/m^3 at 11:00–12:00 LT (Oura *et al.*, 2015). In addition, the air dose rate recorded on the University of Tokyo Kashiwanoha campus (Kashiwa), which was 0.2 μSv/h at 18:00 on 20 March, rose to 0.74 μSv/h when measurements resumed at 9:00 on 21 March, and remained over 0.5 μSv/h through the rest of March (University of Tokyo, 2011). Just after 7:00 on 21 March, rainfall in excess of 0.5 mm/h was recorded at the Bando Automated Meteorological Data Acquisition System (AMeDAS) station operated by the JMA in Bando City, Ibaraki Prefecture, which is located near Kashiwa City. This rainfall is believed to have caused wet deposition of the radioactive substances onto the ground surface in the early morning of 21 March. However, because the formation of

[1] This wet deposition is believed to have caused the contamination of tap water by radioactive substances that occurred at the water purification plant in Tokyo and was reported by newspapers and other media.

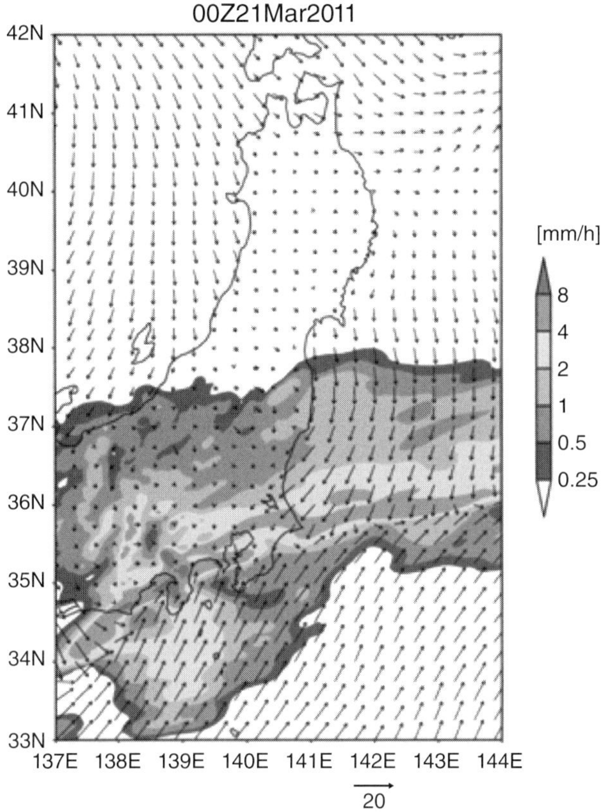

Figure 3.10 Surface wind speed (arrows, m/s) and precipitation rate (shaded areas, mm/h) as of 9:00 JST on 21 March 2011, according to analysis by the JMA. The map shows the heavy rainfall and northeasterly winds that occurred just north of the weather front that migrated southward to the southern coast of the Kanto area.

hotspots depends heavily on local weather conditions and precipitation processes, they cannot be adequately simulated using atmospheric diffusion models. Further verification in this area is needed.

3.3 Atmospheric Dispersion of Releases

ANNE MATHIEU, OLIVIER SAUNIER, DENIS QUÉLO AND DAMIEN DIDIER

3.3.1 Assessment of Atmospheric Releases during the Accident

The accident at the FDNPS in Japan on 11 March 2011 (UTC) resulted in a massive discharge of radionuclides into the atmosphere. Knowledge of the characteristics of atmospheric releases such as the height, the total amounts and the temporal evolution of the release rate of each type of radionuclide released into the atmosphere (source term) is essential to accurately predict environmental impacts and resultant radiological doses received by the public. Following the FDNPS accident, many attempts have been made to estimate the total amounts released into the atmosphere (see Chapters 2 and 4). One of the methods used to estimate source term is a bottom-up approach modelling the processes inside the reactors and the events that led to the radioactive releases. In this respect, in 2012 the NEA (Nuclear Energy Agency) launched the Benchmark Study of the Accident at the Fukushima Daiichi Nuclear Power Station (BSAF) project in order to compute the source term by modelling reactor physics. To date, most of the published estimates have been assessed by coupling observations in the environment and atmospheric transport model (ATM). First, there are reverse methods in which the determination of the release rates is based on a simple comparison between observed and simulated measurements. To estimate the source term of ^{137}Cs and ^{131}I, the JAEA group (Chino *et al.*, 2011; Katata *et al.*, 2012; Terada *et al.*, 2012; Kobayashi *et al.*, 2013) adopted a reverse method and assumed radioactivity ratios for the radionuclides based on available environmental activity concentration measurements (METI, 2011; MEXT, 2011). More recently, Katata *et al.* (2015) revised the JAEA's estimates using observations from both Japan and the Pacific Ocean. They estimated that a total of 151 PBq of ^{131}I and 14.5 PBq of ^{137}Cs have been released into the atmosphere. Mathieu *et al.* (2012) used dose rate observations over Japan and core inventory plant measurement parameters (pressure and temperature measurements in the reactors) to build the source term of 76 radionuclides. Achim *et al.* (2014) used atmospheric transport modelling to assess the arrival time of the plume at International Monitoring System (IMS) stations and to evaluate the quantities of ^{137}Cs, ^{131}I and ^{133}Xe emitted in the atmosphere. Other estimates have been assessed using inverse modelling techniques based on a rigorous mathematical formalism. These are easily automated and are perfectly suited to emergency response-related operational needs. Unlike reverse methods, errors may be explicitly considered in inverse modelling approaches. Stohl *et al.* (2012) estimated the releases of ^{137}Cs and ^{133}Xe using activity concentrations in the air at the Comprehensive Nuclear-Test-Ban Treaty Organization (CTBTO) stations (CTBTO, 2011) on a global scale. Winiarek *et al.* (2012) developed an innovative and efficient method to assess ^{137}Cs and ^{131}I emissions taking into account an evaluation of the instrumental representativeness error together with a fraction of modelling error, as well as the error from the a-priori knowledge on the source term. To refine his previous estimate of ^{137}Cs emissions, Winiarek *et al.* (2014) developed an inversion method using ^{137}Cs deposit measured by aircraft after the end of the releases and air activity concentration simultaneously. Saunier *et al.* (2013) developed an inverse modelling method able to take into account γ dose rate observations; this method does not require using realistic a-priori knowledge of the source term. They estimated the emissions of ^{133}Xe, ^{134}Cs, ^{137}Cs, ^{137m}Ba, ^{131}I, ^{132}I and ^{132}Te and calculated that 105.9 PBq of ^{131}I, 35.8 PBq of ^{132}I, 15.5 PBq of ^{137}Cs and 12 134 PBq of noble gases were emitted. Yumimoto *et al.* (2016) refined ^{137}Cs estimates using the Terada *et al.* (2012) a-priori source term and ^{137}Cs cumulated deposition. It is difficult to compare the ^{137}Cs total amount of all of the source terms since several estimates, only assessed on the basis of environmental measurements in Japan, cannot reproduce the releases transported towards the ocean. Taking into account this fact, Table 3.1 (Mathieu *et al.*, 2018) indicates that there is a relative consensus since the estimation of Stohl *et al.* (2012) found emission rates 2.5 times higher than those of Katata *et al.* (2015). However, the release rates assessed reveal major differences from one estimate to another due to the fact that the source term reconstruction is very sensitive to the type of measurements used and the accuracy of meteorological fields. Saunier *et al.* (2016) assessed the relevance of several source terms by comparing their response to ^{137}Cs atmospheric concentration obtained from the sampling tapes of the Suspended Particulate Matter (SPM) monitoring network (Tsuruta *et al.*, 2014; Oura *et al.*, 2015). The agreement between simulated and observed ^{137}Cs air concentration is not sufficient, thus highlighting the need to further improve the source term reconstruction (Nakajima *et al.*, 2017). Currently, no consensus has emerged that enables one source term to be identified as being more realistic than others (Marzo, 2014; Hirose, 2016; Inomata *et al.*, 2016). The SPM station measurements will inevitably lead to the publication of new source terms (Saunier *et al.*, 2016; Liu *et al.*, 2017; Nakajima *et al.*, 2017).

3.3.2 Modelling the Dispersion of Atmospheric Releases

During the main three weeks of releases, the change of release rates over time as well as the changing weather conditions led to an uneven and complex atmospheric transport of radionuclides. The prevailing weather conditions during the different release phases are summarised in various reports and articles (Kinoshita *et al.*, 2011; Morino *et al.*, 2011; Sugiyama *et al.*, 2012; WMO, 2013; IAEA, 2015; Saito *et al.*, 2015; Nakajima *et al.*, 2017). Since 2011, various numerical studies have been performed, yet those frequently encountered the same difficulties in modelling certain episodes (Terada *et al.*, 2012; Korsakissok *et al.*, 2013; Morino *et al.*, 2013; Saunier *et al.*, 2013; 2016; Draxler *et al.*, 2015; Katata *et al.*, 2015; Nakajima *et al.*, 2017). This results from the complexity of the situation encountered and the continuing uncertainties, linked in particular to the estimation of the source term and the meteorology, as well as the physical and numerical modelling limitations of the ATMs. Incidentally, model-to-measurement discrepancies have

Table 3.1 ^{137}*Cs releases assessed using environmental measurements.*

No.	Source term	Total quantity of ^{137}Cs (PBq)	Method	Observations used
1	Chino *et al.* (2011) JAEA	13	Reverse	Concentrations in the air over Japan. Dose rate close to the facility. Facility events.
2	Stohl *et al.* (2012)	35.7	Inverse	Concentrations in the air over the entire Northern Hemisphere. Facility events.
3	Winiarek *et al.* (2012)	10–19	Inverse	Concentrations in the air over Japan and North America.
4	Terada *et al.* (2012) JAEA	8.7	Reverse	Concentrations in the air over Japan. Dose rates close to the FDNPS. Facility events.
5	Mathieu *et al.* (2012)	20.6	Reverse	Concentrations in the air over Japan. Dose rates close to the FDNPS. Facility events.
6	Saunier *et al.* (2013)	15.5	Inverse	Dose rates in Japan.
7	Kobayashi *et al.* (2013) JAEA	13	Reverse	Depositions over the sea. Concentrations in the air over Japan. Dose rates close to the facility. Facility events.
8	Hirao *et al.* (2013)	9.6	Reverse	Concentrations in the air. Daily depositions by prefecture.
9	Winiarek *et al.* (2014)	11.6–19.3	Inverse	Total deposition over Japan. Concentrations in the air over Japan.
10	Achim *et al.* (2014)	10.8	Reverse	Concentrations in the air of CTBTO stations. Dose rate at the FDNPS. Facility events.
11	Katata *et al.* (2015) JAEA	14.1	Reverse	Concentrations in the air over Japan. Dose rates close to the FDNPS. Depositions over the sea.
12	Yumimoto *et al.* (2016)	8.12	Inverse	Total deposition over Japan.

Table 3.1 (*cont.*)

No.	Source term	Total quantity of ^{137}Cs (PBq)	Method	Observations used
13	Saunier *et al.* (2016)	8.1	Inverse	Concentrations in the air over Japan including SPM data.
14	Liu *et al.* (2017)	7-35	Inverse	Concentrations in the air over Japan including SPM data.

been significantly reduced and the numerical simulations played a key role in the understanding of the different sequences of the FDNPS accident. On 12 March (UTC), the first radioactive plume was detected and measured during its journey along the coast towards the north. This first phase is correctly simulated by ATMs. This episode continued until 13 March (UTC), when monitoring devices in the Sendai region detected the passage of a plume (Tsuruta *et al.*, 2016). It followed a north–northeasterly trajectory away from the coast. Whatever the weather data used, the simulated plume does not move along the coast but is rather transported northeast towards the Pacific Ocean. Figure 3.11(a) illustrates this behaviour. The episode of 14–16 March (UTC) is responsible for the main deposits located to the northwest of the FDNPS and in the Nakadori valley. The turning winds pushed the contamination south–southwest, west, northwest, then south–southwest again, along the coast. The plume first met precipitation into the Nakadori valley. Then it continued towards the southeast, spreading and intensifying. This episode has now been modelled significantly better than in the past. The high values of deposit between the nuclear site and the Fukushima basin have been better simulated, but the location is still slightly off and the scavenging of the plume starts at best 1–2 hours later than in reality (Figure 3.11(b)). The episodes of 18 March (UTC) present similarities with that of 12 March (UTC): the same prevailing wind and an absence of precipitation. The trajectory of the plume over Honshu looks similar, but it appears to be narrower than that of 12 March (UTC). It has now been well modelled (Figure 3.11(c)). Over the episode of 20–21 March (UTC), several plumes were detected north–northwest of the FDNPS, as well as in the Nakadori valley and the Tokyo metropolitan area. Light winds favoured their stagnation, which led, cumulatively, to a high ^{137}Cs concentration and dry deposits of the order of a few kBq/m². It has recently been much better modelled and the various phases of this sequence have now been properly reconstructed. For example, thanks to the weather forecasts provided by the Meteorological Research Institute

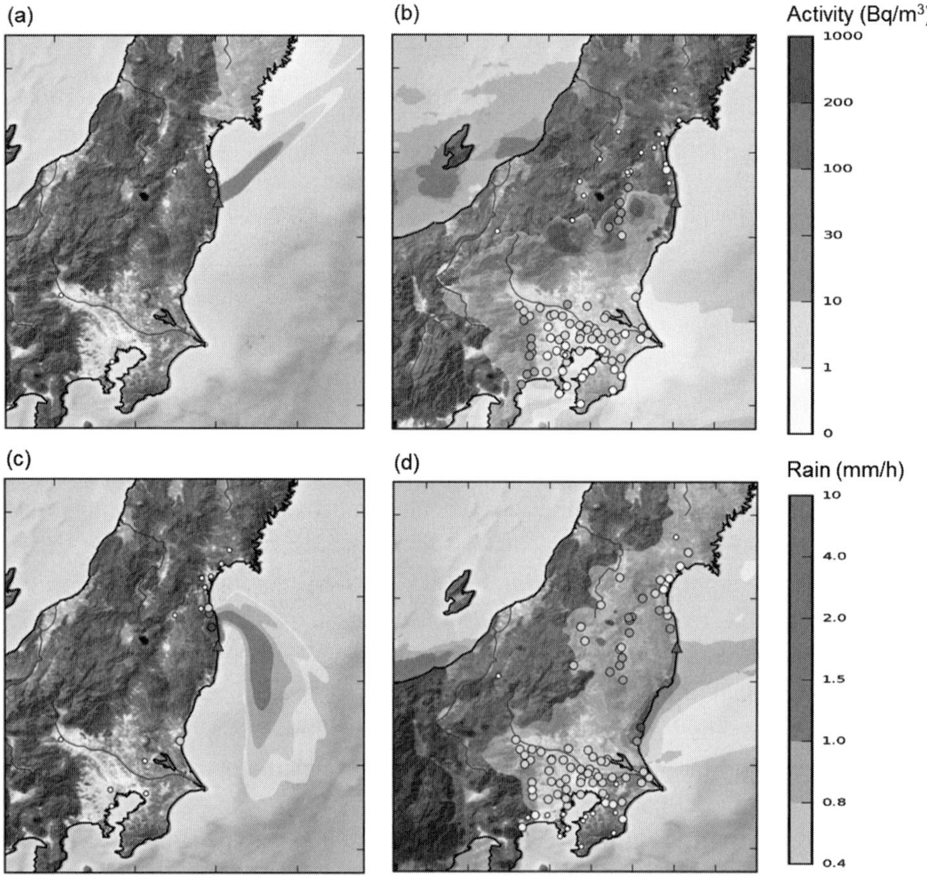

Figure 3.11 Simulated concentrations of ^{137}Cs compared with the measurements (coloured circles). Rainfall is represented by the blue palette. The triangle locates the FDNPS. Each map represents an instant of the main contamination episodes. (a) 12 March at 19:00 (UTC); (b) 15 March at 6:00 (UTC); (c) 18 March at 10:00 (UTC); and (d) 20 March at 20:00 (UTC). A black and white version of this figure will appear in some formats. For the colour version, refer to the plate section.

(MRI) (Sekiyama *et al.*, 2013; 2015), the modelling of the plumes transported southwards has been significantly improved. Even so, a delay in the passage of the plume has not permitted the contamination of the Tokai region to be properly simulated, and then inaccuracy of the simulations in the more distant metropolitan Tokyo area has also not been resolved (Figure 3.11(d)). The modelling studies have singled out the importance of the representation of deposition processes (Morino *et al.*, 2013; Leadbetter *et al.*, 2015; Quérel *et al.*, 2015). The weaknesses of the modelling of the vertical plume distribution and of deposition by light rains also limit the quality of the simulations (Nakajima *et al.*, 2017). Quérel *et al.* (2015; 2016) showed that more complex models do not systematically give a better representation of the total deposition due to the remaining uncertainties. To take

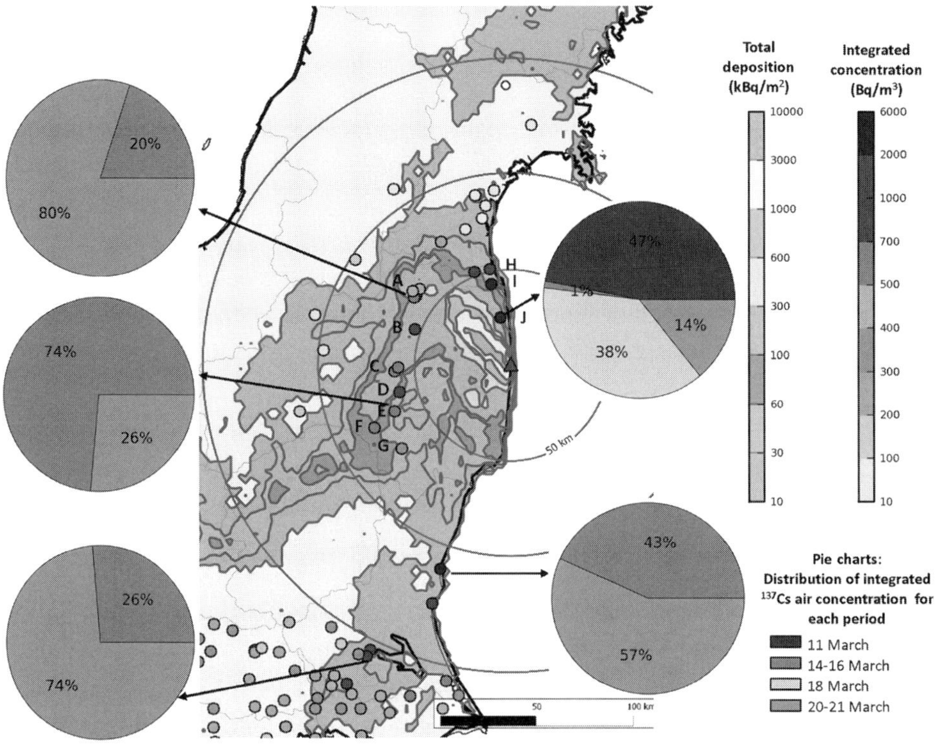

Figure 3.12 The shaded points represent the integrated ^{137}Cs concentration values at the stations between 11 and 24 March (UTC). They are superimposed on the map of total ^{137}Cs deposition. The pie charts show the distribution of observed concentrations in the air for each main contamination period. Reprinted from Appl. Geochem., 91. Mathieu et al (2018). Fukushima Daiichi-derived radionuclides in the atmosphere, transport and deposition in Japan: a review. pp 122–139. Copyright 2018, with permission from Elsevier.

account of uncertainties, ensembles of simulations are produced by varying the configurations of a given ATM, using several ATMs or using several input datasets (Korsakissok *et al.*, 2013; Draxler *et al.*, 2015; Quérel *et al.*, 2015; Solazzo and Galmarini, 2015; Farchi *et al.*, 2016). With regard to the weather, the use of meteorological ensembles as input for the ATMs has been explored (Sørensen *et al.*, 2015; Périllat *et al.*, 2016; Sekiyama *et al.*, 2016). Several ATMs were used by the UNSCEAR task team (Draxler *et al.*, 2015) and by the Science Council of Japan (SCJ) (SCJ, 2014; Nakajima *et al.*, 2017). To provide a ranking of the uncertain parameters, global sensitivity analysis methods were also utilised (Girard *et al.*, 2014; 2016). Figure 3.12 shows the map of total ^{137}Cs deposition. The mapping identified the high-deposition zones to the northwest of the FDNPS and within its vicinity. Distribution of the deposits is greatly influenced by the topography. The mountains would appear to have blocked the transport of the plumes beyond certain regions (IAEA, 2015). The plume exposure of the populations of

Honshu Island is mainly the result of four episodes: that of 12 March (UTC), that of 14–16 March (UTC), that of 18 March (UTC) and that of 20–21 March (UTC) (Terada *et al.*, 2012; UNSCEAR, 2013; IAEA, 2015). Figure 3.12 gathers the analysis of these different episodes and shows the total ^{137}Cs concentrations observed by the SPM stations between 11 and 24 March (UTC). The relative importance of each of the episodes is shown by pie charts for a few representative stations. One of the main findings is that the high-deposition zones do not necessarily match the zones within which the concentrations were the highest. Therefore, when the health impact of the accident is driven by cloud shine exposure, it cannot be assessed only on the basis of total deposition measurements.

3.4 What Would Have Happened if This Accident Had Occurred in a Different Season or at a Different Power Plant?

HISASHI NAKAMURA, YU MORINO AND MASAYUKI TAKIGAWA

The previous analysis clearly shows that the transport of radioactive substances from the FDNPS was strongly influenced by the particular meteorological conditions that occurred during the release of these materials. Let us conduct a thought experiment, as an example, assuming that different meteorological conditions had been realised on 15 March, when the greatest amount of material was released. Assume that the surface cyclone and pressure depression associated with the upper-level pressure trough had passed a day earlier. Then, the near-surface winds over the Fukushima area would have been predominantly northwesterly, transporting most of the radioactive substances offshore towards the Pacific. Under the northwesterlies, contamination over the inland area would have been substantially less, although contamination of the seawater could have been more severe. As another example, assume that the pressure pattern on 15 March had been similar to that actually observed on 21 March; contamination over the Kanto Plain, including the Tokyo metropolitan area, could have been much more severe that it actually was. In March 2011, the influence of cold air masses from the Asian continent was unusually strong. If the cold air masses had already retreated as in a typical year and, thus, migratory cyclones and anticyclones had frequently passed in the vicinity of Japan, the associated southeasterly or southerly winds occurring for one or two days would have likely increased contamination over the Abukuma mountain district and the Sendai Plain.

Let us conduct another thought experiment on what would have occurred if the accident had happened at a different time of the year. If the accident had occurred in mid-winter, prevailing monsoonal northwesterly winds would have enhanced seawater contamination offshore but substantially reduced severe inland contamination. In contrast, if the accident had occurred in mid-summer, monsoonal southerly winds would have transported the radioactive material inland and regional wet deposition, for example, due to afternoon showers, would have enhanced the risk of high concentrations over the inland areas of Fukushima Prefecture and the Sendai Plain. If the accident had occurred during the Baiu rainy season (June to July) or the fall rainy season, the key factor would be the latitude of the seasonal rain front. The monsoonal southerlies towards the rain front located to the north of the power plant could have led to high contamination over the inland Fukushima area and Sendai Plain, as in the mid-summer case. In contrast, the northeasterlies towards the weather front located to the south could have enhanced the risk of high contaminations over the Kanto region, as in the case of 21 March, with greater wet deposition over the extensive rainy area. In particular, the wet, cool northeasterly winds (called 'Yamase' in Japanese) tend to prevail under the influence of the Okhotsk High, which could have enhanced wet deposition of pollutants along the coastal area of Fukushima Prefecture.

Let us suppose further that the accident had occurred in the fall rainy season when a typhoon was approaching. Without the substantial effect from precipitation of wet deposition, the high contamination area could be extended broadly due to strong winds. The direction in which a contaminant plume will spread from the FDNPS depends greatly on the particular path of a typhoon in the warm season or an extratropical cyclone in the cold season relative to the plant. Under a high-pressure system, contaminants tend to accumulate within the vicinity of the power plant, while their distribution can undergo substantial diurnal variations. The local concentrations of the contaminants tend to increase particularly at night and in the early morning under the enhanced stratification near the surface. In the afternoon, the materials tend to be carried inland by sea breeze winds.

Furthermore, what would have happened if the accident occurred at another nuclear power plant in Japan? It is not difficult to conduct thought experiments if the characteristics of the seasonal weather patterns around Japan, as used for the previous thought experiments, are considered. To prepare for the unlikely event of another accident, training should be carried out on a regular basis so that the local government (and, if possible, citizens) can quickly determine whether the particular area is going to be on the downwind or upwind side of the nearby power plant based on the latest weather information and weather forecast charts.

Unless the release of radioactive substances is explosive, as in the case of the Chernobyl accident, the substances released tend to be transported within the PBL and confined up to approximately 1 km above the surface. Because near-surface winds tend to blow towards lower pressure due to surface friction, as a rule of thumb for Japan, inland contamination from a nuclear power plant located on the shore of the Sea of Japan tends to be enhanced under the monsoonal winds in winter or northwesterly winds behind a migratory cyclone. In contrast, inland contamination from a power plant on the shore of the Pacific Ocean will be more likely under the monsoonal southerly winds in summer or easterly winds associated with an approaching cyclone.

If the wind direction changes substantially over time, an accurate assessment of the direction in which materials are being transported is not possible based solely on the latest wind observations. However, useful information on when and in which direction radioactive substances released from a power plant will likely be transported can be obtained from simulations using the SPEEDI under the assumption of a fixed release rate. Even if time series of the amount of radioactive substances released cannot be ascertained, as was the case in the FDNPS accident, and the SPEEDI can only predict concentration distributions relative to the source, the particular prediction of the direction and speed of the transport should be useful in drafting evacuation plans for residents (see Chapter 8).

Lastly, let us consider long-distance transport, called 'transboundary pollution'. Because of the prevailing free-tropospheric westerlies over Japan, regardless of where an accident occurs, there is an extremely high likelihood that radioactive substances lifted up from the PBL into the free atmosphere will be advected eastward to the Pacific Ocean (see Chapter 4). The only situation in which substances might be transported westward can be realised when an anticyclone passes close to or to the north of the accident site. In that case, the radioactive substances may be carried towards the Asian continent by the low-tropospheric easterly winds to the south of the anticyclone. In fact, under such a condition, which occurred in early April 2011, a small amount of radioactive substances was reported in South Korea (Hernandez-Ceballos *et al.*, 2012). Meanwhile, should an accident occur at a nuclear power plant in China or South Korea, radioactive substances could most likely reach Japan by the westerlies, as evident from the transportation of Asian dust to Japan every year. The influence of such an accident on Japan would depend greatly on the amount of radioactive substances released and the meteorological conditions at the time of the accident.

3.5 Factors Contributing to Uncertainty in Atmospheric Diffusion Models

MASAYUKI TAKIGAWA, YU MORINO AND HISASHI NAKAMURA

As mentioned above, atmospheric diffusion models are extremely useful for estimating the regional distribution of radioactive contamination and the formation processes for high-deposition 'hotspots' based on observational data that are both temporally and spatially limited. However, uncertainty exists in model-based simulations, such as the amount of released materials, wind-related transport and diffusion processes and the deposition processes accompanying precipitation resulting in atmospheric removal. Therefore, sufficient care must be exercised when using these simulations.

In the case of the FDNPS accident, estimates by the Emergency Response Support System of the released amounts and measurements of the releases from the reactor vent stacks could not be used due to damage caused by the earthquake. Therefore, to estimate the amount released, we relied on various measurements, including dust sampling and air dose measurements at monitoring posts and other locations (see Chapter 2). In addition, it is believed that up to midday on 14 March, the area near the FDNPS was dominated by westerly winds that carried most of the radioactive substances offshore. Thus, except for temporary spikes in the air dose rate observed at some locations, such as Onagawacho, Miyagi Prefecture, there is little information that can be used to estimate the amounts released during the period in question. Furthermore, as of May 2014, estimates of the released amounts published by different research institutions vary by as much as three orders of magnitude, depending on the estimation period. The reality is that a great deal of uncertainty remains even today regarding the fundamental question of how much was released and over which time period.

Meanwhile, due to the effects of the earthquake and tsunami, there is a substantial gap in the basic meteorological data for a wide region of the Pacific Ocean side of the Tohoku region, including wind speed and precipitation data, which are relevant to the transport and removal processes for radioactive substances. Of the 31 weather monitoring stations operated by the JMA and AMeDAS stations in Fukushima Prefecture, none in the Hamadori region (eastern third of the prefecture) and only a few of the monitoring posts on or near the premises of the NPS were collecting meteorological data on 15 March. This inability to collect data can be attributed to the breakdown of equipment due to direct effects of the earthquake and tsunami or a lack of power due to the malfunction of emergency power sources. Therefore, it is difficult to verify the accuracy of atmospheric models in terms of the meteorological conditions during the time period in question, which is one of the key factors for understanding where the materials were transported and how much was removed from the atmosphere. Although models and other sources suggest that a wide area including Fukushima Prefecture and the surrounding areas experienced very light rainfall (less than 0.5 mm/h) on 15–16 March (Figures 3.6 and 3.7), observational data regarding the temporal and spatial distribution of this rainfall are extremely limited. With regard to the Abukuma mountain district, which lies on the transport path from the FDNPS inland, although it is highly likely that local wind systems and rainfall distribution were affected by the ridges and valleys, models are not able to account for the local effects of terrain whose scale is smaller than the model resolution (mesh size).

The various atmospheric diffusion models such as SPEEDI, used in Japan and in other countries, estimate atmospheric removal and other processes using different calculation modules, which, at their core, comprise formularisations based on historical data and theory. Therefore, even if the same meteorological conditions are used, different models will yield different estimates for deposition distributions. For example, SPEEDI, which accounts for observations from the Chernobyl accident, solves for the removal rate as a function of the precipitation intensity. In contrast, the Models-3 Community Multiscale Air Quality model (Morino *et al.*, 2011; 2013) employed by the National Institute for Environmental Studies, which is normally used to simulate the behaviour of atmospheric pollutants, simulates atmospheric removal of radiocaesium in the same way that it simulates the removal of sulphuric acid aerosols (a cause of acid rain) based on the potential for radiocaesium, given its particle size distribution, to be incorporated into sulphuric acid aerosols (Kaneyasu *et al.*, 2012).

Deposition distribution estimates obtained using atmospheric diffusion models typically deviate from observed data by one order of magnitude or less. It is necessary to advance efforts to verify the reproducibility of atmospheric diffusion model estimates by comparing model estimates to various observational data as well as estimates generated by other models; based on the results, such models should be used to better understand conditions during the accident and to evaluate potential countermeasures.

3.6 Behaviour of Radioactive Substances Based on Atmospheric Monitoring at Fukushima University

AKIRA WATANABE

At Fukushima University, extensive work has been conducted to record changes in radioactive contamination over time. Unrelated to the accident, we began monitoring acid rainfall in April 1988, which we have continued to the present day. During the recent accident, we measured the deposition of radioactive substances using contaminated rainwater samples and particulate deposition samples collected as part of the ongoing sampling. In the midst of our efforts, we were approached by both Vaisala, Inc. and the University of Tokyo Atmosphere and Ocean Research Institute with offers to assist with monitoring. We found volunteers on campus to operate the radiation-monitoring radiosondes and conducted daily monitoring between 15 and 29 April 2011, and once-monthly monitoring up to August of the same year (Figure 3.13). The presence of high concentrations of radioactive substances is consistent with the results of atmospheric models simulating the direct transport of radioactive substances from the accident site. If the dominant radioactive substance in the atmosphere were caesium, we would expect β- and

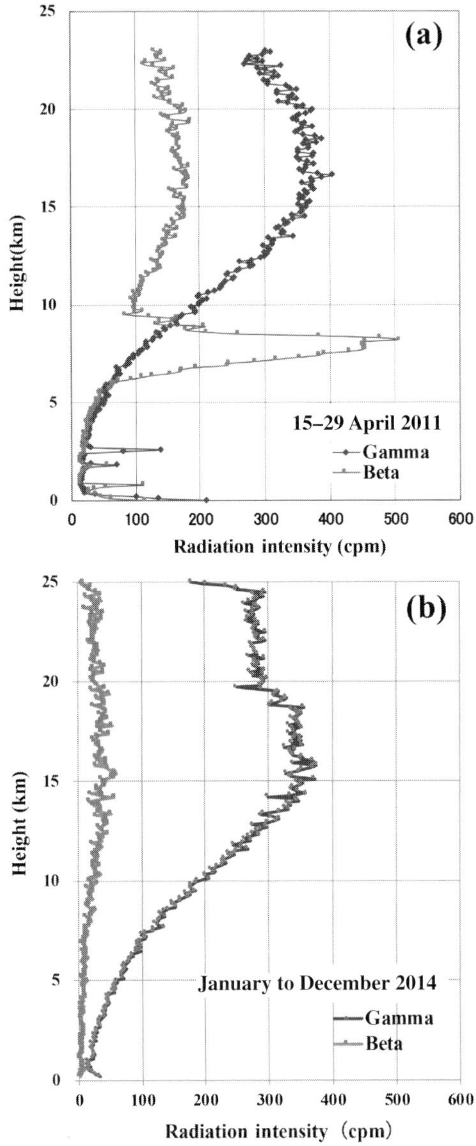

Figure 3.13 (a) Vertical distribution of mean β-rays and γ-ray intensities over the 15 days 15–29 April 2011 and (b) the vertical distribution of annual average β-ray and γ-ray intensity from January to December 2014. Both were averaged every 100 m of altitude. Increases in β- and γ-rays around 15 km altitude are due to cosmic rays.

γ-ray intensities to be similarly elevated in the same locations. However, only elevated β-ray intensity was observed within the troposphere. The levels were particularly high in areas with high humidity, suggesting that the radioactive substances had become concentrated and collected in cloud droplets. The average

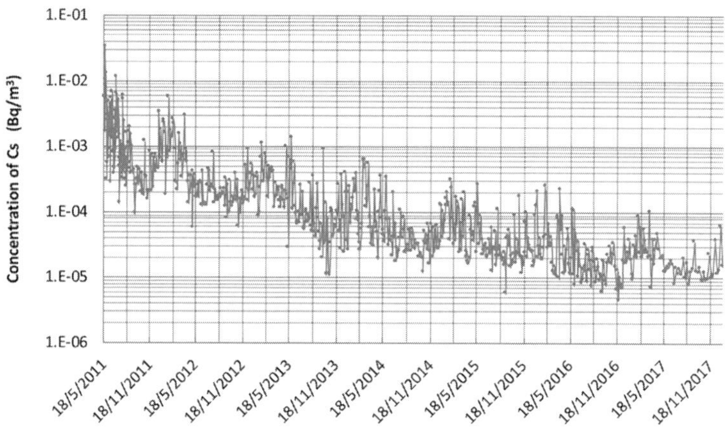

Figure 3.14 Changes in radioactive Cs concentration in the atmosphere with a high-volume sampler (suction volume 700 L/min) installed on a Fukushima University rooftop (24 m above the ground).

values in 2014 show almost the same value except for the peak formed by the radioactive material transported directly from the accident site in 2011. However, it turns out that β-rays at the time of the accident show more than double the strength. It is speculated that the diffusion of water vapour generated from the fuel pool contained a lot of tritium.

The change over time in the atmospheric concentration of the radioactive substances measured on the rooftop of a building on the Fukushima University campus is shown in Figure 3.14. The last day that ^{131}I was detected by the high-volume rooftop sampler at Fukushima University was 15 June 2011. High concentrations of radiocaesium carried directly from the accident site by winds continued to be observed until 27 October 2011. The atmospheric concentration of radioactive substances was expected to simply decline with time; however, it began rising again after reaching a low point in early November 2011. While this increase in concentration could have been due to several factors, including re-entrainment by strong winds, it appears to have occurred on an approximately 40-day cycle. Also, the concentration in the atmosphere is relatively high in the winter and spring season, when the accident site is relatively dry and the wind speed is stronger. This also shows that atmospheric concentrations increase due to rescattering. This last observation suggests the possibility that the radioactive substances in the upper atmosphere, which were monitored using the previously mentioned radiosondes, were gradually descending. Furthermore, until the unit 2 reactor blowout panel opening was closed in March 2013, radioactive substances were being continuously released into the atmosphere at a rate of approximately 10 million Bq/h. Although the rate of release decreased by one order of magnitude after the openings were closed, radioactive substances were still being released at a rate

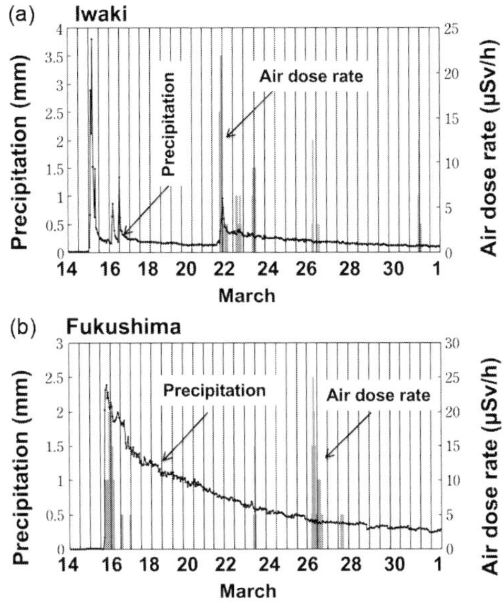

Figure 3.15 Hourly air dose rate in Iwaki City (a) and in Fukushima City (b). The bar graph shows the amount of precipitation, and superimposed line shows the air dose rate.

of 100 000 Bq/h (TEPCO, 2017). This was mainly emitted by the gas exchange work of the reactor to avoid a hydrogen explosion. Analysis of the monitoring data and predictions based on atmospheric diffusion models must be used to improve safety management.

3.6.1 Monitoring of the Air Dose Rate

A large amount of radioactive substances was released into the environment as a result of the FDNPS accident. Most of the legally mandated monitoring posts installed at the FDNPS were unable to fulfil their monitoring function due to damage caused by the earthquake and tsunami or a lack of power. At the time of the accident, the air dose rate was primarily monitored using labour-intensive mobile observations.

Figure 3.15 shows the change over time in the hourly air dose rate after the accident, based on manual measurements collected in Iwaki and Fukushima City in the period 14–31 March 2011, and the change over time in the hourly precipitation rate recorded by the AMeDAS over the same period. On 15 March, an air dose rate of 23.7 µSv/h was recorded at 4:00 in Iwaki, and an air dose rate of 23.9 µSv/h was recorded at 19:00 in Fukushima City. Although essentially the same amounts of radioactive substances were carried to Iwaki and Fukushima, at Iwaki there was little deposition and the atmospheric concentrations declined while undergoing small fluctuations, reaching 2.7 μSv/h by 20:00 on 15 March. In contrast, the air dose at Fukushima City declined following a half-life-like decay curve over a period of eight days after peaking on 15 March. The difference between these two cases is related to the occurrence of rainfall. At Iwaki, the passing of the high-concentration plume did not coincide with a rainfall event, and only dry deposition occurred. For Fukushima, it is believed that the high air dose rate was the result of wet deposition of radioactive substances by rain. Understanding these monitoring results and deposition processes beforehand would have enabled people to remain indoors or to evacuate in an appropriate manner and thus substantially reduce their risk of radiation exposure. Iwaki is an example of a situation in which the risk of exposure could have been reduced by avoiding the contaminated plume as it passed. The problems that occurred in the recent accident underline the importance of independent monitoring systems that do not rely on commercial electrical power. Moreover, I realised that the necessity of the system predicting the passage of the plume and the literacy to be able to understand it are very important.

Taking lessons learned from the FDNPS accident, in 2012 MEXT installed monitoring posts in 2700 schools and other locations throughout Fukushima Prefecture and created a system to monitor the air dose rate at a height of 1 m in real time. However, we are currently facing a situation in which knowing only the air dose rate data is inadequate. One example of this has to do with the fact that the monitoring units are mostly located at schools, and the monitoring is not being conducted in a manner that supports local government restoration and remediation efforts. Another example has to do with the fact that the reasons for fluctuating readings at a given monitoring post, which cause alternating hope and fear, have not yet been clarified. Based on the observations from the two cities, the air dose rates occasionally spiked and fluctuated (Figure 3.15). It is not known why radioactive decay, which is a stochastic process, behaves in this manner. In addition, it has been suggested that the fluctuations are related to the temperature correction procedure for the equipment. Furthermore, the relationship between the air dose rate and temperature rate has not been clarified. If we consider whether or not the amount of radioactive substances in the vicinity of the monitoring units is really changing, it is highly unlikely that there is sufficient movement or accumulation of radioactive substances to change the air dose rate. Therefore, during this period, radioactive substances were transported from the accident site by wind on several occasions (in Fukushima City, this occurred at least five times in March 2011). To understand such changes, it is necessary to install monitoring units that contain equipment for measuring both the air dose rate and meteorological data.

Fukushima University has developed and deployed monitoring posts equipped with solar-powered instruments to measure both meteorological and air dose data. The simultaneous monitoring of meteorological parameters and air dose rates one year after the FDNPS accident revealed that the air dose rate measurements during

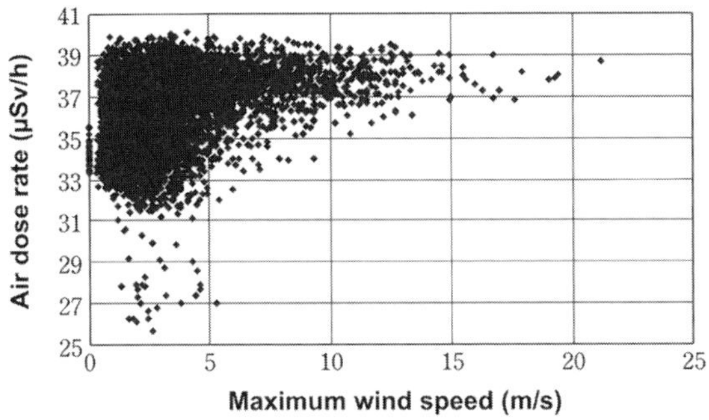

Figure 3.16 Relationship between the maximum hourly wind speed and the dose rate measured by a meteorologist and with a dose rate meter installed at 1 m above the ground at a point 10 km northwest of the accident site.

precipitation events, such as rainfall or snowfall, tend to decrease due to the masking effect of water, which shields the sensors from radioactive rays. In addition, although it appears that high air dose rates tend to occur slightly more frequently when there are stronger winds, no direct relationship was observed between changes in the air dose rate and wind speed (Figure 3.16). Similarly, from the observation of the air dose rate every 10 minutes for six years, it was estimated that it will take 30 years to return to the pre-accident level of 0.04 μSv/h. However, no clear relationship between humidity, temperature or wind direction and the air dose rate was identified in the data. Therefore, with regard to decontamination work that will be carried out in the future, it is necessary to deploy monitoring devices equipped with instruments capable of measuring both meteorological and air dose rate data and to create an internet-based network that provides easy monitoring of temporary and interim storage facilities up to the point that such facilities are no longer necessary. The most important activity is not only to measure the dose rate but also to constantly monitor changes.

3.6.2 Monitoring of Atmospheric Concentrations and Deposition

The air dose rate is not the only parameter that should be monitored. One way of taking responsibility for future nuclear accidents is to ensure that we diligently evaluate exposure doses and prepare data that can be used in epidemiological studies in cases where the radiation exposure is of a magnitude that will impact future generations. Therefore, a wide range of environmental parameters must be monitored. For example, to clarify the behaviour of radioactive substances in the environment, it is necessary to take a mass-balance approach that involves the monitoring of various parameters, including the atmospheric concentration and atmospheric removal by deposition, as well as efflux and movement at the ground surface. However, at present, the only parameter that we have some understanding

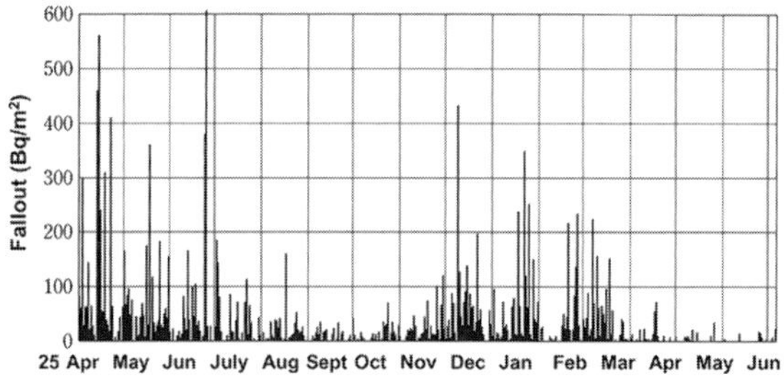

Figure 3.17 Daily radioactive caesium descent amount (Bq/m^2) measured at the Fukushima Prefecture Nuclear Power Center (Fukushima City).

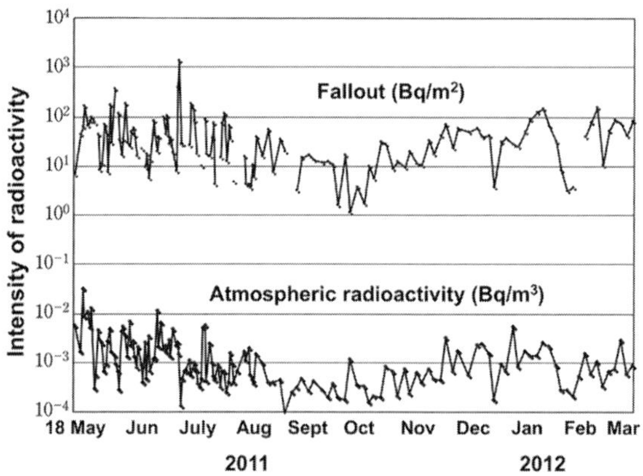

Figure 3.18 The falling amount (Bq/m^2) of radioactive caesium per day measured at the Fukushima Nuclear Power Center (Fukushima City) and the atmospheric concentration (Bq/m^3) of radioactive caesium measured by Fukushima University. Both values are displayed using a logarithmic scale.

of is radiation intensity (dose rate), and we have not yet reached the stage of being able to construct a mass balance.

Figure 3.17 shows the daily radionuclide deposition rate measured by the Environmental Radioactivity Monitoring Center of Fukushima. Considering the assumption that the deposition rate should decline as the amount of radioactive substances released from the accident site declines, we see that the deposition rate actually increased after reaching a low point in early November 2011, with new peaks in January, February and March 2012. This is the same pattern as that observed for the atmospheric concentration. The cumulative amount of radioactive substances deposited between 27 March 2011 and 30 September 2012 was 20 000 Bq/m^2, which is not a trivial amount. At the present time over seven years have passed; the total amount of radioactive caesium in 2016 had fallen to 239 Bq/m^2, but before the accident it was about 10 mBq/m^2. It is also estimated from the attenuation of observation results that to return to the original level will take more than 30 years.

Figure 3.18 shows the atmospheric concentration of radioactive substances measured on the roof of a building on the Fukushima University campus, together with the deposition rate. The concentration exhibits the same trends as the deposition rate. If we ignore the re-entrainment of radioactive substances near the ground surface, deposition essentially represents the settling of radioactive substances suspended in the atmosphere; the deposition speed can be calculated by dividing the deposition rate over a given period of time by the atmospheric concentration. The observed deposition speed (approximately 30 cm/s) is substantially higher than the deposition speeds employed in atmospheric models. In addition, atmospheric concentration measurements can be used to assess internal exposure rates due to inhalation (estimated here to be on the order of 0.17 μSv/year) using the dose conversion factor explained in Chapter 1.

3.7 Atmospheric Radionuclides Concentrations Just After the Fukushima Accident

HARUO TSURUTA, YASUJI OURA, MITSURU EBIHARA AND DAISUKE GOTO

Data on the atmospheric concentrations of radionuclides in Fukushima Prefecture immediately following the FDNPS accident are extremely limited, resulting in a high degree of uncertainty regarding estimates of the early internal exposure rates. From 12 to 13 March 2011, the Ministry of Economy, Trade and Industry (METI) and the Environmental Radioactivity Monitoring Center of Fukushima Prefecture conducted dust sampling in the vicinity of the FDNPS to measure the atmospheric concentrations of radionuclides just after the first release of radionuclides from the FDNPS (METI, 2011). On 18 March, MEXT began to measure radionuclides from a broader area within a 70 km radius of the accident site (MEXT, 2012). TEPCO began to measure radionuclides by using the same type of sampling system as METI and MEXT on the site of the FDNPS on 19 March (TEPCO, 2013).

Meanwhile, in the Kantou area about 10 agencies began to independently conduct atmospheric measurements of radionuclides immediately after the accident (Figure 3.19), and these data have also been made public.

3.7.1 Time Series of Atmospheric Radionuclides in Tokai-mura

The Nuclear Science Research Institute (NSRI), JAEA, measured atmospheric radionuclides for the short sampling time of 20 minutes when the radiation dose rates increased during 15–21 March 2011, and after that measured every 12 hours (Ohkura *et al.*, 2012). They first observed a plume transported to the Kantou area because the site was located on the east coast of Ibaraki Prefecture (Figure 3.19). The time series of atmospheric ^{131}I and ^{137}Cs concentrations during 14–31 March

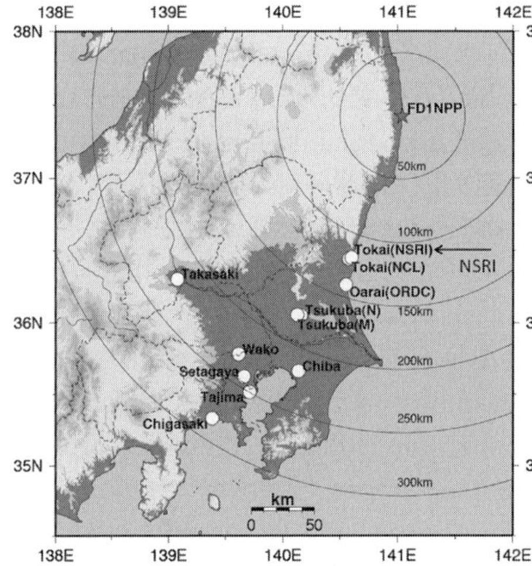

Figure 3.19 Emergent measurement sites of atmospheric radionuclides in the Kantou area (Tsuruta *et al.*, 2012).

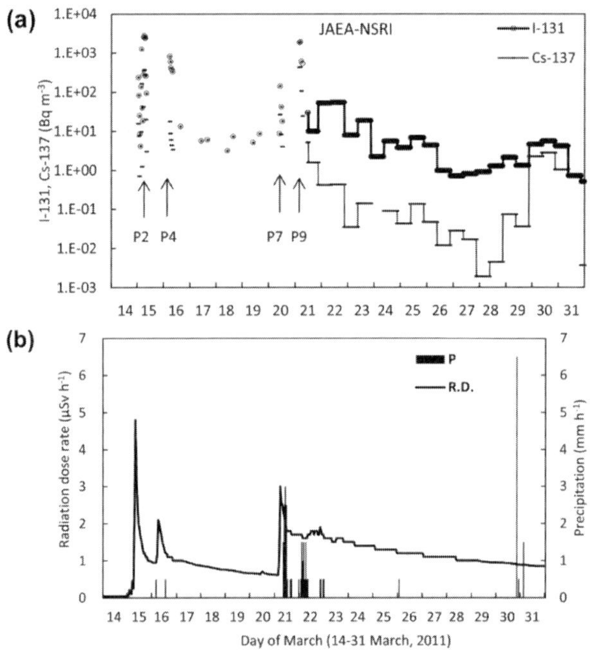

Figure 3.20 Time series of (a) atmospheric ^{131}I and ^{137}Cs concentrations and (b) radiation dose rates (R.D.) and precipitation amounts (P) at NSRI in Tokai-mura, Ibaraki, Japan, during 14–31 March 2011. (Ohkura *et al.*, 2012; Tsuruta *et al.*, 2013). The plumes of P2, P4, P7 and P9 are listed in Table 3.2.

2011 are shown in Figure 3.20. Four plumes (P2, P4, P7 and P9 in Table 3.2) with ^{131}I concentrations higher than 100 Bq/m^3 were transported to the NSRI from the FDNPS on 15, 16, 20 and 21 March, respectively. These plumes were also observed at the other sites in the Kantou area, although some plumes were not observed, depending on the wind directions.

The ^{131}I/^{137}Cs activity ratio at the NSRI was approximately 10 or less in plumes P2, P7 and P9, but much higher (about 75) in plume P4. The value of 10 is close to the inventory ratio of the two radionuclides in reactor units 2 or 3 on 11 March (Nishihara *et al.*, 2012), suggesting the release of radioactive materials from the drywell surrounding the reactor pressure vessel. In contrast, the latter value of about 75 indicates extremely low concentrations of ^{137}Cs and relatively high concentrations of ^{131}I. It still remains unclear whether this reflects a change in the ^{131}I/^{137}Cs ratio of the released material or a relatively high deposition rate of ^{137}Cs by precipitation. ^{131}I existed in particulate and gaseous phases, and they were separately collected and measured by the NSRI (Ohkura *et al.*, 2012). The results indicated that during high-concentration periods, at least half of the ^{131}I existed in the particulate phase, while the gaseous phase was dominant during low-concentration periods. A detailed analysis of all these data will be published elsewhere.

3.7.2 Radionuclides Measurements of Suspended Particulate Matter Collected on Filter-Tapes in SPM Monitors

In Japan, the hourly mass concentrations of suspended particulate matter (SPM: diameter is less than 10 μm) are routinely measured at the air quality monitoring stations operated by local governments under the Air Pollution Control Act. Suspended particulate matter is collected on filter-tapes installed in the SPM monitors. These stations are mainly located in urban or industrial areas to protect human health. One year after the Fukushima accident, the used filter-tapes of more than 400 SPM stations in eastern Japan were offered by many local governments through the Ministry of the Environment, Japan. Since then, we have measured hourly radionuclides of the SPMs at many SPM stations in the Fukushima and Kantou areas during 12–23 March 2011 (Figure 3.21). In the first stage, we identified nine plumes (P1–P9 in Table 3.2) released from the FDNPS just after the accident on 11 March 2011, and analysed their spatiotemporal distributions of atmospheric ^{137}Cs using 40 SPM stations (Tsuruta *et al.*, 2014). In the second stage, we measured 59 SPM stations, and published all the hourly ^{134}Cs and ^{137}Cs concentrations of 99 SPM stations, including the 40 SPM stations in the first stage (Figure 3.21) (Oura *et al.*, 2015). Using these data, we analysed the detailed spatiotemporal distributions of atmospheric ^{137}Cs concentrations (Tsuruta *et al.*, 2017). In the

Table 3.2 *Days and areas of the transported plumes/polluted air masses (p1–P11) (Tsuruta* et al.*, 2018).*

Plume	Day of March, 2011														Range of plumes				Source	Reference
															Hamadori					
No.	12	13	14	15	16	17	18	19	20	21	22	23	24	25	North	South	Nakadori	Kantou	Unit	
p1	○														○				1	3
P1v	●														●				1	3
P1	●	→													●				1	1
P1′		●													●				1	3
p2				○												○		○	1 and/or 3	2
P2				●												●		●	2 and/or 3	1
P3				●	→										●̲		●̲		2 and/or 3	1
P4					●											●̲		●	2 and/or 3	1
P4′					●											●			2 and/or 3	3
P5							●								●				2 and/or 3	1
P5′							●									○			2 and/or 3	3
P6								●							●				2 and/or 3	1
P7									○									○	2 and/or 3	1
P8									●	→					●		●		2 and/or 3	1
P8′									●							●			2 and/or 3	3
P9										●					○	●̲		●̲	2 and/or 3	1
P9′										●						●			2 and/or 3	3
P10													●		●				1	3
P11														●	●				2 and/or 3	3

No., number of major plumes transported from the FDNPS just after the accident.
Hamadori, Nakadori and Kantou: the location of these areas is shown in Figure 3.21.
●: maximum ^{137}Cs concentrations≧100 Bq/m^3.
○: maximum ^{137}Cs concentrations $<$ 100 Bq/m^3.
→: high ^{137}Cs concentrations continued to the following morning.
●̲: precipitation was observed in the area.
References (1) Tsuruta *et al.*, 2014; (2) Tsuruta *et al.*, 2017; (3) Tsuruta *et al.*, 2018.

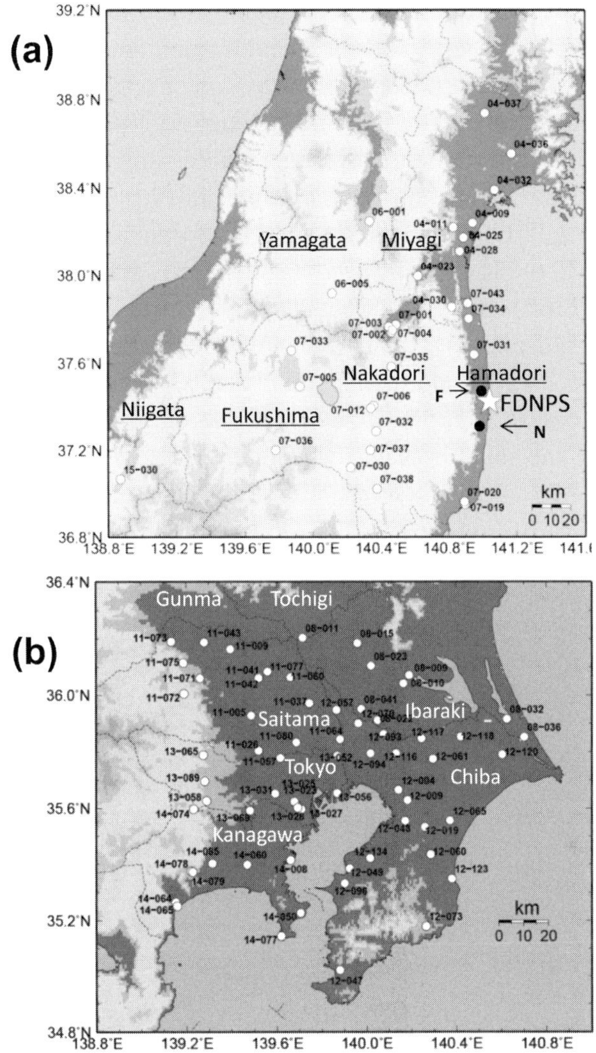

Figure 3.21 SPM monitoring sites (Oura *et al.*, 2015). (a) SPM monitoring sites in the southern Tohoku area, where filter-tapes were used for measurement of radiocaesium. Hamadori and Nakadori are located in the east coast and central area of Fukushima Prefecture, respectively. F and N mean the SPM sites of Futaba and Naraha, respectively. (b) The same as (a) but for the Kantou area.

third stage, by measuring the hourly ^{137}Cs concentrations of the SPMs at two SPM stations (Futaba and Naraha) located within 20 km of the FDNPS (Figure 3.21(a)), the detailed behaviour of radionuclides was revealed near the FDNPS and along Hamadori, the east coast of Fukushima Prefecture (Tsuruta *et al.*, 2018).

Major findings from these papers are summarised as follows. We retrieved 19 plumes transported to the Fukushima and Kantou areas (Table 3.2). At 15:00 JST on 12 March, the highest ^{137}Cs concentration of 13 600 Bq/m^3 was observed at Futaba, 3.2 km northwest of the FDNPS, due to a plume (P1v) from a vent operation of Unit 1 just before the hydrogen explosion at 15:36. On the morning of 15 March, plume P2 was transported to Naraha 17.5 km south of the FDNPS, where a maximum ^{137}Cs concentration of 8300 Bq/m^3 was observed at 3:00, and then transported to the Kantou area by a northeasterly wind following an easterly wind (Figure 3.22(a)). Since the wind direction shifted clockwise on the afternoon of 15 March, plume P3 was transported from the south to the north in Nakadori (Figure 3.22(b)) and transported northwestwards from the FDNPS in the evening. In contrast, on the night of 20 March, plume P8 was transported from the north to the south in Nakadori (Figure 3.22(d)). And on the morning of 21 March, plume P9 was again transported to the Kantou area by a northeasterly wind (Figure 3.22(c)). Figure 3.23 shows time series of related parameters for plumes P3 and P8. The number in Figures 3.22(b) and (d) is the time (hour, JST) of the maximum ^{137}Cs on 15 March and 20–21 March, respectively, during the plume arrival at A to G in Nakadori, and at H to J in northern Hamadori. The wide arrows in Figures 3.22(b) and (d) show typical transport pathways of radioactive materials. Plume P3 transported the materials to Nakadori on the morning of 15 March and to northern Nakadori in the afternoon from the FDNPS by the wind patterns. On the other hand, plume P8 transported materials to northern Nakadori on the afternoon of 21 March and to northern Hamadori in the evening from the FDNPS. The thin arrow in Figure 3.22(d) indicates the transport pathway of the polluted air masses with the maximum ^{137}Cs concentrations when the wind direction shifted from the south to the north in the night. An area of Y in Figure 3.22(c) indicates high ^{137}Cs concentrations of about 10 Bq/m^3 observed at several SPM stations by a northerly wind. Z indicates a wind convergence zone from the east to the west between a northerly wind and a southern wind.

The United Nations Scientific Committee on the Effects of Atomic Radiation (UNSCEAR, 2015; 2016) documented that these papers (Tsuruta *et al.*, 2014; Oura *et al.*, 2015) are considered to make a significant contribution to an identified research need of ‘releases to atmosphere, dispersion and deposition’. In addition to the continuous measurements of radiocaesium of SPM at the other SPM stations, we are now starting to measure ^{129}I (which has a long half-life of 15 700 000 years) of SPMs to estimate atmospheric ^{131}I of SPMs that could not be measured due to a short half-life of eight days.

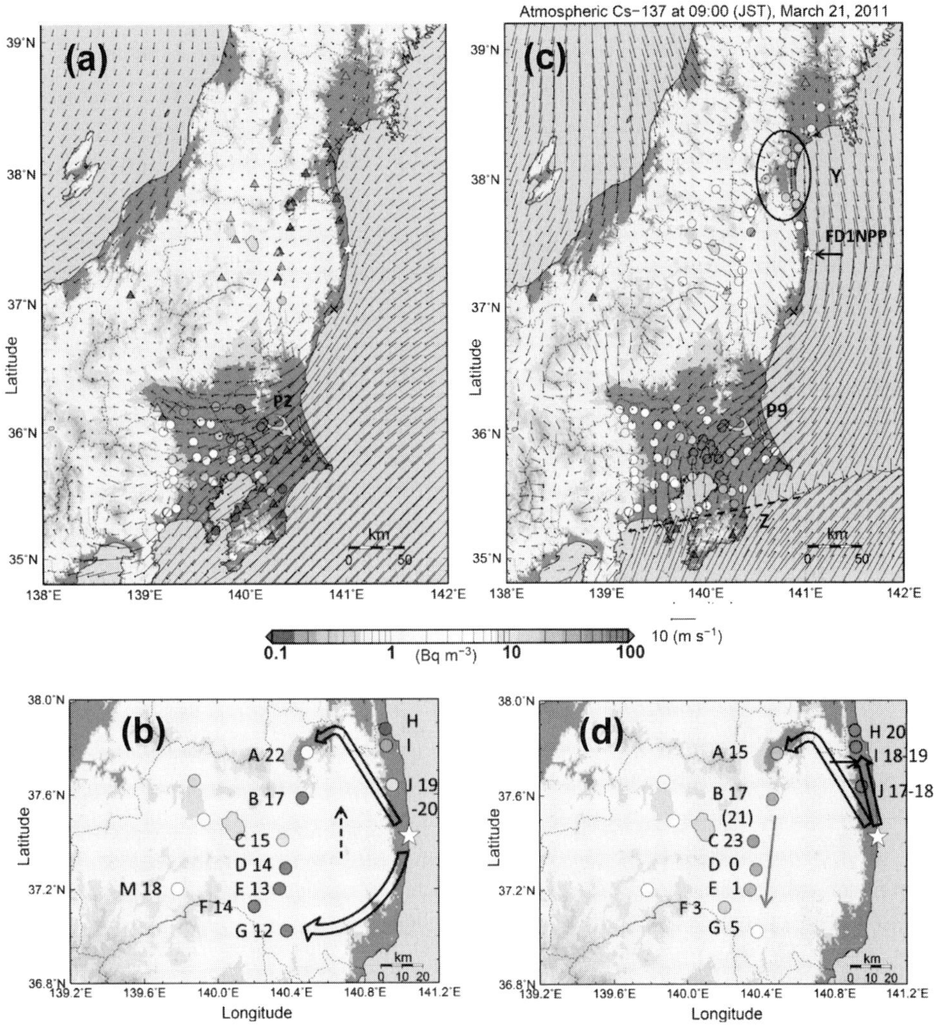

Figure 3.22 (a) Spatial distribution of atmospheric ^{137}Cs concentrations (coloured dot) for plume P2 at the SPM monitoring stations and wind vectors (black arrows) at 1000 hPa by JMA in Kantou and southern Tohoku areas, at 9:00 (JST), 15 March 2011. (b) The maximum ^{137}Cs concentrations with the coloured dot at the SPM stations on 15 March. (c) The same as (a), but at 9:00 (JST), 21 March 2011 for plume P9. (d) The same as (b) but during 20–21 March for plume P8. See the main text for meanings of wide arrows, thin arrow, symbols Y and Z, and the numbers (Tsuruta *et al.*, 2014; 2017). A black and white version of this figure will appear in some formats. For the colour version, refer to the plate section.

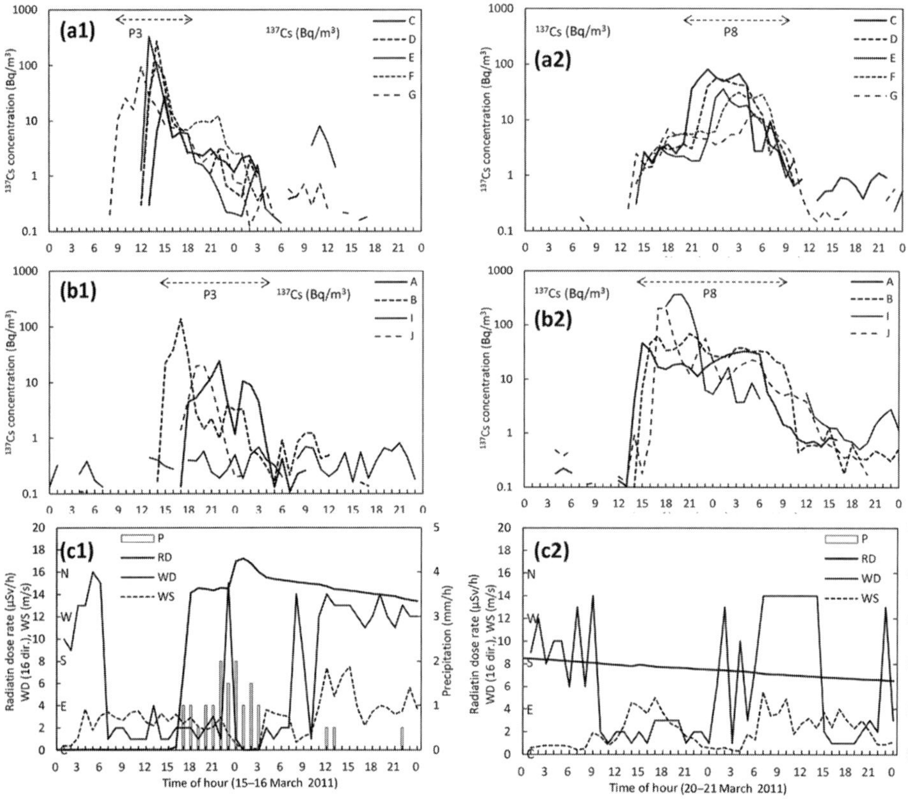

Figure 3.23 Time series of atmospheric ^{137}Cs concentrations at sites shown in Figure 3.22 for plume P3 during 15–16 March (a1, b1) and for plume P8 during 20–21 March 2011 (a2, b2). (c1, c2) indicate time series of radiation dose rates (RD) at Fukushima City monitoring post located near site A, precipitation (P), wind direction (WD) and wind speed (WS) at the Fukushima Local Meteorological Office (Tsuruta *et al.*, 2014; 2017).

3.8 Monitoring the Radioactivity of Atmospheric Aerosols and the Influence of Resuspension from the Ground

KAZUYUKI KITA AND MIZUO KAJINO

Several aspects of the behaviour of anthropogenic radionuclides in the atmosphere following the FDNPS accident have already been discussed in this book. In this section, important data are presented and discussed to better understand the extent of the anthropogenic radionuclides in the atmosphere and mechanisms responsible for spreading them with aerosols over a larger geographic area, which has been revealed by the activities of the Emergency Monitoring Team formed by the Geochemical Society of Japan and other members of the Japan Geoscience Union, as well as the Japan Society of Nuclear and Radiochemical Sciences. Atmospheric aerosol (particulate matter) sampling began immediately following the accident in late March and early April 2011 at more than 20 locations throughout Japan. In May of the same year, the sampling sites were reorganised, and their number was reduced to 11. From September 2011 to December 2015, sampling continued in four locations in Sendai (later Marumori), Fukushima, Koriyama and Hitachi, all surrounding the FDNPS, between 70 and 100 km from its location. Although atmospheric activity concentrations of ^{131}I and radiocaesium at these locations spiked above 1 Bq/m^3 on numerous occasions immediately following the accident in late March and early April, they subsequently declined while undergoing cyclical increases and decreases until September or October 2011. No significant difference was observed in the fluctuation range of the atmospheric activity concentrations at the four sampling locations. The fluctuation range of atmospheric activity concentrations of ^{137}Cs fell to 0.0001–0.02 Bq/m^3 in May, two months after the accident, and to 0.00005–0.005 Bq/m^3 in August, five months after the accident. The atmospheric activity concentration of ^{131}I, which has a shorter half-life than radiocaesium, declined even more rapidly and was mostly below its detection limit from July onward.

At these monitoring locations, the observed atmospheric activity concentrations of ^{131}I and radiocaesium repeatedly varied by one or two orders of magnitude over one- and two-day periods. The changes over time in the atmospheric activity concentration of ^{137}Cs observed at Hitachi, Kashiwa and Yokohama, all of which lie in essentially the same direction from the FDNPS, are shown in Figure 3.24. The same trends can be seen for ^{134}Cs. Their short-term spikes and dips occurred almost simultaneously at each of these monitoring locations. The same figure compares the observed values with estimated values for the atmospheric activity concentrations of ^{137}Cs using an atmospheric transport model (M. Takigawa, personal communication, 2011) and assumed ^{137}Cs release amounts, indicating that the timing of the observed and estimated peaks frequently coincided prior to June 2011. These results strongly suggest that the increased atmospheric radioactivity during this period was due to radionuclides that were continuously released from the FDNPS to the atmosphere, albeit in decreasing quantities, and that were carried to the sampling locations by the wind. Such increases in atmospheric radioactivity are believed to have resulted from the release of radionuclides from the FDNPS and were only rarely observed from October 2011 onward.

As shown in Figure 3.24, during periods when the wind direction largely reduced the transport of radionuclides from the FDNPS to the monitoring locations, the model predicted two to three order of magnitude decreases in the atmospheric activity concentration of ^{137}Cs, whereas the observed values did not decline to the same extent. From October 2011 onward, although the atmospheric concentrations of ^{137}Cs fluctuated repeatedly, no systemic decrease was observed

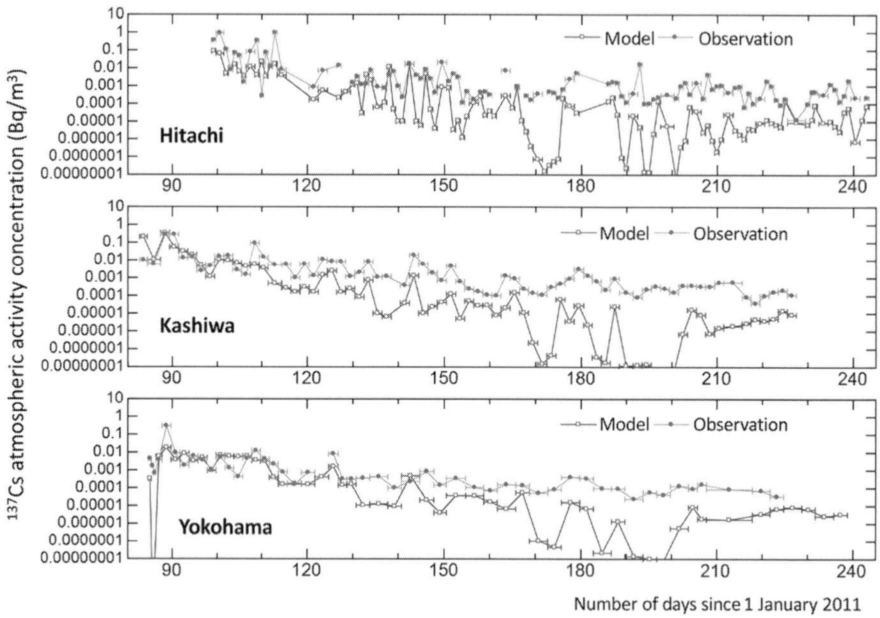

Figure 3.24 Time series of the atmospheric activity concentration of ^{137}Cs observed and simulated by an atmospheric transport model at Hitachi, Kashiwa and Yokohama. (Measurement locations are shown in a glossary map.)

over time. During this period, the concentrations were observed to be as high as 0.01 Bq/m^3, which are three orders of magnitude greater than the values recorded prior to the accident. The concentrations during this period differed systematically among the monitoring locations and appeared to be more related to (i.e. positively correlated with) the radiocaesium density deposited around each monitoring location than to the distance from the FDNPS. From these results, it can be concluded that the atmospheric radiocaesium was significantly supplied from that which had been deposited on the ground (soil and vegetation) near the monitoring locations via resuspension and that the resuspension was the dominant source of the atmospheric radiocaesium beginning in October 2011. Efforts to identify the mechanisms underlying the resuspension of radioactive substances deposited previously over soil and vegetation and to quantify the resuspension flux are being carried out through projects sponsored by MEXT and the NRA and as part of the Grants-in-Aid for Scientific Research in Innovative Areas project, titled 'Interdisciplinary Study on Environmental Transfer of Radionuclides from the Fukushima Daiichi NPP Accident'.

It was an early hypothesis that the main mechanism of the resuspension of radiocaesium from the ground to the atmosphere was the blowing by the wind of soil dust (clay mineral) particles including radiocaesium. However, for the lifting and scattering of soil particles, the ground surface must be dry and ground surface wind speeds have to exceed 5–6 m/s. Despite the fact that such conditions did not exist for most of the sampling period, substantial concentrations of radiocaesium were observed in the atmosphere. Therefore, some mechanisms other than the blowing of soil particles by the wind had to have contributed to the resuspension of radiocaesium, although the specific mechanism has not yet been determined. In addition, the activity concentrations of atmospheric ^{137}Cs were higher at the immature Japanese cedar plantation site than at the other two sites, although the concentration of ^{137}Cs on the ground surface was lower (Figure 3.25(a)) and was not strongly correlated with concentrations recorded at the other two sites (Figure 3.25(c)). These results suggest that different resuspension mechanisms may dominate in forest environments. Efforts are currently underway to elucidate these mechanisms.

For the three years following the accident, there has been no indication that the atmospheric radioactivity concentration of ^{137}Cs has systematically declined in Fukushima. This conclusion indicates that the effects of resuspension will continue. Because the activity concentrations of anthropogenic radionuclides in the atmosphere following the accident are currently lower than those of naturally occurring radionuclides, such as radon, even in Fukushima, it is believed that the health risk from the atmospheric radionuclides is generally low. However, resuspension can cause the diffusion of anthropogenic radionuclides from heavily polluted areas to other areas. Moreover, local increases in atmospheric concentrations of radionuclides could occur due to agricultural work, field burning, forest fires or the incineration of polluted garbage or rubble. Therefore, it will be necessary to continue to diligently monitor changes in atmospheric radioactivity in the future. Further research is also needed to elucidate and quantify the various resuspension mechanisms.

Based on limited but accumulated knowledge of the observations detailed above, Kajino *et al.* (2016) estimated the ^{137}Cs resuspension from contaminated soil and biota due to the Fukushima nuclear accident using three methods: (1) a numerical simulation; (2) a field experiment on dust emission in the contaminated area (the town of Namie) (Ishizuka *et al.*, 2017); and (3) surface air concentration measurements both inside (Namie) and outside (the city of Tsukuba) of the contaminated area. Since the resuspension mechanism remained unknown, the simulation utilised simple assumptions of resuspension flux by adjusting the observed surface concentrations. The simulated and observed concentrations at Namie and Tsukuba are shown in Figure 3.26. Using the dust emission module (Ishizuka *et al.*, 2017), which was developed based on a field experiment, simulated ^{137}Cs resuspension from soil multiplied by five accounted for the observed ^{137}Cs surface air concentration measured at Namie during only the cold season (grey solid lines in the figure). However, the module underestimated the ^{137}Cs concentration by 1–2 orders of magnitude in the warm season. Introducing

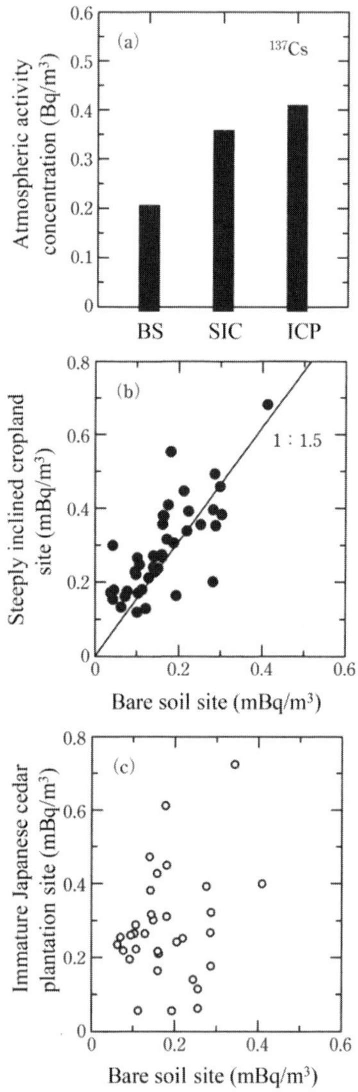

Figure 3.25 Comparisons of the atmospheric activity concentration of ^{137}Cs observed over different land use locations, bare soil (BS), steeply inclined cropland (SIC) and immature Japanese cedar plantation (ICP) sites in Kawamata from January to March 2012.

resuspension from forests using a constant resuspension coefficient of 10^{-7}/h and the monthly green area fraction could quantitatively account for the observed concentration as well as its seasonal variation (grey dashed lines in the figure). The contribution from additional emissions from the reactor buildings of the FDNPS (10^6 Bq/h) was negligible throughout the year (grey dotted lines in the figure). The simulated annual total resuspended amount for the whole region was 1.28 TBq, equivalent to 0.048% of the aircraft-observed total deposited amount of

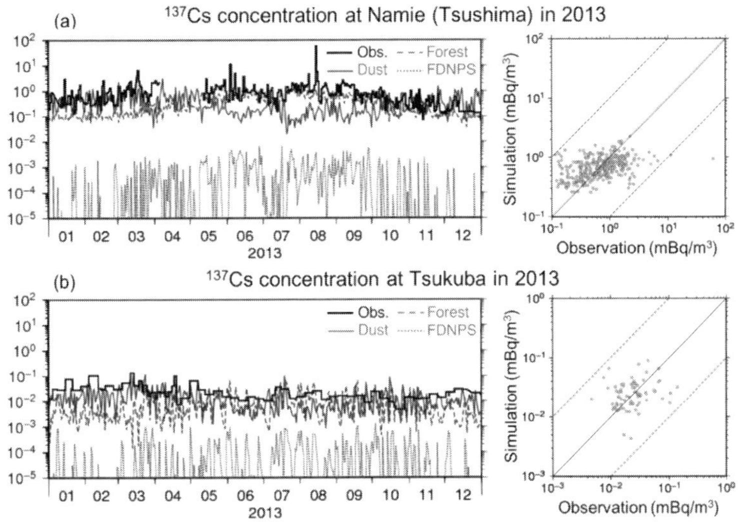

Figure 3.26 Time series of the surface air concentration of observed ^{137}Cs (black) and simulated daily ^{137}Cs levels (greys) on the left and scatterplots between observed and simulated total (i.e. sum of colours) ^{137}Cs at (a) Namie and (b) Tsukuba. The shaded lines indicate simulated ^{137}Cs concentrations due to resuspension from soil (grey solid) using the scheme given by Ishizuka *et al.* (2017) (multiplied by five), resuspension from forests with a resuspension rate of 10^{-7}/h (grey dashed) and emissions from the FDNPS reactor buildings with a constant emission rate of 10^6 Bq/h (grey dotted). Source: Kajino *et al.*, 2016.

2.68 PBq. The total redeposition was 0.18–0.23 TBq, equivalent to 14.1–18.0% of the total resuspended amount. The spatial distribution of the decreasing rate of land surface ^{137}Cs due to resuspension ranged from 2.7×10^{-7}/day to 8.2×10^{-6}/day, which was 2–3 orders of magnitude lower than the first-order rate of decrease of the ambient γ dose rate in Fukushima Prefecture (from 5.2 to 12.1×10^{-4}/day); this resulted from radioactive decay, land surface processes and decontamination. Thus, it appears that resuspension made only a negligible contribution towards reducing ground radioactivity. The resuspension from biota could dominate in the warm season, but the resuspension sources as well as mechanisms remain essentially unknown. Further study based on field experiments and numerical simulations is needed to understand the mechanisms.

3.9 Characteristics of Radioactive Materials in Aerosols

YOSHIO TAKAHASHI AND NAOHIRO YOSHIDA

In this section, the characteristics of aerosols responsible for transporting radionuclides in the atmosphere are discussed. During the FDNPS accident,

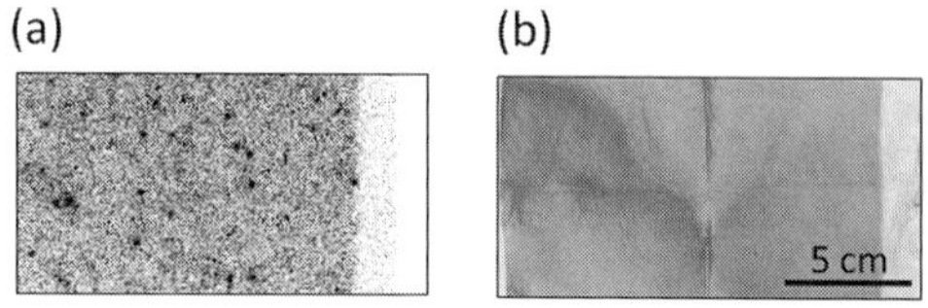

Figure 3.27 Imaging plate image (a) and optical image (b) of aerosol filters collected on 20 March 2011 in Kawasaki City (Tanaka *et al.*, 2013).

radiocaesium and radioiodine travelled as aerosols after their emission from the FDNPS and were dispersed at both regional and global scales. This process, including their behaviour after deposition, is controlled by chemical processes at the molecular scale, which enables us to understand the dispersion process of the radionuclides and to predict their fates after deposition on the ground. However, it is not easy to identify chemical species of radiocaesium and radioiodine in actual aerosol samples based on normal speciation methods because the mole concentrations of radiocaesium and radioiodine are very low, for which spectroscopes and electron microscopes cannot be readily employed. Images from imaging plates of aerosol filters were collected by the Kawasaki Municipal Research Institute for Environmental Protection, including those from the period when radionuclides were emitted from the FDNPS (Figure 3.27; Tanaka *et al.*, 2013). The images of the radionuclides were compared with optical images of the aerosols. The latter images suggested that brown areas showing the presence of aerosols were observed to have characteristics similar to those for normal aerosol particles. However, the former images exhibited black dots that were heterogeneously distributed on the filters. In the samples from five months after the accident, radioiodine (^{131}I) must have completely decayed. Thus, the main radionuclide in the black spot was radiocaesium. If the number of aerosol particles that contained high concentrations of radiocaesium was similar to that of normal radionuclide-free aerosols, heterogeneous distribution of radioactivity would not be observed. Hence, this heterogeneity suggests that there were far fewer aerosol particles that contained very large amounts of radiocaesium. Leaching experiments of the filters by water showed that 50–70% of radiocaesium was found to be leached, suggesting that most of the radiocaesium was water-soluble species, such as chloride, sulphate or nitrate. This result suggests that radiocaesium initially deposited via dry deposition (aerosols) can be leached into water and immediately adsorbed on mineral particles in soil. However, this also indicates that the other radiocaesium fraction is water-insoluble. Adachi *et al.* (2013) found insoluble caesium-enriched particles (Cs ball), which were observed as black spots detected by the imaging plates, as shown in Figure 3.27(a). Further study is strongly needed to quantitatively determine the ratio of insoluble radiocaesium in Cs balls to soluble Cs in the aerosols emitted from the FDNPS accident. Kaneyasu *et al.* (2012) determined particle size distributions for radiocaesium (concentration peak in the 0.49–0.70 μm fraction) in aerosols collected in April and May 2011, which indicated that the distribution of radiocaesium was similar to that of sulphate in aerosols. This correlation showed that the radiocaesium fraction was once incorporated into droplets with sulphate ions in the atmosphere and subsequently evaporated in the atmosphere to form the particle size distribution known as the accumulation or droplet mode. This component is generally water soluble, although the samples in this study were collected one month after the accident, which may not contain the Cs ball fraction emitted in the initial stage. Thus, further survey of aerosols, including discoveries of aerosol filter samples collected during the FDNPS accident (March 2011), is important to understand the initial chemical form of radiocaesium (Chapter 12), which is closely related to the water-soluble nature of radiocaesium emitted from the FDNPS. This chemical state of radiocaesium in turn controls the secondary migration of radiocaesium after the deposition on land surfaces or in seawater.

3.10 Sizes and Distributions of Metallic Particles Caused by Burning or Explosion

ISAO TANIHATA AND MAMORU FUJIWARA

Radioactive materials caused by burning or explosion are emitted in the gaseous or solid phases. In the case of the solid phase, knowledge of the sizes of the solid phase (particle size) and their distributions is particularly important for evaluating processes related to the release and movement of materials into the environment. The particle size distribution of heavy metal particles caused by burning or explosions depends substantially on the specific conditions. In general, the number of particles dramatically decreases with increasing particle size. There is a good experimental example for the case of burning Mg. This trend is well illustrated by a logarithmic function with a parameter of the particle size emitted by burning Mg (Figure 3.28; Yashima *et al.*, 2010). In this case, the particle density distribution, $d(x)$, is approximately described as a function of the form $d(x) \propto x^{-1.9}$, where x is the particle size in microns.

The number of a given atomic element forming a particle is proportional to the cube of the particle diameter. Therefore, the number of large particles decreases because the amount of a given element forming large particles is larger than that of those forming smaller particles. For example, let us imagine that there is a particle A with a certain size. The number of another particle B with twice the size is expected to be one-quarter of that of particle A. However, the total amount of the atomic element forming particle B is eight times larger than that forming particle A, with half its size.

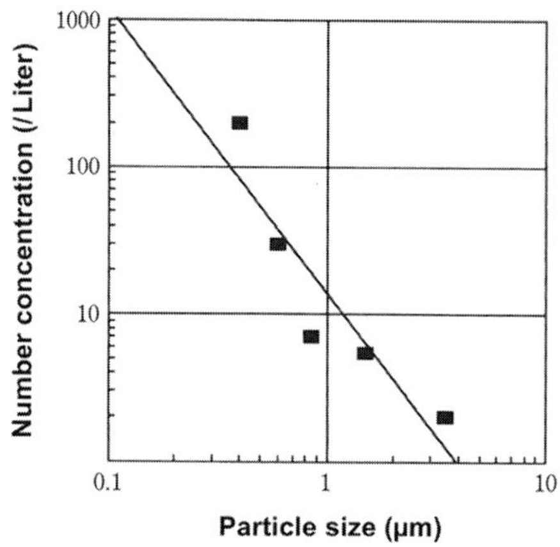

Figure 3.28 Relation between size and number concentration of particles released by a burning process (Yashima *et al.*, 2010).

If the sizes of the deposited particles are identical, the particle size distribution can be used as a proxy for the radioactive material distribution. If the number of particles deposited per unit area (N) increases, the fluctuation of N is simply described as $1/\sqrt{N}$. However, the presence of even a few large particles containing large amounts of radioactive material in the deposited material results in substantial fluctuations in the deposited amounts of radioactive material. If there is substantial fluctuation in the size distribution of the deposited particles, substantial fluctuations in the deposited amounts of radioactive material occur. Such a condition is always satisfied when the particle density decreases at a rate less than x^{-3}.

References

Achim, P., M. Monfort, G. Le Petit, *et al.* (2014). Analysis of radionuclide releases from the Fukushima Dai-ichi Nuclear Power Plant accident part II. *Pure Appl. Geophys.*, **171**, 645–67, doi:10.1007/s00024-012-0578-1.

Adachi, K., M. Kajino, Y. Zaizen and Y. Igarashi (2013). Emission of spherical cesium-bearing particles from an early stage of the Fukushima nuclear accident. *Sci. Rep.*, **3**, 2554.

Chino, M., H. Nakayama, H. Nagai, *et al.* (2011). Preliminary estimation of release amounts of ^{131}I and ^{137}Cs accidentally discharged from the Fukushima Daiichi nuclear power plant into the atmosphere. *J. Nucl. Sci. Technol.*, **48**(7), 1129–34, doi:10.1080/18811248.2011.9711799.

CTBTO (2011). Fukushima-related measurements by CTBTO. Comprehensive Nuclear-Test-Ban Treaty Organization, Preparatory Commission.

Draxler, R. R., D. Arnold, M. Chino, *et al.* (2015). World Meteorological Organization's model simulations of the radionuclide dispersion and deposition from the Fukushima Daiichi nuclear power plant accident. *J. Environ. Radioact.*, **139**, 172–84.

Farchi, A., M. Bocquet, Y. Roustan, A. Mathieu and A. Quérel (2016). Using the Wasserstein distance to compare fields of pollutants: application to the radionuclide atmospheric dispersion of the Fukushima-Daiichi accident. *Tellus B*, **68**, 31682, doi:10.3402/tellusb.v68.31682.

Girard, S., I. Korsakissok and V. Mallet (2014). Screening sensitivity analysis of a radionuclides atmospheric dispersion model applied to the Fukushima disaster. *Atmos. Environ.*, **95**, 490–500, doi:10.1016/j.atmosenv.2014.07.010.

Girard, S., V. Mallet, I. Korsakissok and A. Mathieu (2016). Emulation and Sobol' sensitivity analysis of an atmospheric dispersion model applied to the Fukushima nuclear accident. *J. Geophys. Res. Atmospheres*, **121**, 3484–96, doi:10.1002/2015JD023993.

Hernández-Ceballos, M. A., G. H. Hong, R. L. Lozano, *et al.* (2012). Tracking the complete revolution of surface westerlies over Northern Hemisphere using radionuclides emitted from Fukushima. *Sci. Total Environ.*, **438**, 80–5.

Hirao, S., H. Yamazawa and T. Nagae (2013). Estimation of release rate of iodine-131 and cesium-137 from the Fukushima Daiichi nuclear power plant: Fukushima NPP accident related. *J. Nucl. Sci. Technol.* **50**, 139–147.

Hirose, K. (2016). Fukushima Daiichi Nuclear Plant accident: atmospheric and oceanic impacts over the five years. *J. Environ. Radioact.*, **157**, 113–30, doi:10.1016/j.jenvrad.2016.01.011.

IAEA (ed.) (2015). *The Fukushima Daiichi Accident*, Vol. 4, *Radiological Consequences*. Vienna: International Atomic Energy Agency.

Inomata, Y., M. Aoyama, T. Tsubono, D. Tsumune and K. Hirose (2016). Spatial and temporal distributions of Cs-134 and Cs-137 derived from the TEPCO Fukushima Daiichi Nuclear Power Plant accident in the North Pacific Ocean by using optimal interpolation analysis. *Env. Sci. Process. Impacts*, **18**, 126–36, doi:10.1039/C5EM00324E.

Ishizuka, M., M. Mikami, T. Y. Tanaka, *et al.* (2017). Use of a size-resolved 1-D resuspension scheme to evaluate resuspended radioactive material associated with mineral dust particles from the ground surface. *J. Environ. Radioact.*, **116**, 436–48, doi:10.1016/j.envrad.2015.12.023.

Kajino, M., M. Ishizuka, Y. Igarashi, *et al.* (2016). Long-term assessment of airborne radiocesium after the Fukushima nuclear accident: re-suspension from bare soil and forest ecosystems. *Atmos. Chem. Phys.*, **16**, 13149–72, doi:10.5194/acp-16-13149-2016.

Kaneyasu, N, H. Ohashi, F. Suzuki, T. Okuda and F. Ikemori (2012). Sulfate aerosol as a potential transport medium of radiocesium from the Fukushima nuclear accident. *Environ. Sci. Technol.*, **46**(11), 5720–6, doi:10.1021/es204667h.

Katata, G., M. Ota, H. Terada, M. Chino and H. Nagai (2012). Atmospheric discharge and dispersion of radionuclides during the Fukushima Dai-ichi Nuclear Power Plant accident. Part I: source term estimation and local-scale atmospheric dispersion in early phase of the accident. *J. Environ. Radioact.*, **109**, 103–13, doi:10.1016/j.jenvrad.2012.02.006.

Katata, G., M. Chino, T. Kobayashi, *et al.* (2015). Detailed source term estimation of the atmospheric release for the Fukushima Daiichi Nuclear Power Station accident by coupling simulations of atmospheric dispersion model with improved deposition scheme and oceanic dispersion model. *Atmos. Chem. Phys.*, **15**, 1029–70, doi:10.5194/acp-15-1029-2015.

Kinoshita, N., K. Sueki, K. Sasa, *et al.* (2011). Assessment of individual radionuclide distributions from the Fukushima nuclear accident covering central-east Japan. *Proc. Natl. Acad. Sci.*, **108**, 19526–9, doi:10.1073/pnas.111172410.

Kitayama, K., Y. Morino, M. Takigawa, *et al.* (2018). Model inter-comparison of transport and deposition of atmospheric ^{137}Cs from the Fukushima Dai-ichi Nuclear Power Plant. *J. Geophys. Res. Atmos.*, **123**, 7754–70, doi:10.1038/2017JD028230.

Kobayashi, T., H. Nagai, M. Chino and H. Kawamura (2013). Source term estimation of atmospheric release due to the Fukushima Dai-ichi Nuclear Power Plant accident by atmospheric and oceanic dispersion simulations: Fukushima NPP accident related. *J. Nucl. Sci. Technol.*, **50**, 255–64, doi:10.1080/00223131.2013.772449.

Korsakissok, I., A. Mathieu and D. Didier (2013). Atmospheric dispersion and ground deposition induced by the Fukushima Nuclear power plant accident: a local-scale simulation and sensitivity study. *Atmos. Environ.*, **70**, 267–79, doi:10.1016/j.atmosenv.2013.01.002.

Leadbetter, S. J., M. C. Hort, A. R. Jones, H. N. Webster and R. R. Draxler (2015). Sensitivity of the modelled deposition of caesium-137 from the Fukushima Dai-ichi nuclear power plant to the wet deposition parameterisation in NAME. *J. Environ. Radioact.*, **139**, 200–11, doi:10.1016/j.jenvrad.2014.03.018.

Liu, Y., J.-M. Haussaire, M. Bocquet, *et al.* (2017). Uncertainty quantification of pollutant source retrieval: comparison of Bayesian methods with application to the Chernobyl and Fukushima Daiichi accidental releases of radionuclides. *Q. J. R. Meteorol. Soc.*, **143**, 2886–90, doi:10.1002/qj.3138.

Marzo, G. A. (2014). Atmospheric transport and deposition of radionuclides released after the Fukushima Dai-chi accident and resulting effective dose. *Atmos. Environ.*, **94**, 709–22, doi:10.1016/j.atmosenv.2014.06.009.

Mathieu, A., I. Korsakissok, D. Quélo, *et al.* (2012). Fukushima Daiichi: atmospheric dispersion and deposition of radionuclides from the Fukushima Daiichi Nuclear Power Plant accident. *Elements*, **8**, 195–200, doi:10.2113/gselements.8.3.195.

Mathieu, A., M. Kajino, I. Korsakissok, *et al.* (2018). Fukushima Daiichi-derived radionuclides in the atmosphere, transport and deposition in Japan: a review. *Appl. Geochem.*, **91**, 122–39.

METI (2011). Results of urgent monitoring of radioactive materials around Fukushima Daiich and Daini Nuclear Power Stations. http://bit.ly/2VpUGXJ (accessed 19 September 2018) (in Japanese).

MEXT (2011) Results of the fourth airborne monitoring survey by MEXT. http://bit.ly/2VlOIag (accessed 19 September 2018).

MEXT (2012) Readings of dust sampling by MEXT, as of Sep. 12, 2012. http://bit.ly/2VqRjjy (accessed 19 September 2018) (in Japanese).

Morino, Y., T. Ohara and M. Nishizawa (2011). Atmospheric behavior, deposition, and budget of radioactive materials from the Fukushima Daiichi nuclear power plant in March 2011. *Geophys. Res. Lett.*, **38**, L00G11, doi:10.1029/2011gl048689.

Morino, Y., T. Ohara, M. Watanabe, S. Hayashi and M. Nishizawa (2013). Episode analysis of deposition of radiocesium from the Fukushima Daiichi nuclear power plant accident. *Environ. Sci. Technol.*, **47**, 2314–22, doi:10.1021/es304620x.

Nakajima, T., S. Misawa, Y. Morino, *et al.* (2017). Model depiction of the atmospheric flows of radioactive cesium emitted from the Fukushima Daiichi Nuclear Power Station accident. *Prog. Earth Planet. Sci.*, **4**, 18, doi:10.1186/s40645-017-0117-x.

Nishihara, K., H. Iwamoto and K. Suyama (2012). Estimation of fuel compositions in Fukushima-Daiichi nuclear power plant. JAEA-Data/Code 2012-018, Japan Atomic Energy Agency.

Nuclear Regulation Authority (2011) Reading of radioactivity level in fallout by prefecture. https://radioactivity.nsr.go.jp/en/list/194/list-1.html (accessed 22 March 2019).

Ohkura, T., T. Oishi, M. Taki, *et al.* (2012). Emergency monitoring of environmental radiation and atmospheric radionuclides at Nuclear Science Research Institute, JAEA following the accident of Fukushima Daiichi Nuclear Power Plant. JAEA/Data Code 2012-010.

Oura, Y., M. Ebihara, H. Tsuruta, *et al.* (2015). A database of hourly atmospheric concentrations of radiocesium (^{134}Cs and ^{137}Cs) in suspended particulate matter collected in March 2011 at 99 air pollution monitoring stations in eastern Japan. *J. Nucl. Radiochem. Sci.,* **15**, 1–12.

Périllat, R., I. Korsakissok, V. Mallet, *et al.* (2016). Using meteorological ensembles for atmospheric dispersion modeling of the Fukushima nuclear accident. Presented at the 17th International Conference on Harmonisation within Atmospheric Dispersion Modelling for Regulatory Purposes, Budapest, Hungary, 9 May 2016.

Quérel, A., Y. Roustan, D. Quélo and J.-P. Benoit (2015). Hints to discriminate the choice of wet deposition models applied to an accidental radioactive release. *Int. J. Environ. Pollut.*, **58**, 268–79.

Quérel, A., D. Quélo, Y. Roustan, *et al.* (2016). Impact of changing the wet deposition schemes in ldX on 137-Cs atmospheric deposits after the Fukushima accident. Presented at the 17th International Conference on Harmonisation within Atmospheric Dispersion Modelling for Regulatory Purposes, Budapest, Hungary, 9 May 2016.

Saito, K., T. Shimbori and R. Draxler (2015). JMA's regional atmospheric transport model calculations for the WMO technical task team on meteorological analyses for Fukushima Daiichi Nuclear Power Plant accident. *J. Environ. Radioact.*, **139**, 185–99, doi:10.1016/j.jenvrad.2014.02.007.

Saunier, O., A. Mathieu, D. Didier, *et al.* (2013). An inverse modelling method to assess the source term of the Fukushima Nuclear Power Plant accident using gamma dose rate observations. *Atmos. Chem. Phys.*, **13**, 11403–21, doi:10.5194/acp-13-11403-2013.

Saunier, O., A. Mathieu, T. T. Sekiyama, *et al.* (2016). A new perspective on the Fukushima releases brought by newly available ^{137}Cs air concentration observations and reliable meteorological fields. Presented at the 17th International Conference on Harmonisation within Atmospheric Dispersion Modelling for Regulatory Purposes, Budapest, Hungary, 9 May 2016.

SCJ (2014). A review of the model comparison of transportation and deposition of radioactive materials released to the environment as a result of the Tokyo Electric Power Company's Fukushima Daiichi Nuclear Power Plant accident, Committee on Comprehensive Synthetic Engineering, Science Council of Japan. Science Council of Japan.

Sekiyama, T., M. Kajino, and M. Kunii. (2013). *Ensemble Simulation of the Atmospheric Radionuclides Discharged by the Fukushima Nuclear Accident.* Vienna: EGU General Assembly.

Sekiyama, T. T., M. Kunii, M. Kajino and T. Shimbori (2015). Horizontal resolution dependence of atmospheric simulations of the Fukushima nuclear accident using 15 km, 3 km, and 500m grid models, *J. Meteor. Soc. Japan* **93**, 49–64.

Solazzo, E. and S. Galmarini (2015). The Fukushima-^{137}Cs deposition case study: properties of the multi-model ensemble. *J. Environ. Radioact.*, **139**, 226–33, doi:10.1016/j.jenvrad.2014.02.017.

Sørensen, J. H., B. Amstrup, H. Feddersen, *et al.* (2015). Fukushima accident: uncertainty of atmospheric dispersion modelling (FAUNA). Nordic Nuclear Safety Research.

Stohl, A., P. Seibert, G. Wotawa, *et al.* (2012). Xenon-133 and caesium-137 releases into the atmosphere from the Fukushima Dai-ichi nuclear power plant: determination of the source term, atmospheric dispersion, and deposition. *Atmos. Chem. Phys.*, **12**, 2313–43, doi:10.5194/acp-12-2313-2012.

Sugiyama, G., J. Nasstrom, B. Pobanz, *et al.* (2012). Atmospheric dispersion modeling: challenges of the Fukushima Daiichi response. *Health Phys.*, **102**, 493–508, doi:10.1097/HP.0b013e31824c7bc9.

Takemura, T., H. Nakamura, M. Takigawa, *et al.* (2011). A numerical simulation of global transport of atmospheric particles emitted from the Fukushima Daiichi Nuclear Power Plant. *SOLA*, **7**, 101–4, doi:10.2151/sola.2011-026.

Tanaka, K., A. Sakaguchi, Y. Kanai, *et al.* (2013). Heterogeneous distribution of radiocesium in aerosols, soil and particulate matters emitted by the Fukushima Daiichi Nuclear Power Plant accident: retention of micro-scale heterogeneity during the migration of radiocesium from the air into ground and river systems. *J. Radioanal. Nucl. Chem.*, **295**, 1927–37.

TEPCO (2013). Influence to surrounding environment | Archives.Air, 2011, March. The results of nuclide analyses of radioactive materials in the air at the site of Fukushima Daiichi Nuclear Power Station (from the first to the tenth releases). www.tepco.co.jp/en/nu/fukushima-np/f1/index2-e.html (accessed 19 September 2018).

TEPCO (2017). Evaluation of additional release from nuclear reactor building www.tepco.co.jp/nu/fukushima-np/handouts/2017/index-j.html (accessed 19 September 2018) (in Japanese).

Terada, H., G. Katata, M. Chino and H. Nagai (2012). Atmospheric discharge and dispersion of radionuclides during the Fukushima Dai-ichi Nuclear Power Plant accident. Part II: verification of the source term and analysis of regional-scale atmospheric dispersion. *J. Environ. Radioact.*, **112**, 141–54, doi:10.1016/j.envrad.2012.05.023.

Tsuruta, H., M. Takigawa, and T. Nakajima (2012). 'Summary of atmospheric measurements and transport pathways of radioactive materials released by the Fukushima Daiichi Nuclear Power Plant accident'. *Proceedings of The First NIRS Symposium on Reconstruction of Early Internal Dose the TEPCO Fukushima Daiichi Nuclear Power Station Accident*, pp. 101–111, National Institute of Radiological Sciences, Japan, 10–11 July 2012.

Tsuruta, H., T. Arai, K. Shiba, *et al.* (2013). Atmospheric ^{131}I and ^{137}Cs measured in the east coast of Kanto area and the east Fukushima area in an early period after the Fukushima Dai-ichi Nuclear Power Plant accident. In *Proceedings of the 14th Workshop on Environmental Radioactivity*. High Energy Accelerator Research Organization, pp. 90–8 (in Japanese).

Tsuruta, H., Y. Oura, M. Ebihara, T. Ohara and T. Nakajima (2014). First retrieval of hourly atmospheric radionuclides just after the Fukushima accident by analyzing filter-tapes of operational air pollution monitoring stations. *Sci. Rep.,* **4**, 6717, doi:10.1038/srep06717.

Tsuruta, H., Y. Oura, M. Ebihara, Y. Moriguchi and T. Nakajima (2016). Comprehensive retrieval of spatio-temporal distribution of atmospheric radionuclides just after the Fukushima accident by analyzing filter-tapes of operational air pollution monitoring stations. Abstract, Goldschmidt Abstracts, p. 3198.

Tsuruta, H., Y. Oura, M. Ebihara, *et al.* (2017) Spatio-temporal distribution of atmospheric radiocesium in eastern Japan just after the TEPCO Fukushima Daiichi Nuclear Power Plant accident: analysis of used filter-tapes of SPM monitors in air quality monitoring stations. *Earozoru kenkyu*, **32**, 244–54 (in Japanese).

Tsuruta, H., Y. Oura, M. Ebihara, *et al.* (2018). Time-series analysis of atmospheric radiocesium at two SPM monitoring sites near the Fukushima Daiichi Nuclear Power

Plant just after the Fukushima accident on March 11, 2011. *Geochemical J.*, **52**(2), 103–21, doi:10.2343/geochemj.2.0520.

University of Tokyo (2011). The University of Tokyo environmental radiation info (provisional values). http://bit.ly/2UeTgze (accessed 19 September 2018).

UNSCEAR (2013). Sources, effects and risks of ionizing radiation. In *UNSCEAR 2013 Report to the General Assembly with Scientific Annexes*, Vol. **2**, *Scientific Annex B*. New York: United Nations.

UNSCEAR (2015). Developments since the 2013 UNSCEAR report on the levels and effects of radiation exposure to the nuclear accident following the great east-Japan earthquake and tsunami: a 2015 white paper to guide the Scientific Committee's future programme of work.

UNSCEAR (2016). Developments since the 2013 UNSCEAR report on the levels and effects of radiation exposure to the nuclear accident following the great east-Japan earthquake and tsunami: a 2016 white paper to guide the Scientific Committee's future programme of work.

Winiarek, V., M. Bocquet, O. Saunier and A. Mathieu (2012). Estimation of errors in the inverse modeling of accidental release of atmospheric pollutant: application to the reconstruction of the cesium-137 and iodine-131 source terms from the Fukushima Daiichi power plant. *J. Geophys. Res.*, **117**, D05122, doi:10.1029/2011JD016932.

Winiarek, V., M. Bocquet, N. Duhanyan, *et al.* (2014). Estimation of the caesium-137 source term from the Fukushima Daiichi nuclear power plant using a consistent joint assimilation of air concentration and deposition observations. *Atmos. Environ.*, **82**, 268–79, doi:10.1016/j.atmosenv.2013.10.017.

WMO (2013). Third meeting of WMO task team on meteorological analyses for Fukushima-Daiichi nuclear power plant accident. http://bit.ly/2ECN7HS.

Yashima, M., M. Ono, M. Takaya and T. Ashida (2010). *JNIOSH-SRR*, **40**, 19–26.

Yumimoto, K., Y. Morino, T. Ohara, *et al.* (2016). Inverse modeling of the ^{137}Cs source term of the Fukushima Dai-ichi Nuclear Power Plant accident constrained by a deposition map monitored by aircraft. *J. Environ. Radioact.*, **164**, 1–12, doi:10.1016/j.jenvrad.2016.06.018.

4

Global Transport of Radioactive Materials

The radioactive materials that were released into the atmosphere due to the Fukushima Daiichi Nuclear Power Station (FDNPS) accident not only spread within Japan but also dispersed over the entire globe through atmospheric flows. As described in Chapter 3, there were strong westerlies and a low pressure that passed over the Tohoku region when the accident occurred, which transported most of the released radioactive materials towards the east in the form of gas and/or aerosol particles. Previous studies on atmospheric trace materials (e.g. Okada *et al.*, 1992; Husar *et al.*, 2001; Uno *et al.*, 2009) have shown that aerosol particles can be transported over long distances. A good example is the air pollution originating over East Asia, such as the Asian dust that arises from the dry land of China and Mongolia, which has been identified in the USA as well as over the Pacific Ocean. Therefore, to understand the entire picture of the radioactive pollution caused by the accident, it is necessary to clarify how the radioactive materials were transported, as well as deposited, over Japan and around the world. The radioactive materials that were produced by the FDNPS accident have been detected throughout the world. To date, numerical simulations of the transport of radioactive materials over large areas have been carried out by various organisations. In this chapter, we will discuss the global transport of the radioactive materials caused by the accident by examining their detection around the world, the characteristics of the atmospheric transport of radioactive materials using global numerical simulations, and estimations of the release of the radioactive materials using observations and numerical simulations.

4.1 Global Observation of Radioactive Material

TAICHU YASUMICHI TANAKA, TOSHIHIKO TAKEMURA AND MICHIO AOYAMA

Observation of airborne radioactive materials began around the world after the FDNPS accident to evaluate the potential risks to the environment and the human body through exposure to the radioactive materials. Observations of global-scale radioactive material transport have indicated the presence of radioactive xenon (^{133}Xe), as well as caesium 137 (^{137}Cs) and iodine 131 (^{131}I).

^{133}Xe is used as an indicator for underground nuclear tests or leakage from nuclear fuel containers because its half-life is approximately 5.2 days and its chemically inactive and non-soluble characteristics enable it to survive atmospheric removal processes such as precipitation and chemical reactions, thus allowing it to be readily detected in observations.

To monitor the manufacture and experiments of nuclear weapons as well as the operation of nuclear facilities, the Comprehensive Nuclear Test Ban Treaty Organization (CTBTO) will operate a worldwide monitoring system of radionuclides called the International Monitoring System (IMS). As of 2017, the Comprehensive Nuclear Test Ban Treaty (CTBT) had not entered into force; however, the Preparatory Commission for the CTBTO, which was founded in 1996, has been building global systems for the detection of nuclear activities. The monitoring systems are not complete, but are largely operational. When it is complete, the CTBTO IMS will set up 80 observatories for particulate radioactive material, as well as 40 observatories for radioactive xenon. Among the CTBTO IMS stations, 64 stations measured radioactive aerosol particles and 27 stations measured radioactive xenon after the FDNPS accident (Medici, 2001; Yonezawa and Yamamoto, 2011; Comprehensive Nuclear-Test-Ban Treaty Organization Preparatory Commission, 2011).

The CTBTO stations in Japan, which are in Takasaki (Gunma) and Okinawa, released their measurements immediately following the accident (Yonezawa and Yamamoto, 2011). The CTBTO observatories around the world have continuously measured radioactive materials and have reported the detection of radioactive substances from the accident. The radioactive materials have been detected at most observatories in the Northern Hemisphere, although the concentrations outside Japan tend to be low and are thus thought to be at a level that does not impact the human body due to dilution during transport. Based on the CTBTO observations, the radioactive materials that were released by the FDNPS accident were detected in Eastern Russia on 14 March as well as in the USA and Canada after 16 March. Other than the CTBTO, several institutes have also reported observations of radioactive materials. Beginning on 16 March, the Pacific Northwest National Laboratory (PNNL) in Richland detected ^{133}Xe that was released during the accident (Bowyer *et al.*, 2011; Figure 4.1(a)). Leon *et al.* (2011) detected ^{131}I, ^{134}Cs, ^{137}Cs and ^{132}Te in Seattle on 17–18 March, indicating that the transport from Japan to Seattle took 5–6 days. The detected radioactive material had a short lifetime, indicating that it was not from spent nuclear fuel but possibly from recently active fuel elements (Leon *et al.*, 2011). The observations indicated that ^{133}Xe reached North America faster than ^{137}Cs, meaning that ^{133}Xe was released relatively early in the accident (e.g. Stohl *et al.*, 2012).

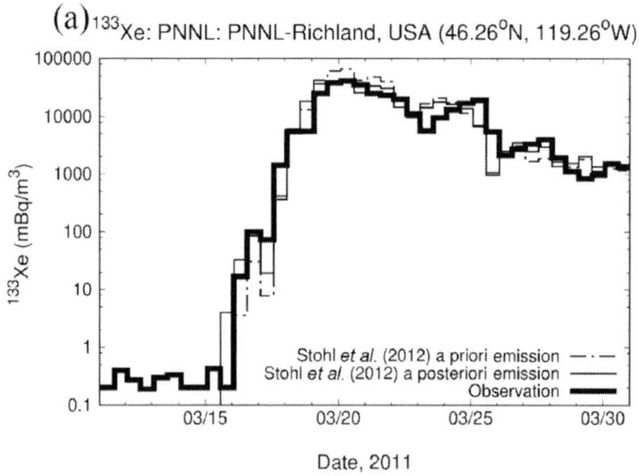

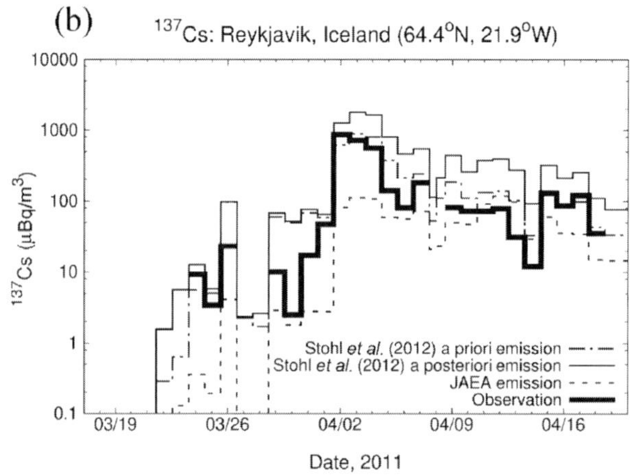

Figure 4.1 Comparisons of time series of observed and simulated radioactive materials, (a) ^{133}Xe at Richland (observation: Bowyer *et al.*, 2011), USA; and (b) ^{137}Cs at Reykjavik, Iceland (observation: Icelandic Radiation Safety Authority, 2011). The simulations are conducted using a global model of the Meteorological Research Institute (MRI) that uses a-priori (dot-dash line) and a-posteriori (solid thin line) estimation of Stohl *et al.* (2012), and estimation by the Japan Atomic Energy Agency (JAEA) (dotted line).

The Environmental Protection Agency (EPA) of the USA reported on 22 March that RadNet had detected radioactive materials on 18 March. The deposition of ^{137}Cs, ^{134}Cs and ^{131}I has been evaluated based on the EPA RadNet and the National Atmospheric Deposition Program (NADP) observations (Wetherbee *et al.*, 2012). The RadNet observations indicate that the ^{131}I observed in the USA had the typical characteristics of a material transported over a long distance, and deposited ^{131}I decreased during the transport across the USA from west to east. Moreover, the fallout of the radioactive materials from the FDNPS accident found in the USA was reported to be higher than that of the Chernobyl accident. Additionally, a radioisotope of sulphur, ^{35}S, presumed to be neutrons that leaked from the nuclear reactors, was detected in La Jolla, California, on 28 March (Priyadarshi *et al.*, 2011) in concentrations exceeding the natural levels.

The radioactive material was transported across the Atlantic to Scandinavia and to Europe. More than 150 European observatories measured radioactive materials from the accident. Most of the observatories are participating in a network called 'Ring of 5' (Ro5), which started in 1983 and exchanges data on occasional enhanced concentrations of anthropogenic radioactive materials (Masson *et al.*, 2011). On 19–20 March, an observatory in Iceland detected ^{131}I and ^{137}Cs, and other radioactive materials were observed on 23 March (Icelandic Radiation Safety Authority, 2011; Figure 4.1(b)). Elsewhere in Europe – in Norway, Germany, France and Greece – similar observations occurred. Masson *et al.* (2011) estimated that the ^{131}I from the accident that passed over Europe was approximately 1 PBq, which was less than 1% of the total emission (approximately 150 PBq).

After passing over Europe, the radioactive materials were detected at the CTBTO observatories in Russia, Mongolia and China. In South Korea, ^{133}Xe was detected during 23–27 March, and ^{131}I was found on 28 March (Kim *et al.*, 2012). Relatively high concentrations of ^{131}I (3.12 mBq/m^3), ^{137}Cs (1.19 mBq/m^3) and ^{134}Cs (1.25 mBq/m^3) were detected on 6–7 April in South Korea. The observed high concentrations in South Korea on April were, however, not estimated to be transported westward. From a trajectory analysis, these radioactive materials were presumed to have reached the East China Sea after circling over the southern part of the Sea of Japan (Kim *et al.*, 2012).

The radioactive materials that were transported eastward by the westerlies spread throughout the entire Northern Hemisphere, and the concentrations increased above detectable levels by 27–28 March. Furthermore, in the tropical Pacific Ocean region, observatories in Hawaii detected radioactive materials that were separated by low-pressure systems that developed in the northern Pacific during March and April. Additionally, the radioactive materials that were transported to the southwest from Japan were detected in Okinawa, Taiwan (Hsu *et al.*, 2012; Huh *et al.*, 2012) and southeastern Asia. ^{131}I, ^{137}Cs and ^{134}Cs were detected in the Philippines on 23 March and in late April (Philippine Nuclear Research Institute, 2011), and ^{131}I, ^{137}Cs and ^{134}Cs were detected from 27 March to 22 April in Dalat, Hanoi and Ho Chi Minh City, Vietnam (Long *et al.*, 2012). The CTBTO observatories reported that the radioactive materials had reached the Southern Hemisphere in April based on detections in Australia, Malaysia and Papua New Guinea (Comprehensive Nuclear-Test-Ban Treaty Organization Preparatory Commission, 2011).

4.2 Simulations of the Long-Range Transport of Radioactive Materials after the Accident

TAICHU YASUMICHI TANAKA, TOSHIHIKO TAKEMURA AND MICHIO AOYAMA

Atmospheric long-range transport simulations of radioactive materials from the FDNPS were conducted after the accident caused by the earthquake and tsunami. For tracking and predicting the spread of airborne hazardous substances, such as volcanic eruptions and nuclear accidents, the World Meteorological Organization (WMO) has established the Emergency Response Activities (ERA) programme. The WMO responded to the FDNPS accident immediately (World Meteorological Organization, 2011). At the request of the International Atomic Energy Agency (IAEA), the WMO requested information related to the dispersion of radioactive materials from the Environmental Emergency Response (EER) Regional Specialized Meteorological Center (RSMC). The Japan Meteorological Agency (JMA) is specified as the Asian node of the EER RSMC by the WMO and provides dispersion prediction information of harmful substances released into the atmosphere in the case of a nuclear accident. Therefore, the JMA provided atmospheric dispersion simulation information to the IAEA based on a global Lagrangian environment emergency response model. The simulation used the JMA's meteorological analysis and the IAEA's specified conditions that included a constant emission rate. The calculated output included the near-surface (up to an altitude of 500 m) concentrations, fallout to the ground and trajectory analysis of the radioactive materials. The spatial resolution of the model was approximately 100 km, and the observed radioactive concentrations were not reflected in the calculation. Thus, the JMA stated that it was not suitable for a domestic emergency evacuation plan (Japan Meteorological Agency, 2011). Zentralanstalt für Meteorologie und Geodynamik (ZAMG) in Austria published estimated dispersion information of the emitted radioactive materials on 12 March (Wotawa, 2011) and provided detailed meteorological and dispersion information to the Emergency Response Center (IEC) of the IAEA at the request of the WMO on 15 March. The simulated results by ZAMG were received in Japan via the internet after first being featured in the German magazine *Spiegel* (Becker, 2011) on 15 March. The Institute for Radiological Protection and Nuclear Safety (IRSN) of France, the Norwegian Meteorological Institute (Norsk institutt for luftforskning; NILU) and the Germany Weather Service (Deutscher Wetterdienst; DWD) provided long-range dispersion predictions for radioactive materials.

These large-scale dispersion predictions from institutions outside of Japan drew attention from people in Japan, especially because the operational status of SPEEDI was not known immediately following the accident, and the SPEEDI predictions were not available to the public until 25 April. The JMA was also criticised because it provided the dispersion simulation information of the EER to the IAEA but not to the Japanese public. After the chief cabinet secretary held a press conference on 4 April (Cabinet Office, Government of Japan, 2011), the EER dispersion information from the JMA was made available to the public via its website. The problems regarding the role of public information in an emergency can be addressed when the information is quickly shared. A research group from universities in Japan also conducted a large-scale transport experiment using a global aerosol transport model and published a research paper in June 2011 (Takemura *et al.*, 2011). Transport simulations of radioactive material from the accident have been conducted by several research groups inside and outside of Japan (Ten Hoeve and Jacobson, 2012; Winiarek *et al.*, 2012; Christoudias and Lelieveld, 2013; Kristiansen *et al.*, 2016; Mészáros *et al.*, 2016; Sarkar *et al.*, 2017). In the next section, we describe the transport pathway of the radioactive material estimated by these simulations.

4.3 Estimation of the Transport Pathway and Simulation of Radioactive Materials Using Global-Scale Models

TAICHU YASUMICHI TANAKA, TOSHIHIKO TAKEMURA AND MICHIO AOYAMA

The radioactive materials from the FDNPS accident were transported in the atmosphere in the form of gas and aerosol particles, as described in Chapter 3. Currently, many large-scale numerical atmospheric models include aerosol transport processes because they are important to the climate. This makes it relatively easy to conduct large-scale simulations of the transport of radioactive materials using existing numerical models. Takemura *et al.* (2011) used the SPRINTARS (Spectral Radiation-Transport Model for Aerosol Species) global aerosol model with a horizontal resolution of 0.56 degrees (approximately 55 km). They assumed a constant emission rate from the FDNPS and aerosol particles with a diameter of 10 μm as a tracer. They estimated that the aerosol particles were transported over the Pacific Ocean and reached the western coast of the USA in four days, and the concentrations rapidly decreased to 10^{-8} of the initial level in the area surrounding the power plant (Figure 4.2).

The particles reached Iceland after six days and Europe shortly thereafter (Figure 4.1). The results are largely consistent with the detected timing of the observations by the CTBTO and other institutions. This simulation reproduced the transport of radioactive materials well throughout the Northern Hemisphere, although the assumed emission rate was constant. Numerical simulations of the transport of radioactive materials under more realistic conditions became possible after the estimated temporal variations in the emission fluxes of radioactive materials were published by the JAEA and other researchers (Chino *et al.*, 2011; Stohl *et al.*, 2012). The MRI of the JMA performed a numerical simulation of the global transport of radioactive materials from the FDNPS using estimated emission fluxes (Tanaka *et al.*, 2012). Their global simulation was conducted with their global aerosol model, called MASINGAR mk-2, which is coupled with

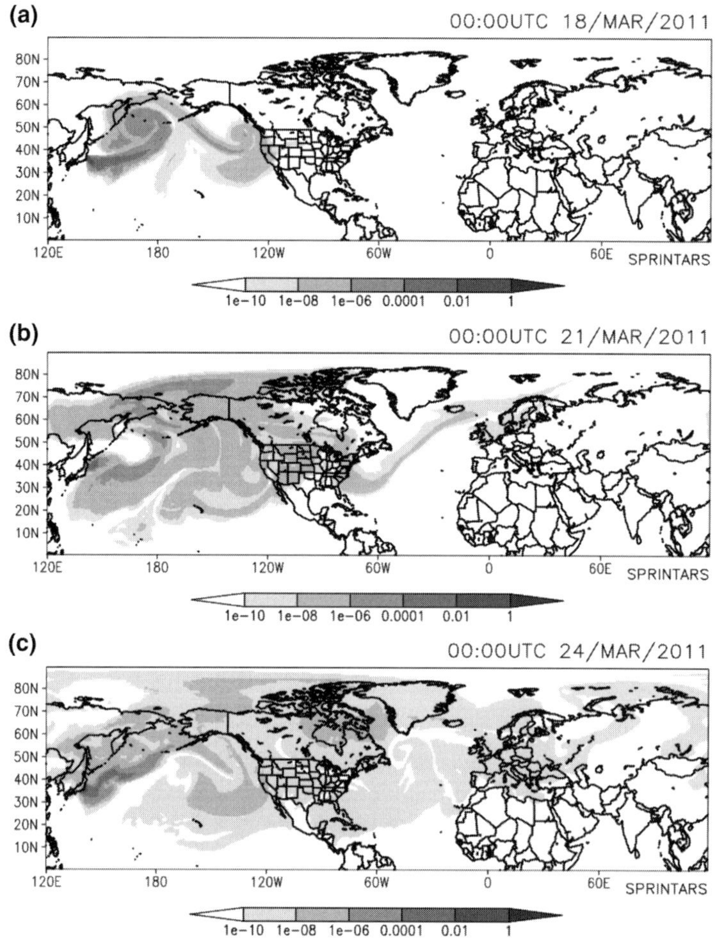

Figure 4.2 Numerically simulated near-surface particle concentration emitted continuously from the FDNPS since 12:00 UTC, 14 March 2011, using the global aerosol model SPRINTARS. Each figure shows 00:00 UTC on (a) 18, (b) 21 and (c) 24 March 2011. The concentration indicated is relative to that within a few tens of kilometres around the power plant. Each range of shading contours corresponds to one order of magnitude (Takemura *et al.*, 2011).

the MRI-AGCM3 general circulation model. This numerical model is essentially the same model as that used by the JMA's Aeolian dust prediction, but with finer grid resolutions (60 km in the horizontal and 48 layers in the vertical) compared with that used for the JMA's Aeolian dust prediction (110 km and 40 layers). In the simulation, ^{131}I, ^{132}I, ^{132}Te, ^{133}Xe, ^{134}Cs and ^{137}Cs were treated as radioactive materials. In the numerical model, ^{133}Xe was treated as a noble gas and removed only by radioactive decay. The other substances were assumed to be attached to sulphate aerosol particles and removed from the atmosphere by dry/wet deposition

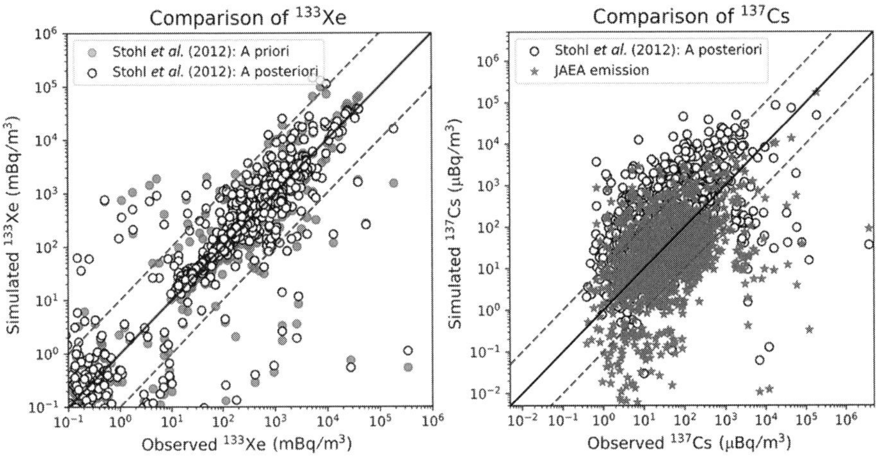

Figure 4.3 Comparisons of the observed and simulated concentrations of (a) ^{133}Xe and (b) ^{137}Cs using a global simulation model by the MRI. The period is from 10 March 2011 to 30 April 2011. In (a), the grey circles indicate the simulated results using the first guess (a priori) of ^{133}Xe emission by Stohl *et al.* (2012), and the white circles indicate the results using the inversely analysed (a posteriori) ^{133}Xe emission. The solid line in the centre represents the 1:1 value of the simulation and the observation value. The upper and lower dashed lines show a factor 10, which represents the over-/underestimations.

and radioactive decay. The MRI used the emission fluxes of radioactive materials estimated by the JAEA (Chino *et al.*, 2011; Terada *et al.*, 2012) for ^{131}I, ^{132}I, ^{134}Cs and ^{137}Cs, and by Stohl *et al.* (2012) for ^{133}Xe and ^{137}Cs. The emission fluxes for ^{137}Cs differed greatly, with the total emission amount through late April 2011 being 8.8 PBq according to the JAEA and 36.6 PBq based on Stohl *et al.* (2012), differing by a factor of four. The total emission flux of ^{133}Xe was 15.3 EBq according to Stohl *et al.* (2012), slightly larger than the a-priori estimate of 12.4 EBq.

The reported results of the simulation reproduced well the observed arrival time in the Pacific Ocean, North America and Europe (Figures 4.1 and 4.3). The arrival time and the concentration of ^{133}Xe that reached North America exhibited very good agreement. The simulated concentration of ^{137}Cs generally correlated with the observed value, although the numerical model of the MRI underestimated the concentration that reached Europe, while it overestimated the concentration that reached southeastern Asia and the tropical Pacific Ocean. The cause of the discrepancies is thought to be uncertainties during the process of removal via precipitation.

The simulated concentrations of radioactive materials tended to be underestimated when the emission fluxes estimated by the JAEA were used, while the simulated results tended to be overestimated when the emission estimates of Stohl *et al.* (2012) were used (Figure 4.3). The simulated patterns and the order of

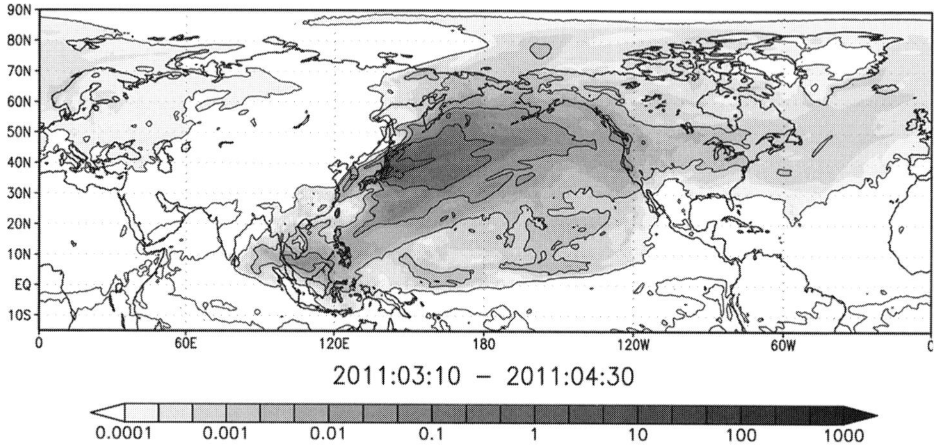

Figure 4.4 Simulated distribution of the accumulated deposition of ^{137}Cs from March to April 2011 (unit: kBq/m^2) by the global transport model of the MRI. Meteorological fields are assimilated with a nudging scheme using the global analysis of the JMA as a reference. The estimated emission flux of ^{137}Cs by the JAEA is used.

magnitude of the deposition of ^{137}Cs were comparable to the aircraft observations by the Ministry of International Trade and Industry (Torii *et al.*, 2012) and *in-situ* observations, although the detailed structures were not reproduced by the global model due to the insufficient grid resolution. The simulated results showed that most of the radioactive materials were transported eastward and deposited in the northern Pacific Ocean (Figure 4.4). The simulation also suggested that 70% of the released ^{137}Cs was deposited in the Pacific Ocean when the emission flux estimates of the JAEA were used, while 82% was deposited when the Stohl *et al.* (2012) emission flux estimates were used.

4.4 Inverse Estimation of Emission Fluxes Based on Global Observations and Numerical Simulations

TAICHU YASUMICHI TANAKA, TOSHIHIKO TAKEMURA AND MICHIO AOYAMA

Table 4.1 shows several estimates of ^{137}Cs emissions calculated by various numerical models. The first emission estimate of the JAEA was calculated by comparing the model simulation and observed concentrations within Japan (Chino *et al.*, 2011; Terada *et al.*, 2012; see Chapter 2). The JAEA has been continuously updating their emission estimates of ^{137}Cs and ^{131}I (Terada *et al.*, 2012; Kobayashi *et al.*, 2013; Katata *et al.*, 2015). The limitation in the number of observations may lead to large uncertainties over the Pacific Ocean, although the radioactive

Table 4.1 *Estimated total emission of ^{137}Cs to the atmosphere from the power plant due to the FDNPS accident, using a numerical model.*

Reference	Total ^{137}Cs emission	Integration period	Remarks
Chino *et al.* (2011)	13 PBq	12 March to 6 April	Reverse analysis
Terada *et al.* (2012)	8.8 PBq	12 March to 30 April	Reverse analysis
Stohl *et al.* (2012)	36.6 PBq (20.1–53.1)	10 March to 20 April	Inversion analysis
Winiarek *et al.* (2012)	10–19 PBq	11 March to 26 March	Inversion analysis
Ten Hoeve and Jacobson (2012)	17.0 PBq	12 March to 12 April	
Kobayashi *et al.* (2013)	13.0 PBq	12 March to 1 April	Reverse analysis with atmospheric and ocean models
Saunier *et al.* (2013)	15.5 PBq	12 March to 27 March	
Winiarek *et al.* (2014)	11.6–19.3 PBq	11 March to 1 April	Inversion analysis
Maki (2015)	19.4 ± 3 PBq	10 March to 20 April	Inversion analysis
Katata *et al.* (2015)	14.5 PBq	12 March to 1 May	Reverse analysis with both land and ocean observation and models
Aoyama *et al.* (2016)	15–20 PBq	11 March to 31 May	From observed concentrations in seawater and numerical models

materials transported within Japan may be estimated with relatively high accuracy. This may be a reason that the total emission estimates by the JAEA (Terada *et al.*, 2012) are relatively small (8.8 PBq). The JAEA used both land and ocean observation data in their recent estimate of radionuclides with their atmospheric (WSPEEDI-II) and oceanic (SEA-GEARN-FDM) models (Kobayashi *et al.*, 2013; Katata *et al.*, 2015). The emissions estimated by Stohl *et al.* (2012) were derived using an inversion method with the FLEXPART atmospheric transport model. The inversion method used the primary fluxes from the nuclear reactor and observed airborne concentrations and deposition fluxes from worldwide observatories, including those in Japan. Their method reflected the observed radioactive materials transported across the Pacific Ocean based on worldwide observations; however, as the transport distance increased, uncertainty in the processes in the numerical model, such as wet deposition, may be greater. The estimated total ^{137}Cs emission based on the inversion analysis of Stohl *et al.* (2012) was larger (36.6 PBq) than the other emission estimates. Stohl *et al.* (2012) also conducted an inversion analysis of ^{133}Xe and argued that the emissions of ^{133}Xe occurred earlier than the assumed initial guess, suggesting that the leak of ^{133}Xe possibly began before the vent rupture immediately following the earthquake on 11 March. When the numerical simulation of the MRI used the ^{137}Cs emissions estimated by Stohl *et al.* (2012), the ^{137}Cs was overestimated, suggesting that the estimated emission fluxes based on the inversion analysis strongly depend on the transport model and the inverse analytical method. Maki (2015) estimated the emission flux of ^{137}Cs using a global aerosol model of the MRI and an inversion method similar to that of Stohl *et al.* (2012), and reported a total emission of ^{137}Cs of 13–24 PBq, which is between the results of the JAEA and Stohl *et al.* (2012). Winiarek *et al.* (2012) noted that the estimated emission fluxes of the radioactive materials strongly depended on the assumed initial guess conditions and the solutions of the inversion analysis method. Winiarek *et al.* (2014) developed a joint assimilation of the ^{137}Cs source term using all available data such as the atmospheric concentration, the daily fallout and the total cumulated deposition, and estimated the total ^{137}Cs emission of 11.6–19.3 PBq, with standard deviation of 15–20%. A reason for the large uncertainty in the emission estimates of the radioactive materials by the inversion analyses is the insufficient number of observations over the Pacific Ocean, which is where most of the ^{137}Cs was deposited. To overcome this problem, Aoyama *et al.* (2016) estimated the emissions based on the concentrations of ^{137}Cs and ^{134}Cs measured in seawater samples, as outlined in Section 5.4. Aoyama *et al.* (2016) created an inventory of the concentrations of radioactive caesium in the surface seawater based on observations. They also calculated the concentrations of ^{137}Cs in seawater using the ocean model of the Central Research Institute of Electric Power Industry, coupled with the ^{137}Cs atmospheric deposition fluxes from three large-scale atmospheric transport models of the MRI. Based on a comparison of the inventory and calculated concentrations of ^{137}Cs, they estimated the total emission of ^{137}Cs into the atmosphere from the accident to be 15–20 PBq. The inversion analyses of emission fluxes using observations and simulations still have challenges, such as the utilisation of Kalman filter/four-dimensional data assimilation methods for numerical weather prediction, the inclusion of additional high-quality observations, and improvement in the accuracy of the numerical models. It is expected that inversion analysis using the γ dose rate will provide more detailed information about the emissions of radioactive materials as dose rate measurements become more widespread (Saunier *et al.*, 2013).

4.5 Future Issues in the Global Simulation of Radioactive Materials

TAICHU YASUMICHI TANAKA, TOSHIHIKO TAKEMURA AND MICHIO AOYAMA

As discussed in this chapter, the global transport of radioactive materials has been increasingly verified by observations from around the world and global-scale numerical simulations. In this section, we describe the challenges of reducing uncertainties in global atmospheric numerical simulations of radioactive materials.

The challenges plaguing global atmospheric numerical models are similar to those of regional models. As described by Ohara and Morino (2012), the following points are especially important:

- reducing the uncertainty in the emission estimations of radioactive materials;
- determining the state of radioactive materials in the atmosphere, such as gases and aerosols;
- accurately representing meteorological fields and refining the dry and wet deposition processes related to radioactive materials;
- ensuring a sufficient spatial resolution that can adequately represent the terrain around the FDNPS – to accomplish this requirement, nested structure models or variable spatial resolution grid models will be required;
- understanding the ground surface and vegetation conditions near the power plant and numerical modelling of the resuspension of radioactive materials from the ground; and
- quantitatively analysing the uncertainties in numerical models via intercomparisons between models and/or multiple model ensemble analysis.

As mentioned earlier, research on the emission estimation using observations and model simulations is particularly important for evaluating the impacts of nuclear accidents. The most uncertain process in numerical models is the removal from the atmosphere by precipitation. The wet removal process includes two issues: how the particles or gases are captured by the precipitating water drops and the accuracy of the precipitation process. Previous studies have indicated that the results of radioactive transport simulations can be considerably different when using different meteorological analysis fields. Stohl *et al.* (2012) showed that the inversely estimated emission fluxes of radioactive materials can be different when using different meteorological analysis fields. The spatial resolution of numerical models should be fine enough to represent the point source of a power plant and its complicated surrounding topography; however, the resolutions currently used in global simulations of radioactive material, which are at most several tens of kilometres in the horizontal, are not adequate. *In-situ* and aircraft observations have indicated that there are sharp gradients in the deposited radioactive materials on the order of several to several tens of kilometres and ‘hotspots’ of high radiation levels over the Kanto Plain. The current global models cannot represent these distributions. We should carefully investigate the following points. Can we discuss the global transport of radioactive materials using large-scale numerical models that cannot adequately express the topography around the FDNPS? Is it appropriate to estimate the emission fluxes with such large-scale numerical models? As discussed by Stohl *et al.* (2012) and Ohara and Morino (2012), it is necessary to consider using nested global and regional atmospheric models or using a high-resolution Lagrangian transport model and a large-scale Eulerian transport model. Because the atmospheric transport models used in different research organisations have different characteristics, it is necessary to evaluate the uncertainties of the simulated distributions of radioactive materials via intercomparisons. After the Chernobyl Nuclear Power Plant accident, the WMO, the Commission of European Communities (CEC) and the IAEA jointly conducted an atmospheric transport model evaluation study called ATMES (Atmospheric Transport Model Evaluation Study), which was followed by ETEX (European Tracer Experiments). In order to assess the uncertainties of the simulated distributions of the radioactive materials from the FDNPS, the environmental modelling working group of the Science Council of Japan (SCJ) has conducted an intercomparison experiment using regional and large-scale atmospheric models (Sectional Committee on Nuclear Accident Committee on Comprehensive Synthetic Engineering, Science Council of Japan, 2014). Observations and simulations of the global transport of the radionuclides from the accident were used for the study of the atmospheric lifetime of aerosols. Kristiansen *et al.* (2012) analysed the ratios of aerosol-bound radionuclides ^{137}Cs and ^{131}I, and ^{133}Xe as a passive tracer. They found the *e*-folding lifetimes of ^{137}Cs and ^{131}I to be 10.0–13.9 days and 17.1–24.2 days, respectively, during April and May 2011, suggesting that the typical aerosol lifetime of the numerical models (3–7 days) may be too fast. Kristiansen *et al.* (2016) compared the ^{133}Xe and ^{137}Cs transport experiment with 19 global models and observations at CTBTO stations. They found that the modelled median lifetime of the ^{137}Cs was 9.4 ± 2.3 days, which suggests the removal of aerosols was too fast in most models. Several studies have been published that use the simulated results of the Fukushima accident using large-scale atmospheric transport models that evaluate the risk of health damage caused by the radioactive materials (e.g. Ten Hoeve and Jacobson, 2012; Christoudias and Lelieveld, 2013) and assess the risk of contamination by the nuclear accident (Lelieveld *et al.*, 2012; Christoudias *et al.*, 2014). The information obtained from the studies with the observations and simulations of the accident can be utilised for making hazard maps. The way to utilise the information obtained from the studies should be investigated (e.g. for making hazard maps). Since the atmospheric transport is important not only for nuclear accidents but also for other disasters such as volcanic eruptions and vegetation fires, the risk evaluation studies with the observations and simulations are beneficial for a wide range of public safety aspects.

References

Aoyama, M., M. Kajino, T. Y. Tanaka, *et al.* (2016). ^{134}Cs and ^{137}Cs in the North Pacific Ocean derived from the March 2011 TEPCO Fukushima Dai-ichi Nuclear Power Plant accident, Japan. Part two: estimation of ^{134}Cs and ^{137}Cs inventories in the North Pacific Ocean. *J. Oceanogr.*, **72**, 67–76, doi:10.1007/s10872-015-0332-2.

Becker, M. (2011). Wind bläst radioaktive Wolke nach Tokio. *Spiegel Online*, 14 March.

Bowyer, T. W., S. R. Biegalski, M. Cooper, *et al.* (2011). Elevated radioxenon detected remotely following the Fukushima nuclear accident. *J. Environ. Radioact.*, **102**, 681–7, doi:10.1016/j.jenvrad.2011.04.009.

Cabinet Office, Government of Japan (2011). Press conference by the chief cabinet secretary (4 April at 16:02). http://nettv.gov-online.go.jp/eng/prg/prg2065.html (accessed 19 September 2018).

Chino, M., H. Nakayama, H. Nagai, *et al.* (2011). Preliminary estimation of release amounts of ^{131}I and ^{137}Cs accidentally discharged from the Fukushima Daiichi Nuclear Power Plant into the atmosphere. *J. Nucl. Sci. Technol.*, **48**, 1129–34, doi:10.1080/18811248.2011.9711799.

Christoudias, T. and J. Lelieveld (2013). Modelling the global atmospheric transport and deposition of radionuclides from the Fukushima Dai-ichi nuclear accident. *Atmos. Chem. Phys.*, **13**, 1425–38, doi:10.5194/acp-13-1425-2013.

Christoudias, T., Y. Proestos and J. Lelieveld (2014). Global risk from the atmospheric dispersion of radionuclides by nuclear power plant accidents in the coming decades. *Atmos. Chem. Phys.*, **14**, 4607–16, doi:10.5194/acp-14-4607-2014.

Comprehensive Nuclear-Test-Ban Treaty Organization Preparatory Commission (2011). The 11 March Japan disaster. http://bit.ly/2NxZ5W2 (accessed 19 September 2018).

Hsu, S.-C., C.-A. Huh, C.-Y. Chan, *et al.* (2012). Hemispheric dispersion of radioactive plume laced with fission nuclides from the Fukushima nuclear event. *Geophys. Res. Lett.*, **39**, doi:10.1029/2011GL049986.

Huh, C.-A., S.-C. Hsu and C.-Y. Lin (2012). Fukushima-derived fission nuclides monitored around Taiwan: free tropospheric versus boundary layer transport. *Earth Planet. Sci. Lett.*, **319-320**, 9–14, doi:10.1016/j.epsl.2011.12.004.

Husar, R. B., D. M. Tratt, B. A. Schichtel, *et al.* (2001). Asian dust events of April 1998. *J. Geophys. Res. Atmospheres*, **106**, 18317–30, doi:10.1029/2000JD900788.

Icelandic Radiation Safety Authority (2011). Summary of radionuclide concentrations in air: March 19th–April 13th, Reykjavik, Iceland. http://bit.ly/2Vxhy83

Japan Meteorological Agency (2011). On WMO GDPFS RSMC on the provision of transport model products for Environmental Emergency Response. www.jma.go.jp/jma/kokusai/kokusai_eer.html (accessed 19 September 2018) (in Japanese).

Katata, G., M. Chino, T. Kobayashi, *et al.* (2015). Detailed source term estimation of the atmospheric release for the Fukushima Daiichi Nuclear Power Station accident by coupling simulations of an atmospheric dispersion model with an improved deposition scheme and oceanic dispersion model. *Atmos. Chem. Phys.*, **15**, 1029–70, doi:10.5194/acp-15-1029-2015.

Kim, C. K., J. I. Byun, J. S. Chae, *et al.* (2012). Radiological impact in Korea following the Fukushima nuclear accident. *J. Environ. Radioact.*, **111**, 70–82, doi:10.1016/j.jenvrad.2011.10.018.

Kobayashi, T., H. Nagai, M. Chino and H. Kawamura (2013). Source term estimation of atmospheric release due to the Fukushima Dai-ichi Nuclear Power Plant accident by atmospheric and oceanic dispersion simulations. *J. Nucl. Sci. Technol.*, **50**, 255–64, doi:10.1080/00223131.2013.772449.

Kristiansen, N. I., A. Stohl and G. Wotawa (2012). Atmospheric removal times of the aerosol-bound radionuclides ^{137}Cs and ^{131}I measured after the Fukushima Dai-ichi nuclear accident: a constraint for air quality and climate models. *Atmos. Chem. Phys.*, **12**, 10759–69, doi:10.5194/acp-12-10759-2012.

Kristiansen, N. I., A. Stohl, D. J. L. Olivié, *et al.* (2016). Evaluation of observed and modelled aerosol lifetimes using radioactive tracers of opportunity and an ensemble of 19 global models. *Atmos. Chem. Phys.*, **16**, 3525–61, doi:10.5194/acp-16-3525-2016.

Lelieveld, J., D. Kunkel and M. G. Lawrence (2012). Global risk of radioactive fallout after major nuclear reactor accidents. *Atmos. Chem. Phys.*, **12**, 4245–58, doi:10.5194/acp-12-4245-2012.

Leon, J. D., D. A. Jaffe, J. Kaspar, *et al.* (2011). Arrival time and magnitude of airborne fission products from the Fukushima, Japan, reactor incident as measured in Seattle, WA, USA. *J. Environ. Radioact.*, **102**, 1032–8, doi:10.1016/j.jenvrad.2011.06.005.

Long, N. Q., Y. Truong, P. D. Hien, *et al.* (2012). Atmospheric radionuclides from the Fukushima Dai-ichi nuclear reactor accident observed in Vietnam. *J. Environ. Radioact.*, **111**, 53–8, doi:10.1016/j.jenvrad.2011.11.018.

Maki, T. (2015). Emission source estimation by an inverse model. In *Contribution of JMA to the WMO Technical Task Team on Meteorological Analyses for Fukushima Daiichi Nuclear Power Plant Accident and Relevant Atmospheric Transport Modeling at MRI.* Tsukuba: Meteorological Research Institute, pp. 150–3.

Masson, O., A. Baeza, J. Bieringer, *et al.* (2011). Tracking of airborne radionuclides from the damaged Fukushima Dai-Ichi nuclear reactors by European networks. *Environ. Sci. Technol.*, **45** (18), 7670–7, doi:10.1021/es2017158.

Medici, F. (2001). The IMS radionuclide network of the CTBT. *Rad. Phys. Chem.*, **61**, 689–90, doi:10.1016/S0969-806X(01)00375-9.

Mészáros, R., Á. Leelőssy, T. Kovács and I. Lagzi (2016). Predictability of the dispersion of Fukushima-derived radionuclides and their homogenization in the atmosphere. *Sci. Rep.*, **6**, 19915, doi:10.1038/srep19915.

Ohara, T. and Y. Morino (2012). Current status and issues of the atmospheric transport and deposition models of radioactive substances. In Abstracts of the symposium of the 2012 spring meeting of Meteorological Society of Japan, 2012, 2–8.

Okada, K., M. Ikegami, O. Uchino, *et al.* (1992). Kuwaiti soot over Japan. *Nature*, **355**, 120.

Philippine Nuclear Research Institute (2011). Fukushima-Daiichi Nuclear Power Plant accident in Japan. Information Bulletin No. 29 (19 May 2011 update as of 11:00 AM). http://bit.ly/2VmSpMX (accessed 19 September 2018).

Priyadarshi, A., G. Dominguez and M. H. Thiemens (2011). Evidence of neutron leakage at the Fukushima nuclear plant from measurements of radioactive ^{35}S in California. *Proc. Natl. Acad. Sci. USA*, **108**, 14422–5, doi:10.1073/pnas.1109449108.

Sarkar, T., S. Anand, K. D. Singh, *et al.* (2017). Simulating long range transport of radioactive aerosols using a global aerosol transport model. *Aerosol and Air Quality Research*, **17**, 2631–42, doi:10.4209/aaqr.2017.01.0049.

Saunier, O., A. Mathieu, D. Didier, *et al.* (2013). An inverse modeling method to assess the source term of the Fukushima Nuclear Power Plant accident using gamma dose rate observations. *Atmos. Chem. Phys.*, **13**, 11403–21, doi:10.5194/acp-13-11403-2013.

Sectional Committee on Nuclear Accident Committee on Comprehensive Synthetic Engineering, Science Council of Japan (2014). A review of the model comparison of transportation and deposition of radioactive materials released to the environment as a result of the Tokyo Electric Power Company's Fukushima Daiichi Nuclear Power Plant Accident. Report, 2 September.

Stohl, A., P. Seibert, G. Wotawa, *et al.* (2012). Xenon-133 and caesium-137 releases into the atmosphere from the Fukushima Dai-ichi nuclear power plant: determination of the source term, atmospheric dispersion, and deposition. *Atmos. Chem. Phys.*, **12**, 2313–43, doi:10.5194/acp-12-2313-2012.

Takemura, T., H. Nakamura, M. Takigawa, *et al.* (2011). A numerical simulation of global transport of atmospheric particles emitted from the Fukushima Daiichi Nuclear Power Plant. *SOLA*, **7**, 101–4, doi:10.2151/sola.2011-026.

Tanaka, T. Y., Y. Inomata, Y. Igarashi, *et al.* (2012). Current status and issues of the atmospheric transport simulation of radioactive materials in Meteorological Research Institute. *Tenki*, **59**, 239–50 (in Japanese).

Ten Hoeve, J. E. and M. Z. Jacobson (2012). Worldwide health effects of the Fukushima Daiichi nuclear accident. *Energy Environ. Sci.*, **5**, 8743, doi:10.1039/c2ee22019a.

Terada, H., G. Katata, M. Chino and H. Nagai (2012). Atmospheric discharge and dispersion of radionuclides during the Fukushima Dai-ichi Nuclear Power Plant accident. Part II: verification of the source term and analysis of regional-scale atmospheric dispersion. *J. Environ. Radioact.*, **112**, 141–54, doi:10.1016/j.jenvrad.2012.05.023.

Torii, T., Y. Sanada, T. Sugita, *et al.* (2012). Investigation of radionuclide distribution using aircraft for surrounding environmental survey from Fukushima Dai-ichi Nuclear Power Plant. *JAEA-Technology*, **36**, doi:10.11484/jaea-technology-2012-036.

Uno, I., K. Eguchi, K. Yumimoto, *et al.* (2009). Asian dust transported one full circuit around the globe. *Nature Geosci.*, **2**, 557–60, doi:10.1038/ngeo583.

Wetherbee, G. A., D. A. Gay, T. M. Debey, C. M. Lehmann and M. A. Nilles (2012). Wet deposition of fission-product isotopes to North America from the Fukushima Dai-ichi incident, March 2011. *Environ. Sci. Technol.*, **46**, 2574–82, doi:10.1021/es203217u.

Winiarek, V., M. Bocquet, O. Saunier and A. Mathieu (2012). Estimation of errors in the inverse modeling of accidental release of atmospheric pollutant: application to the reconstruction of the cesium-137 and iodine-131 source terms from the Fukushima Daiichi power plant. *J. Geophys. Res.*, **117**, doi:10.1029/2011jd016932.

Winiarek, V., M. Bocquet, N. Duhanyan, *et al.* (2014). Estimation of the caesium-137 source term from the Fukushima Daiichi nuclear power plant using a consistent joint assimilation of air concentration and deposition observations. *Atmos. Environ.*, **82**, 268–79, doi:10.1016/j.atmosenv.2013.10.017.

World Meteorological Organization (2011). No. 909: WMO monitoring meteorological conditions in quake-hit area. http://bit.ly/2GMBoIP (accessed 19 September 2018).

Wotawa, G. (2011). Accident in the Japanese NPP Fukushima: spread of radioactivity/first source estimates from CTBTO data show large source terms at the beginning of the accident/weather currently not favourable/low level radioactivity meanwhile observed over U.S. East Coast and Hawaii (update: 22 March 2011 15:00). www.zamg.ac.at/docs/aktuell/Japan2011-03-22_1500_E.pdf (accessed 19 September 2018)

Yonezawa, C. and Y. Yamamoto (2011). Measurements of artificial radionuclides in the atmosphere by nuclear test monitoring for radioactive nuclides networks. *Bunseki*, **440**, 451–8 (in Japanese).

5

Ocean Transport of Radioactive Materials

5.1 Introduction

MICHIO AOYAMA, MITSUO UEMATSU, SEIYA NAGAO,
TAKASHI ISHIMARU, JOTA KANDA, TATSUO AONO,
YUKIO MASUMOTO AND DAISUKE TSUMUNE

Radioactive substances were released from the TEPCO Fukushima Daiichi Nuclear Power Station (FDNPS) accident into the environment, beginning on 11 March 2011. A large amount of radioactive material was released into the atmosphere from the three damaged cores and 80% of it was deposited into the ocean. Radioactive materials also discharged directly into the ocean as leaked stagnant water from the reactor housing. River runoff and groundwater discharge can also be considered as minor sources of the FDNPS-derived radioactivity in the ocean.

Because a large portion of released radioactivity from the FDNPS, which included both radiocaesium and radioiodine, might be soluble, the movement of water can be responsible for large-scale transport. Because the FDNPS is located in a coastal region with strong coastal currents along the coast in a north–south direction, radioactivity discharged or deposited in the coastal region can be advected by these strong coastal currents and is less affected by diffusion.

In this region, the Kuroshio Current comes from the south, originating far from Japan, and the Oyashio Current comes from the north, originating near the Aleutian Islands and Kuril Islands. Therefore, it is a sea area at the crossroad of two current systems. Initial ocean transport studies on FDNPS radioactivity along the coast suggested that transport to the south was initially dominant before eastward propagation resulting from the Kuroshio and Kuroshio Extension became dominant. Some portion of the atmospheric release from the FDNPS might have been deposited over a wide area of the northwestern North Pacific Ocean, and a

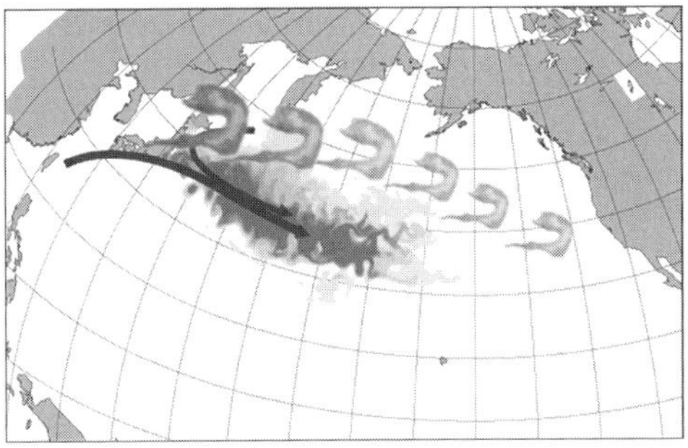

Figure 5.1 Schematic diagram of transport of FDNPS-derived radionuclides.

few hotspots have been identified in observations (Figure 5.1). We also need to pay attention to the particulate matter and marine sediments that react with radionuclides in seawater.

In this chapter, we present research results obtained from coastal areas and an analysis of diffusion and advection in the open water based on observations and numerical model simulations.

5.2 Measurement of Radioactive Materials Over the Marine Atmosphere

MITSUO UEMATSU

The number of atmospheric measurements of radioactive materials over the sea are very limited compared to over land. In this section, temporal variations in the atmospheric concentrations of ^{131}I and ^{137}Cs in marine aerosols are summarised based on published data measured at observation points within 30 km of the FDNPS, which were collected by the Ministry of Education, Culture, Sports, Science and Technology (MEXT) and during cruises by research vessels and a US research vessel.

5.2.1 Marine Aerosol Sampling

As soon as the Sea Area Monitoring Plan was proposed by MEXT, R/V *Hakuho Maru* left port on 22 March and started monitoring on 23 March 2011, at eight locations within 30 km of Fukushima. The monitoring was continued by other research vessels, including R/Vs *Mirai*, *Kairei* and *Yokosuka*, belonging to the Japan Agency for Marine-Earth Science and Technology (JAMSTEC). Aerosol samples in the marine atmosphere were collected at every station, and a quantitative analysis of the radioactive nuclides was performed immediately. The Sea Area Monitoring Plan was later strengthened, with more observation stations added and increased sampling frequency for the precise mapping of a wider area and temporal changes of radioactive nuclide concentrations.

5.2.2 Temporal and Spatial Variation in Atmospheric ^{131}I and ^{137}Cs at the Fixed Marine Stations along a 30-km Line Off the Coast of Fukushima

Beginning on 23 March, atmospheric filter sampling and seawater sampling were conducted at 13 stations of the Sea Area Monitoring network by MEXT (Figure 5.2). The network consisted of eight stations along the line of 141°24′E, 30 km east of the FDNPS, and five stations from the northernmost end and the southernmost end of the north–south line to land.

Time variations at each fixed station are shown for the atmospheric concentration of ^{131}I (half-life: 8.02 days) and ^{137}Cs (half-life: 30.2 years) obtained from 98 filter samples measured over one month (Figure 5.3). While the ^{131}I concentration was high in late March, ^{137}Cs showed high concentrations in mid-April. From the FDNPS, radiogenic caesium was emitted as particulate matter, and most of the iodine was released in its gaseous phase (Ohara *et al.*, 2011). This difference was reflected in the emission rates of each radionuclide and the physical and chemical transformation during transport in the atmosphere.

The maximum concentration of ^{131}I was 23.5 Bq/m^3 at station 9 on 28 March, and the ^{137}Cs concentration in the same sample was 3.3 Bq/m^3. The ^{137}Cs concentration was 14.9 Bq/m^3 at station B on the north side on 11 April 2011, and the ^{131}I concentration was 13.3 Bq/m^3. On the same day, 8.4 Bq/m^3 of ^{137}Cs and 5.1 Bq/m^3 of ^{131}I were observed at the southernmost station 10 due to the release of radioactive material and the changes in wind direction from the FDNPS.

According to the estimated release rates of ^{131}I and ^{137}Cs emitted from the FDNPS in March following the accident (Chino *et al.*, 2011), the ratio of the

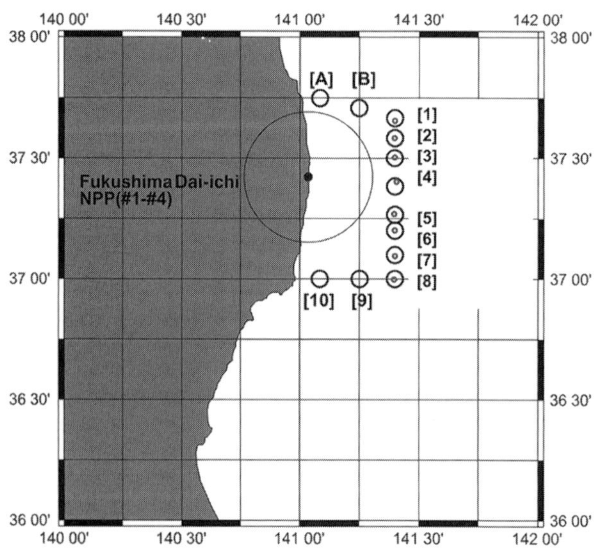

Figure 5.2 Initial monitoring stations of the FDNPS.

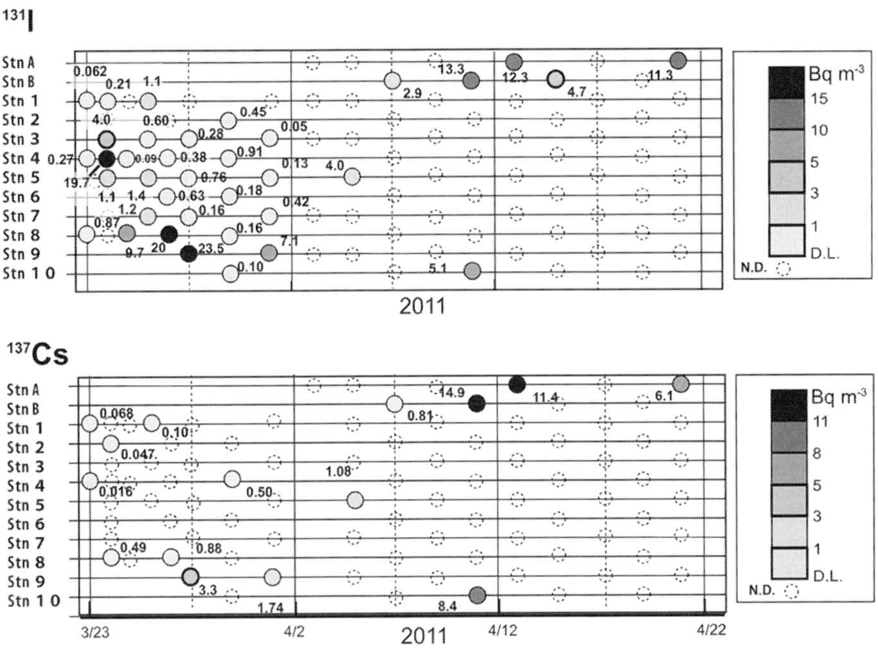

Figure 5.3 Atmospheric concentration change of ^{131}I (half-life: 8.02 days) and ^{137}Cs (half-life: 30.2 years) over the coastal region off the FDNPS.

release rates of ^{131}I and ^{137}Cs was large at the beginning, and ^{137}Cs became more prominent with time, which resembles the tendency of the observed values over the ocean. The release of radioactive materials from the FDNPS into the atmosphere was generally intermittent in April.

5.3 Behaviour of Radiocaesium from Rivers to the Coastal Marine Environment

SEIYA NAGAO

5.3.1 Export of Radiocaesium from Rivers to the Coastal Area

The Low Level Radioactivity Laboratory of Kanazawa University conducted river research to understand the transport of ^{134}Cs and ^{137}Cs in the upper and lower reaches of the Abukuma River, the Uta River and the Niida River, beginning in May 2011. The activity concentration of ^{137}Cs in the rivers on 20 May 2011 ranged from 230 to 4170 Bq/m^3, which was approximately three orders of magnitude higher than that in the Kuji River and Tone River before the FDNPS accident (Nagao, 2015; Nagao *et al.*, 2015). The activity ratio ^{134}Cs/^{137}Cs for the river water samples was almost 1.0 and coincided with that of the surface soil in Fukushima Prefecture after the accident. The results suggest that a large portion of the radiocaesium in the river waters originated from the FDNPS accident. The activity concentration of ^{137}Cs showed a decreasing trend from 64–1540 Bq/m^3 on 12–13 July 2011 to 19–790 Bq/m^3 on 12–13 September and 11–190 Bq/m^3 on 7–8 December 2011 (Nagao, 2015; Nagao *et al.*, 2015). The results suggest that the supply of ^{137}Cs from the watershed to the river systems decreased over time following the accident. Therefore, the export of ^{137}Cs from the river systems to the coastal marine area in December 2011 was significantly lower than that in May and September 2011.

The flux of ^{137}Cs from land to ocean are estimated as 2.9~5.2 $\times$ 10^{10} Bq/y for the Natsui River and Same River in 2011 (Nagao *et al.*, 2013) and 354 $\times$ 10^{10} Bq for the Abukuma River during August 2011 to May 2012 (Yamashiki *et al.*, 2014). The export flux of ^{137}Cs attributable to the heavy rain from Typhoon Roke in September 2011 accounted for 30–50% of the annual radiocaesium flux for the Natsui River (Nagao *et al.*, 2013) and 61% for the Abukuma River in 2011 (Yamashiki *et al.*, 2014).

5.3.2 Dynamics of Radiocaesium in the Coastal Marine Environment

Figure 5.4 shows the temporal variations of the radioactivity of ^{134}Cs in the surface seawater from the coastal area in December 2011. Inoue *et al.* (2012a) investigated the dispersion of ^{134}Cs radioactivity in surface seawater at 10 sites from near

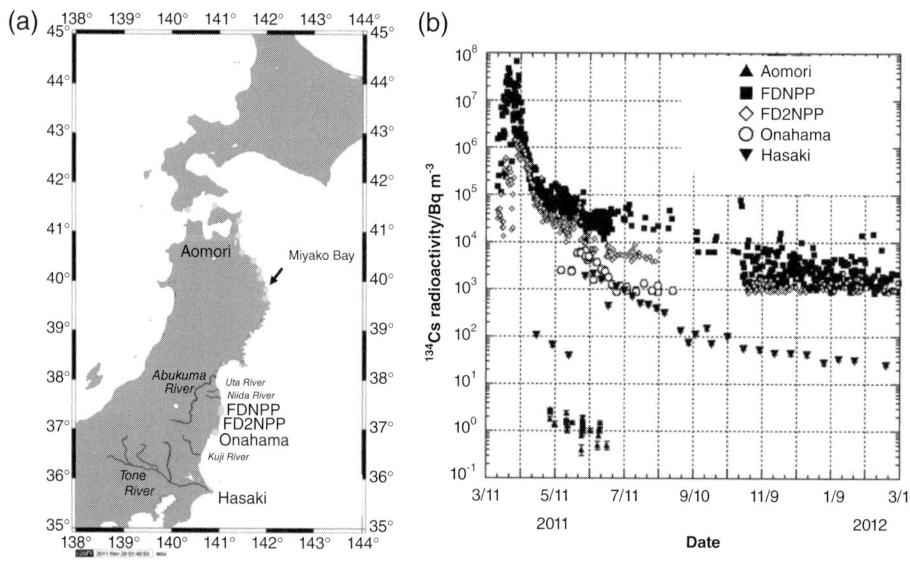

Figure 5.4 Sampling location map of the coastal area of the Pacific Ocean (a) and variations of ^{134}Cs radioactivity in the surface seawaters (b). FDNPP is the Fukushima Daiichi Nuclear Power Plant and FD2NPP is the Fukushima Daini Nuclear Power Plant. (Aoyama *et al.*, 2012b: Inoue *et al.*, 2012a).

Miyako Bay in Iwate Prefecture to Ooma in Aomori Prefecture, located 250–450 km from the FDNPS, once every two weeks from early May to late June 2011. The ^{134}Cs radioactivity in the surface seawater samples was 2–3 Bq/m^3 in early May 2011 and decreased to 0.5 Bq/m^3 after one month. The results indicated that radiocaesium was transported from the coastal area to the pelagic ocean or deeper layers of the ocean.

Furthermore, the ^{134}Cs radioactivity in the surface seawater at Onahama significantly decreased from 7000 Bq/m^3 in early June 2011 to 1000 Bq/m^3 at the end of June, remaining nearly constant thereafter (Aoyama *et al.*, 2012b). The monitoring site at Hasaki in Chiba Prefecture showed ^{134}Cs radioactivity of 40–110 Bq/m^3 until May 2011, although it suddenly increased to 2080 Bq/m^3 on 6 June 2011. The radioactivity decreased towards the end of August 2011 and then remained constant (40–50 Bq/m^3) until December 2011 (Aoyama *et al.*, 2012b). Water current data around Japan indicated the presence of warm eddies through the end of May 2011. After the disappearance of the eddies, the ^{134}Cs radioactivity increased suddenly. Therefore, the coastal area near Fukushima and Ibaraki Prefectures saw complicated dispersion abetted by coastal currents and eddies separated from the Kuroshio Current during this time.

To understand the transport behaviour of radiocaesium in the coastal area near Ibaraki and Fukushima, a research expedition was conducted during the KT-11-22 cruise of R/V *Tansei Maru*. Surface seawater samples were collected at three or four sites in each line from four transects on 7–12 September 2011. Higher radioactivity was detected near the coastline off Ibaraki compared to that near the FDNPS. Along the transect line, higher radioactivity was detected closer to the coast. These results suggested that (1) radiocaesium directly released from the FDNPS was transported to the coastal area off Ibaraki; and (2) radiocaesium was transported from the contaminated land surface through the river systems to the coastal area. In early September 2011, higher radioactivity may have contributed by the coastal ocean current from north to south because of lower radiocaesium concentration in river water than the surface seawater.

5.3.3 Dynamics of Radiocaesium in the Okhotsk Sea and Japan Sea

Inoue *et al.* (2012b) reported the spatial variations in ^{134}Cs radioactivity in the surface seawater based on six transects from south to north in the Japan Sea and one transect in the Okhotsk Sea during the OS-229 cruise of T/S *Oshoro-Maru* (Figure 5.5). As shown in Figure 5.5, the ^{134}Cs radioactivity showed a maximum (~1 Bq/m^3) off the Tsugaru Peninsula in Aomori Prefecture and near Oshima and Ishikari in Hokkaido. However, the maximum fallout deposition was observed at Akita after the accident and the fallout accumulation decreased in the following

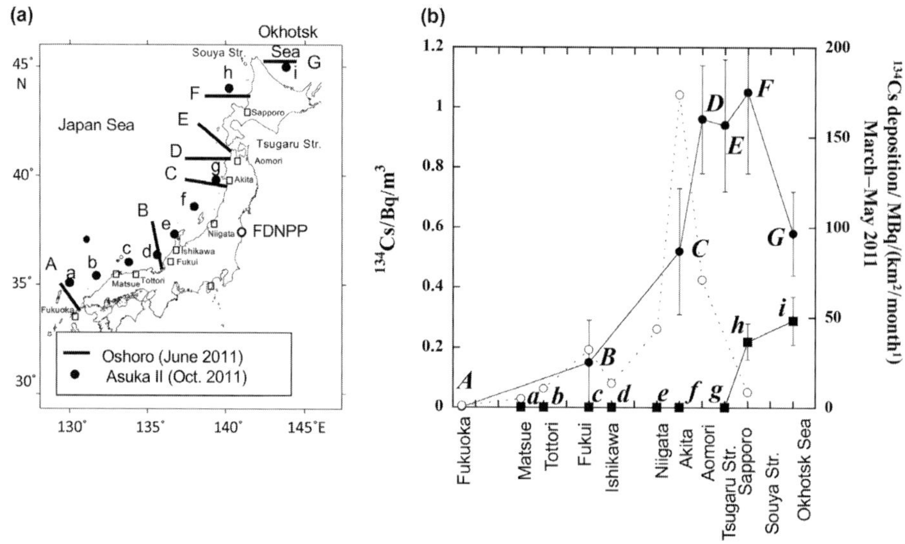

Figure 5.5 Sampling location map of Okhotsk Sea and the Sea of Japan (a), and variations of ^{134}Cs radioactivity in the surface seawaters (b) (Inoue *et al.*, 2012a; 2012b).

order: Akita > Aomori > Niigata > Fukui > Sapporo. The fallout deposition shows different spatial variations for the surface seawater samples shortly after the accident, over May to September 2011.

In October 2011, surface seawater samples were collected near the transect lines in the Japan Sea and Okhotsk Sea during the cruise of *Asuka II*. The radioactivity of ^{134}Cs was less than the detection limit, except for the samples in the Ishikari Bay and Okhotsk Sea. The decrease in the ^{134}Cs radioactivity during the four months after the OS-229 cruise was controlled by the lateral transport of water mass from south to north in the Japan Sea via the Tsushima Current. However, in February to March of 2016, 0.12–0.17 Bq/m^3 of ^{134}Cs was detected in surface seawater from the eastern part of the Japan Sea (Aoyama *et al.*, 2017). These results suggest that a small part of the FDNPS-derived radiocaesium had already recirculated and reached the Japan Sea coast.

5.4 Transport of Radiocaesium in the North Pacific Ocean

MICHIO AOYAMA

We collected surface seawater samples at 220 stations in the North Pacific Ocean (Figures 5.6 and 5.7) from March 2011 to March 2012. The sample volume of 220 stations ranged from 2 to 10 L. The sampling details for the

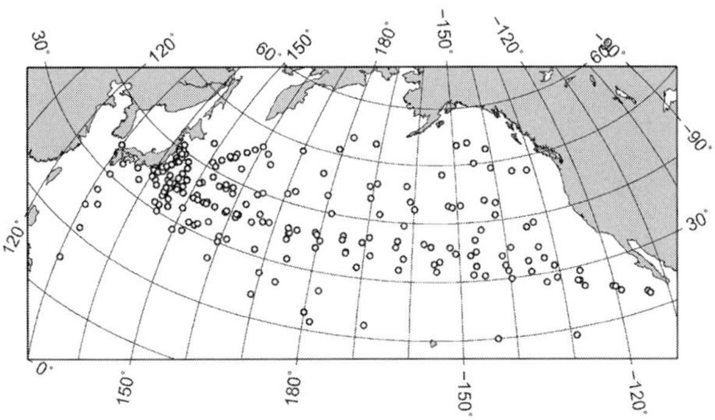

Figure 5.6 Sampling locations of surface water during the period from March 2011 to March 2012.

cargo ships in the North Pacific Ocean are described elsewhere (Aoyama *et al.*, 2012a; 2013).

During the first month after the accident, the measured $^{134}Cs/^{137}Cs$ activity ratios were very close to 1 (0.99 $\pm$ 0.03 in the FDNPS north and south discharge channels) and extremely uniform (Buesseler *et al.*, 2011). Therefore, the ^{134}Cs activity trends from the FDNPS accident can be considered the same as the ^{137}Cs activity trends from the FDNPS accident. In the ocean, the behaviour of caesium is considered to be conservative: it is soluble in seawater and less than 1% of radiocaesium adsorbs into marine particles (Aoyama and Hirose, 1995). Thus, the presence of ^{134}Cs is a unique isotopic signature of FDNPS-derived radiocaesium, and ^{134}Cs can be used for tracking radiocaesium in ocean waters.

In addition, for the ^{137}Cs before the FDNPS accident in 2011, ^{137}Cs derived from atmospheric nuclear weapons tests was detected in the oceans. In the western North Pacific Ocean, the ^{137}Cs concentration in the surface water ranged from 10 to 100 Bq/m^3 in the 1960s; the ^{137}Cs concentration in the surface water decreased gradually from the 1970s to the 2000s. Thereafter, the ^{137}Cs concentration in the surface water was approximately 1–2 Bq/m^3 (Inomata *et al.*, 2009; Aoyama *et al.*, 2011). In addition, the ^{137}Cs activity in the Pacific Ocean in the 2000s showed a maximum in the mid-latitudes of the western North Pacific Ocean and showed a minimum in the subarctic ocean in the North Pacific. This pattern may have resulted from the basin-scale transport in this region.

^{137}Cs observed after the FDNPS accident might originate from both atmospheric nuclear weapons tests and the FDNPS accident. In fact, using the ^{134}Cs versus ^{137}Cs radioactivity ratio of 1, we can estimate the ^{137}Cs activity from the FDNPS accident by subtracting the ^{134}Cs activity from the observed ^{137}Cs activity (Aoyama *et al.*, 2013). In April to June 2011, radiocaesium released from the

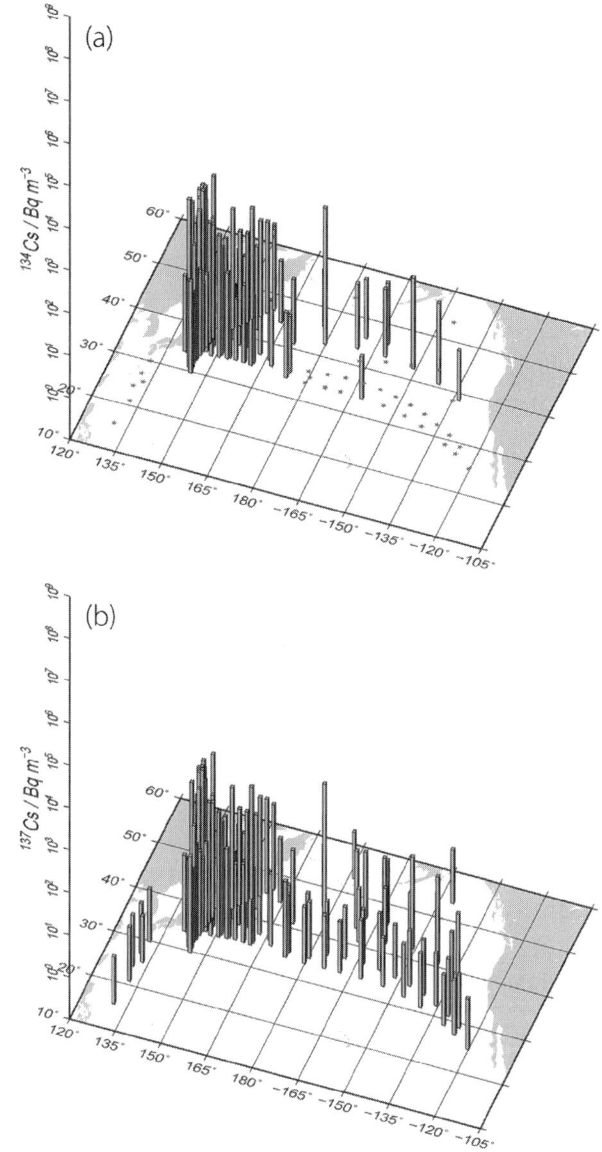

Figure 5.7 ^{134}Cs (a) and ^{137}Cs (b) activity in surface water during the period from March 2011 to June 2011.

FDNPS to the atmosphere was transported northeast from Japan. Therefore, the radiocaesium activity in the surface water was high at high latitudes in the western North Pacific Ocean (Figure 5.7). In addition, relatively high radiocaesium activity was observed due to locally deposited radiocaesium from the atmosphere at several places in the eastern North Pacific Ocean (Figure 5.8).

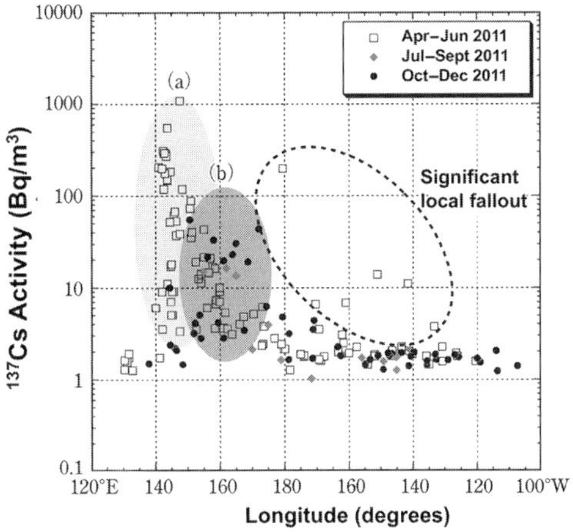

Figure 5.8 Eastern spread and local fallout of radiocaesium in surface water in the North Pacific Ocean originated from the Fukushima accident. High activity concentration region of ^{137}Cs in April to June 2011 and high activity concentration region of ^{137}Cs in October to December 2011. A significant local fallout region in the eastern North Pacific Ocean due to atmospheric transport of radiocaesium that originated from the Fukushima accident is also shown.

We also examined the eastward transport of radiocaesium from the FDNPS accident. The region where the ^{137}Cs activity in the surface water exceeded 10 Bq/m^3 extended to 160°E in June 2011 before reaching 170°E in October to December 2011, as shown in Figure 5.7 and described elsewhere (Aoyama *et al.*, 2013).[1]

The radiocaesium from the FDNPS accident spread eastward in the surface water across the mid-latitude North Pacific at a speed of 7 km/day (8 cm/s) until March 2012, and 3 km/day (3.5 cm/s) from March 2012 through August 2014. Thereafter the radioactive material reached the western coast of North America in 2015. To present the general trends of FDNPS-derived radiocaesium in the North Pacific Ocean, we can use the result of a model simulation of the horizontal distribution of ^{134}Cs activity concentrations in surface water, based on Tsubono *et al.*'s study (Tsubono *et al.*, 2016) for October 2011 (Figure 5.9), for example. Results of the model simulation, shown in Figure 5.9, successfully represented the general trend of surface transport of FDNPS-derived radiocaesium concentrations, especially eastward transport of the main body of FDNPS-derived radiocaesium

[1] The Japanese version of this article was written in the summer of 2012 and published in 2014. Since that time, much progress has been made, and a review article (Buesseler *et al.*, 2017) and many articles have been published regarding the transport processes of FDNPS radiocaesium in surface layers and the ocean interior (Kaeriyama *et al.*, 2013; 2016; Kumamoto *et al.*, 2014; Aoyama *et al.*, 2016a; 2016b; 2017; Tsubono *et al.*, 2016). New information regarding the transport processes of the FDNPS-derived radiocaesium is described in the text.

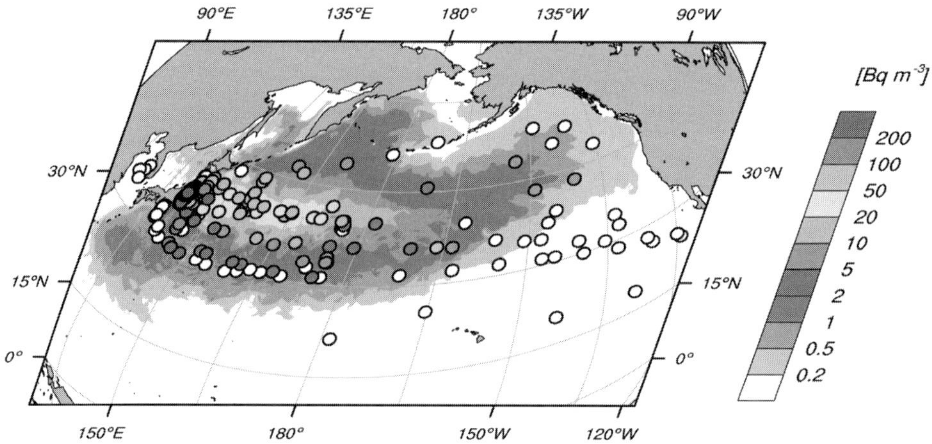

Figure 5.9 Horizontal distribution of ^{134}Cs activity concentrations in surface water based on the study of Tsubono *et al.* (2016) and observations of ^{134}Cs activity concentrations in October 2011, reproduced from figure 3.1 in Aoyama *et al.* (2018). Open circles with colour scale: observations of ^{134}Cs activity concentrations; colour shading: model results. A black and white version of this figure will appear in some formats. For the colour version, refer to the plate section.

with less northward and southward transport in mid-latitude in the North Pacific Ocean, as discussed here and in the published articles cited in this chapter.

Beneath the ocean's surface, in June 2012, the ^{134}Cs activity reached a maximum of 6.12 ± 0.50 Bq/m^3 at a depth of 151 m (potential density, $\sigma_\theta = 25.3$ kg/m^3) at 29°N, 165°E. This subsurface maximum, which was also observed along 149°E, might reflect the southward transport of the FDNPS radiocaesium in association with the formation and subduction of subtropical mode water (STMW). In June 2012, between 34°N and 39°N along 165°E, the ^{134}Cs activity showed a maximum of approximately $\sigma_\theta = 26.3$ kg/m^3, corresponding to the central mode water (CMW). The ^{134}Cs activity was higher in the CMW than in any of the surrounding waters, including the STMW. These observations also indicated that the most efficient pathway by which FDNPS radiocaesium was introduced into the ocean on a one-year timescale was through CMW formation and subduction.

5.5 Dispersion Simulation and Estimation of the Total Amount of ^{137}Cs Directly Discharged into the Ocean

YUKIO MASUMOTO AND DAISUKE TSUMUNE

After the accident at the FDNPS, how much of the radionuclides originating from the FDNPS entered the ocean? How did the discharged radionuclides spread in the ocean? Seeking answers to these and other related questions is essential for a better understanding of the environmental impacts of radionuclides discharged from the

FDNPS. Efforts using a multiple modelling approach are critical to obtaining key answers to the above questions.

5.5.1 Estimation of the Discharged Amount and a Possible Discharge Scenario

TEPCO reported that the estimated amount of ^{137}Cs in highly contaminated water discharged into the ocean in the period 1–6 April 2011 was 0.94 PBq (TEPCO, 2011c). However, in the case of an unexpected accident, including the present case, it is almost impossible to obtain accurate and timely information from observed values regarding the timing and amount of the radionuclides released into the ocean. Efforts have been made to estimate the total amount of radionuclides directly discharged for the Fukushima case using numerical models together with observed data.

Kawamura *et al.* (2011) created a discharge scenario for ^{137}Cs by assuming that the discharged ^{137}Cs was distributed evenly within a 1.5 km^2 region in front of the FDNPS in the upper 1 m, with the same value as that observed at the FDNPS discharge channels. After adjusting the values with the above estimate by TEPCO for early April, the total amount of discharged ^{137}Cs was estimated to be 3.6 PBq from 21 March to the end of April 2011. On the other hand, Tsumune *et al.* (2012) estimated the total amount of discharged ^{137}Cs by adjusting the results from a numerical model with a unit release rate of radionuclides to the observed monitoring data and obtained a value of 3.5 PBq of ^{137}Cs from 26 March to the end of May 2011. They also conducted an analysis of the ^{131}I/^{137}Cs ratio, which led to the conclusion that the direct release of contaminated water into the ocean started on 26 March 2011. The simulated results of Tsumune *et al.* (2012) reproduce the ^{137}Cs distribution in the coastal region near Fukushima, including the maximum values in late March and early April 2011.

There are several other models that have been used to estimate the total amount of discharged ^{137}Cs that are similar to the approach of Tsumune *et al.* (2012). However, due to differences in model settings, such as boundary conditions and surface mixed-layer parameterisations, the results range from 2.3 to 27 PBq. After detailed evaluations of the model results and readjustment of several parameters, recent estimates of the total amount of ^{137}Cs that was directly discharged range from 3 to 6 PBq. There is a certain degree of uncertainty in these estimated values due to a strong bias towards monitoring data taken by TEPCO near the FDNPS and a lack of information on the vertical distribution of radionuclides originating from the FDNPS.

5.5.2 Distribution and Movement of ^{137}Cs a Few Months after the Accident

To understand the evolution of the ^{137}Cs distribution and its dispersion processes in the ocean near Fukushima, a model intercomparison study was conducted for the directly discharged component of ^{137}Cs (e.g. Masumoto *et al.*, 2012; Science Council of Japan, 2014).

In late March 2011, highly contaminated water was observed along the coast of Fukushima near the FDNPS. All models reasonably reproduced the southward expansion of the high-concentration area associated with weak southward flow along the coast. However, the relatively high concentration values obtained in the monitoring data 30 km offshore were not reproduced by any of the models. This suggested that the relatively high concentrations in the offshore region were caused by fallout of ^{137}Cs from the atmosphere, released just after the accident. This result is consistent with that of Tsumune *et al.* (2012), in which the observed ^{131}I/^{137}Cs ratio was used to distinguish between the direct and indirect discharge components.

One month later, in late April 2011 (see Figure 5.10), the region affected by ^{137}Cs spread farther southward and southeastward along the coast. It also

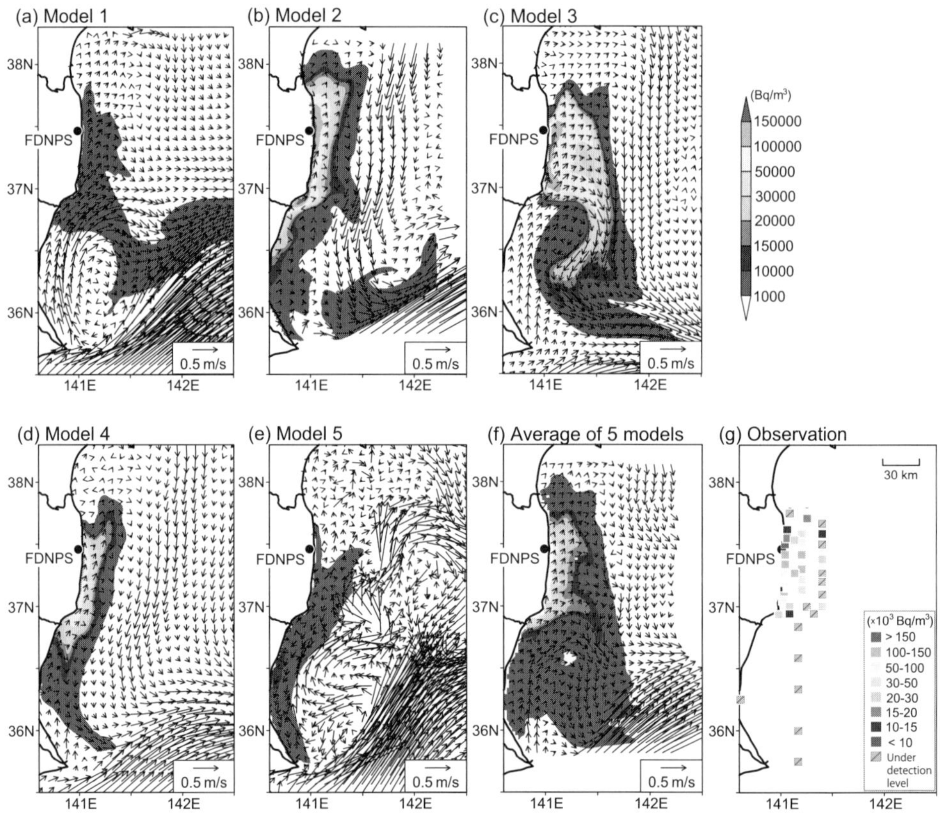

Figure 5.10 Horizontal distributions of the surface flow field and ^{137}Cs concentration in late April 2011. Panels (a) to (e) show results from five different models for the same period, while panel (f) indicates the average of the five model results. (g) Observed distribution of ^{137}Cs concentration during the period 21–30 April 2011 (Masumoto *et al.*, 2012). A black and white version of this figure will appear in some formats. For the colour version, refer to the plate section.

expanded eastward in the region off Fukushima due to weak northeastward surface flow. These tendencies are consistent with the monitoring observations. However, much monitoring data from the eastern and southern stations indicated values below the detection level. This makes a detailed evaluation of the simulated evolution of the ^{137}Cs distribution very difficult. In addition, another uncertainty comes from the mesoscale anticyclonic eddy off the Ibaraki coast, which was observed in satellite sea surface temperature and ocean colour images in mid-April 2011. While some models demonstrated the southeastward penetration of the ^{137}Cs distribution associated with this clockwise eddy circulation, others indicated a southward expansion of highly contaminated water along the Ibaraki coast. These discrepancies among the model results demonstrated that the surface flow field in the region east of Tohoku was very complex and difficult to reproduce, especially the mesoscale eddies and location and magnitude of the Kuroshio Current.

After May 2011, the ^{137}Cs-contaminated water expanded rapidly to the east along the northern flank of the Kuroshio Current and gradually dispersed into a wide area due to large temporal variability in the flow field and strong mesoscale eddy activities in the Kuroshio Extension region.

5.6 Investigation of Radioactive Contamination of Marine Biota: A Chronicle

TAKASHI ISHIMARU

On 15 March 2011, the author (T. Ishimaru) first realised the effects of the FDNPS accident in Tokyo when the gas-monitoring alarm suddenly sounded in the Radioisotope Facility of the Tokyo University of Marine Science and Technology (Minato-ku, Tokyo; hereafter the RI facility). The RI facility has an air circulation system that brings air into the facility, and the air is eventually discharged from the facility through a high-efficiency filter unit. An alarm device for monitoring radioactivity in the discharged air works continuously in order to not release significant radioactive substances into the environment. On that day, no one was using the facility; there was no reason for any radioactive substances to be released from the facility. The alarm actually caught the radioactive material from Fukushima that arrived in Tokyo; we found a value 10 times higher than the normal air dose measured using a radiation survey meter.

Meanwhile, units 1, 2, 3 and 4 of the FDNPS lost their cooling capability due to the earthquake and tsunami, and seawater and freshwater were injected for the purpose of cooling. The injected water was contaminated by the radioactive materials in the reactors, and then flowed out to the sea. On 25 March, seawater collected in the vicinity of the south discharge gate of the power plant showed

50 MBq/m^3 of ^{131}I (1250 times higher than the guideline level of waters outside the facility regulatory zone) (Tokyo Electric Power Company (TEPCO), 2011a).

Under these circumstances, many ocean scientists, including the author, believed that as long as the leakage of radioactive material was transient, it would not be a big problem because of the expected rapid diffusion in the ocean. The author himself answered queries from the media with this response. TEPCO measured the seawater radioactivity at four points on the coast of Fukushima; the author plotted the data (TEPCO, 2011a) in a graph and found that the radioactive water that had been observed at the south discharge gate moved southwards. Surprisingly, at a monitoring point 16 km south of the power plant, a radioactive caesium concentration of 2.4 MBq/m^3 was detected in the seawater on 28 March; the radioactivity had been diluted to only one-twentieth of the 25 March measurement. Therefore, the contaminated water containing radioactive material moved along the coast with only moderate dispersion in the offshore direction.

On 4 April, as will be discussed later, the kounago fish (the juvenile of a sand lance species) from the Kitaibaraki area, located south of Fukushima Prefecture, showed caesium radioactivity that exceeded the provisional regulation guideline of 500 Bq/g-wet. In addition, although the provisional regulation guideline of radioactive iodine for fish had not yet been set, an ^{131}I radioactivity value of 4080 Bq/kg-wet, a value more than twice the guideline level for leaf-type vegetables (2000 Bq/kg-wet), was found (Fisheries Agency, 2012). The issue raised enormous public concern.

Tokyo University of Marine Science and Technology (TUMSAT) operates four training vessels, including the *Umitaka-maru* and the *Shinyo-maru*, and immediately after the earthquake these vessels were standing by for a possible emergency operation in response to demands for transportation of rescue supplies. Because the ports in the Tohoku region were damaged and more damage from the tsunami was imminent, the transportation operation by the university vessels was considered unrealistic and never occurred. However, university colleagues wished to contribute to the recovery process of the Tohoku disasters in other ways. Meanwhile, in response to a call by the Japan Association of National Universities, TUMSAT listed possible ways that it could contribute to the Tohoku area; among them was radioactivity observation in the marine environment, using the university vessels. For radioactivity observations in the marine environment, MEXT monitored seawater and seafloor sediments, and the Fisheries Agency observed edible fish species. Irrespective of the importance, the radioactivity transferred to fish through the marine food chain was not investigated. Thus, TUMSAT decided to investigate the remaining components of the marine ecosystem, including plankton and benthic organisms.

It is essential for a successful ship survey to gain cooperation from local communities. The collection of fish and shellfish for research must be conducted under a special permit from the local government, which is issued only under the condition that the local fishing cooperatives agree to the research. At that time, fishing communities in eastern Japan were facing the so-called 'reputation damage' for fishery products from the Pacific coasts, and they were expected to oppose the radioactivity survey. Professor Midori Kawabe of TUMSAT interviewed fishermen in Onahama, Fukushima, in mid-May. She found that they were actually in favour of the radioactivity survey. They considered that the radioactivity contamination of their fishing grounds should be thoroughly disclosed. Professor Kawabe also learned that Dr Satoshi Igarashi, director of the Fukushima Prefectural Fisheries Experimental Station, was insisting that a long-term, widespread and comprehensive ecosystem survey should be conducted by the responsible national governmental agencies. The national government seemed to be quite reluctant to conduct such an ecosystem survey. Therefore, we undertook an independent survey using the vessels of TUMSAT in the Iwaki area. Based on consultations with the fishermen and the Fisheries Experimental Station, and with the approval of university executives, we developed a survey plan that would use the *Umitaka-maru*. The survey was conducted from 1 July through 8 July, immediately after the completion of the regular dock examination of the *Umitaka-maru*.

The Oceanographic Society of Japan (JOS) organised a working group for the earthquake disaster (hereafter, JOS-WG) in mid-April. The missions of the JOS-WG include mediation of ship opportunities that can be used for research of radioactive material in marine environments and a provision of analytical resources for radioactivity samples. For the special cruise of the *Umitaka-maru* (UT1107), we received the support of this working group. Several faculty members, students, staff of the Fukushima Prefectural Fisheries Experimental Station, experts sent from private companies and other volunteers gave support. The investigation cruises by the vessels of TUMSAT continued; the second cruise was in October 2011, with the *Shinyo-maru* (SY1110), and the third cruise was in May 2012 with the *Umitaka-maru* (UT1205). Successive cruises were conducted twice per year through 2016. Radioactivity data on marine organisms other than edible fish and shellfish are very limited, and the investigation by the TUMSAT vessels are critical for studying the transfer mechanism of radioactive materials in marine ecosystems.

In the sea area within 20 km from the FDNPS (the 20 km zone), the only radioactivity survey conducted at that time was the monitoring by TEPCO of seawater and sediments; researchers not associated with TEPCO had no access to the zone, and research on biological samples was not conducted at all. The 20 km zone was thus a large blank area of radioactivity data. In response to a proposal from NHK, the national broadcasting corporation of Japan, the JOS-WG carried out a joint investigation of the 20 km zone from late November to December 2011. The JOS-WG participants included T. Ishimaru, J. Kanda (TUMSAT), A. Tsuda (Atmosphere and Ocean Research Institute, University of Tokyo) and Y. Kato (Tokai University). Mapping of sediment radioactivity was conducted at 5 km intervals, and seawater, plankton, benthos and fish samples were collected within and around the zone. The results of the radioactivity analysis showed that the Japanese rockfish, the greenling and the slime flounder collected with gill nets near the 20 km zone showed higher caesium radioactivity levels, although the values were not significantly higher than those in other sea areas around Fukushima. In March 2012, TEPCO began sampling fish within the 20 km zone. The joint investigation by the JOS-WG and NHK likely resulted in the implementation of TEPCO's biological survey.

Among the plankton samples collected during the *Umitaka-maru* cruise in July 2011, several samples collected in the coastal areas near Iwaki City showed high levels of caesium radioactivity up to 700 Bq/kg-wet, and later examinations suggested that at least some portions of the radioactivity were derived from highly radioactive caesium-bearing particles similar to those found in atmospheric and soil samples (Ishimaru *et al.*, 2016). These samples and data obtained in the early phase of the contamination will be important for studying the distribution and transition of radioactivity within marine ecosystems in the future.

5.7 Pollution in Coastal Environments: Seawater and Sediment

JOTA KANDA

5.7.1 Pollution of Coastal Seawater

^{137}Cs is an artificially produced isotope and was dispersed globally by atomic bomb testing. ^{137}Cs radioactivity was detected in coastal seawater at a level of 1–2.5 Bq/m^3 before the FDNPS accident (MEXT, 2010). ^{134}Cs, which is also an artificial isotope, has a half-life of only approximately two years and was not detected in seawater until after the accident.

In response to the accident, MEXT began offshore monitoring on 23 March 2011 (Nuclear Regulation Authority, 2017a). The monitoring results showed that ^{131}I, ^{134}Cs and ^{137}Cs were detected in surface seawater up to a level of several tens of kBq/m^3. Radioactive material in seawater during this period was considered to be derived from atmospheric deposition and was detected at a similar level in a wide area extending to relatively distant locations from the FDNPS.

At the end of March, radioactive substances with very high concentrations were detected in seawater in the immediate vicinity of the power plant, and the radioactivity was considered to be due to the direct outflow of contaminated water from the FDNPS. From seawater outside the plant harbour, 47 MBq/m^3 (30 March, near the south discharge gate) and 68 MBq/m^3 (7 April, near the north discharge gate) of ^{137}Cs were detected. Inside the plant harbour, the value reached 120 000 MBq/m^3 (2 April, the unit 2 screen) (TEPCO, 2011a). While the discharged volume of contaminated water was relatively small, the ^{137}Cs radioactivity of the water was as high as 1 800 000 MBq/m^3 (TEPCO, 2011b).

As mentioned in Section 5.6, there was a peak in ^{137}Cs in excess of 1 MBq/m^3, with a delay of several days in the seawater near the coast, specifically 10–16 km south of the FDNPS. A radioactivity peak of 200–300 kBq/m^3 was also observed at the same time or slightly later in the eastward offshore area at 15 km from the FDNPS (TEPCO, 2011a). While the distance from the power plant was similar at these locations, the radioactivity exhibited a large difference: a southward extension of the high-radioactivity area was a distinct feature of the seawater contamination due to the accident. Based on these observations, we may assume that there was a peak of a few hundred kBq/m^3 after 2–3 weeks in the coastal area off Iwaki. Radioactivity in the seawater of the Fukushima area fell sharply in May at a rate of one-several-hundredths per month (Kanda, 2013).

After June 2011, the direct analysis of seawater samples by a γ-ray spectrometer, which is commonly used in national emergency monitoring, frequently showed readings below the detection limit; these cases constituted the majority of the seawater monitoring data and they were reported as 'not detected' (ND). The radioactivity of caesium isotopes in seawater can be determined via a sensitive analytical method using adsorption concentration by ammonium phosphomolybdate (AMP); with this method, the ^{137}Cs radioactivity in the seawater was determined before the Fukushima accident (Aoyama *et al.*, 2000). According to the high-sensitivity analysis, many of the ^{137}Cs values in the seawater off the coast of Fukushima Prefecture in autumn 2011 were 10–100 Bq/m^3, which were still 1–2 orders of magnitude higher than those found before the accident but were substantially lower than the values in April and May 2011 (TEPCO, 2011a; Nuclear Regulation Authority, 2017a). The results of continued monitoring showed that seawater radioactivity in much of the areas off Fukushima had returned to the pre-accident levels, except for in the immediate vicinity of the power plant facility (Nuclear Regulation Authority, 2017a).

The rapid decrease in seawater radioactivity reflected the characteristics of the Fukushima accident, in which a large amount of radioactivity was released into the environment in a relatively short period of time (Yoshida and Kanda, 2012). However, the TEPCO monitoring data indicated ^{137}Cs of several tens to several hundreds of kBq/m^3 in and around the harbour of the power plant through 2012 (Kanda, 2013). The values obtained from the following monitoring were still higher than the values obtained before the accident (Nuclear Regulation Authority,

2017a). If the release of radioactive material from the plant to the sea had completely stopped, the values would have dropped accordingly. Thus, the release is likely still occurring from the power plant site, although the amount is much smaller than that in the weeks following the accident (Kanda, 2013).

5.7.2 Pollution of Coastal Sediments

All of the materials delivered to the sea eventually arrive at the seafloor by deposition. The processes related to this deposition generally involve 'settling particles'. Various suspended particles aggregate in seawater, which increases the settling rate; the formation is mediated by organic material, such as detritus, plankton and its remains and faecal pellets. Radioactive materials are incorporated by plankton cells or adsorbed by various particles, and they reach the sediment surface along with the particles. An alternative process of radioactivity transfer is adsorption by the sediment particles during direct contact with the radioactive seawater. Regardless of the major process for the concentration of sediment radioactivity, we cannot expect a rapid decrease as observed in seawater.

Radioactivity transfer to sediments and the behaviour of the radioactive material within the sediment vary greatly depending on the radionuclide. A quantitative indicator of radionuclide association with sediments is the distribution coefficient, Kd. This coefficient is defined as the ratio of the concentration or radioactivity of an element in sediments and in seawater (IAEA, 2004; Takata *et al.*, 2010a).

$$\begin{aligned}\mathrm{Kd} = {} & (\text{radioactivity or concentration in sediments } (\mathrm{Bq/kg}))/ \\ & (\text{radioactivity or concentration in seawater } (\mathrm{Bq/L})).\end{aligned}$$

The radionuclide concentration in sediments is usually represented on a dry weight (kg-dry) basis. The recommended value for the Kd of caesium by the IAEA is 2000 for the open ocean (deep-sea sediments) and 4000 in coastal sediments (IAEA, 2004; Takata *et al.*, 2010a). Thus, the ^{137}Cs concentration in sediments is expected to be several thousand times that in seawater. However, this value is achieved when the levels of radionuclides in sediments and seawater are in equilibrium. For the Fukushima accident, the respective levels of ^{131}I, ^{134}Cs and ^{137}Cs in seawater increased and then decreased rapidly, which illustrates that predicting the radionuclide concentration in sediments using this coefficient is extremely difficult.

Monitoring of the radioactivity in marine sediments was initiated at the end of April 2011 by TEPCO, MEXT and the Fukushima Prefectural government. The number of monitoring points and data were far fewer than those of the seawater monitoring efforts. In July 2011, a ^{137}Cs value of 150 000 Bq/kg was reported

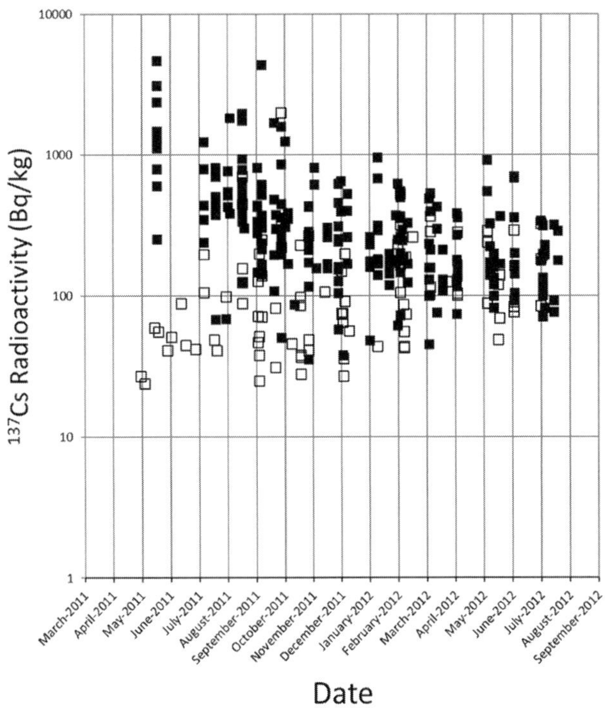

Figure 5.11 ^{137}Cs radioactivity data for sediments of the sea area off Iwaki City (Bq/kg-dry). Black squares show radioactivity data from shallower coastal areas (bottom depth <70 m) and white squares show data from deeper pelagic areas (bottom depth >70 m) (data compiled by the Nuclear Regulation Authority).

from the sediments in the power plant harbour.[2] Except for these values, the ^{137}Cs radioactivity values from coastal sediments rarely exceed a thousand Bq/kg-dry and were typically observed at several tens to several hundreds of Bq/kg-dry (TEPCO, 2011a; Nuclear Regulation Authority, 2017b).

Figure 5.11 summarises the ^{137}Cs radioactivity data for sediments of the sea area off Iwaki City; the figure shows radioactivity data from both shallower coastal areas (bottom depth <70 m) and deeper pelagic areas (bottom depth >70 m) from May 2011 to July 2012. Compared with the data obtained in the initial period from May through July 2011, when sediment monitoring began, the sediment radioactivity decreased with time in the coastal area. However, the radioactivity increased slightly over time in the pelagic area. Along with the slower decrease in sediment radioactivity compared with seawater, the variance in the radioactivity

[2] Radioactivity of sediments is generally reported as a value per dry weight (kg), although the measurements by TEPCO until the summer of 2012 was reported using a wet weight (kg) basis. For the same sample, the wet weight basis value was lower than that of the dry weight basis. The value for July 2011 in the plant harbour was on a wet weight basis.

was large. The radioactivity values in sediments, even if obtained at the same sampling location, may vary significantly. This large variation reflects the non-uniformity of the seafloor environment in coastal areas; the chemical composition, particle size distribution, and other characteristics of sediments are highly variable even within a very small area. In addition, the highly radioactive caesium-bearing particles were also found in sediment samples, and the contribution of those particles might be significant (Ishimaru *et al.*, 2016). Interpreting time series data of sediment radioactivity, even if taken at the same location, should include a consideration of these variations.

Particulate material in sediment is highly mobile, and radioactive material can move together with particulate matter. In addition, soil particles and suspended solids of higher radioactivity from the terrestrial environment can be transported by rivers and will eventually accumulate in marine sediments. Off the coast of Ibaraki Prefecture, a few sampling locations showed an increasing trend in sediment radioactivity in June and July 2011 (Nuclear Regulation Authority, 2017b), and these cases were considered to be due to the effect of terrestrial supply. The overall levels of sediment radioactivity decreased gradually (Nuclear Regulation Authority, 2017b). Kusakabe *et al.* (2017) reported a comprehensive summary of distribution and temporal variation of radioactivity in coastal sediment.

Research results from Kindai University (Nakagawa *et al.*, 2012) and others have suggested that sediment ^{137}Cs radioactivity increased during 2012 in the area around the mouth of the Arakawa River in Tokyo Bay; this increase was assumed to be due to radioactive material deposited on the Kanto Plain and then supplied through the solid transfer process in rivers. The values, however, continued to decrease later (Nuclear Regulation Authority, 2017a).

5.8 Pollution of Marine Fish and Shellfish

TAKASHI ISHIMARU AND TATSUO AONO

5.8.1 Radioactive Caesium

In early April 2011, radioactive iodine and radioactive caesium exceeding the provisional regulation guidelines for food at that time were detected in kounago (juveniles of a sand lance species). In the period following the detection, radioactive caesium exceeding the guideline was detected in other fish species, such as the shirasu (larvae of the Japanese sardine), and some seaweed species (Fisheries Agency, 2012). Juvenile and larvae fish such as the kounago and the shirasu grow relatively quickly and migrate in surface waters; thus, radioactive materials in seawater and plankton are considered to transfer rapidly. The provisional regulation guideline in food was set for radioactive caesium (total of ^{134}Cs and ^{137}Cs), and the value was 500 Bq/kg-wet. In response to public concern about food safety and public demand for stricter regulations, the Japanese national government reduced the guideline value to 100 Bq/kg-wet in April 2012, following an assessment of radiation dose due to food consumption among the Japanese public (Department of Food Safety, Pharmaceutical and Food Safety Bureau, Ministry of Health, Labour and Welfare, 2011b). If a food material exceeded the guideline, shipment of the material from the place of production should be halted.

Beginning in June 2011, an increasing number of fish samples showed caesium radioactivity exceeding the provisional regulation guidelines in the coastal sea off Fukushima Prefecture (Fisheries Agency, 2012). The caesium radioactivity of some fish species is shown in Figure 5.12. Radioactivity in shirasu decreased rapidly following the rapid decrease in seawater radioactivity. On the other hand,

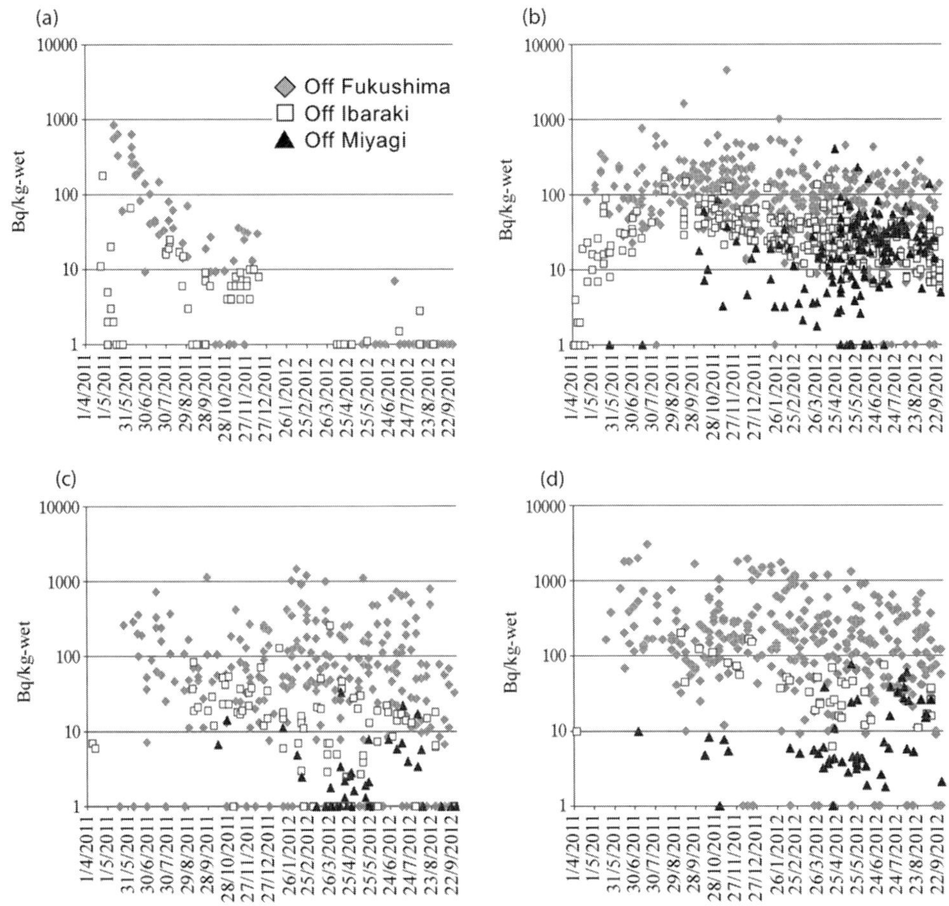

Figure 5.12 Caesium radioactivity ($^{134}Cs + ^{137}Cs$) data on some fish species during the period from April 2011 to September 2012. (a) shirasu (larvae of Japanese sardine), (b) olive flounder, (c) slime flounder, (d) greenling, (e) Japanese white rockfish, (f) Japanese sea bass, (g) black porgy (data compiled by the Fisheries Agency).

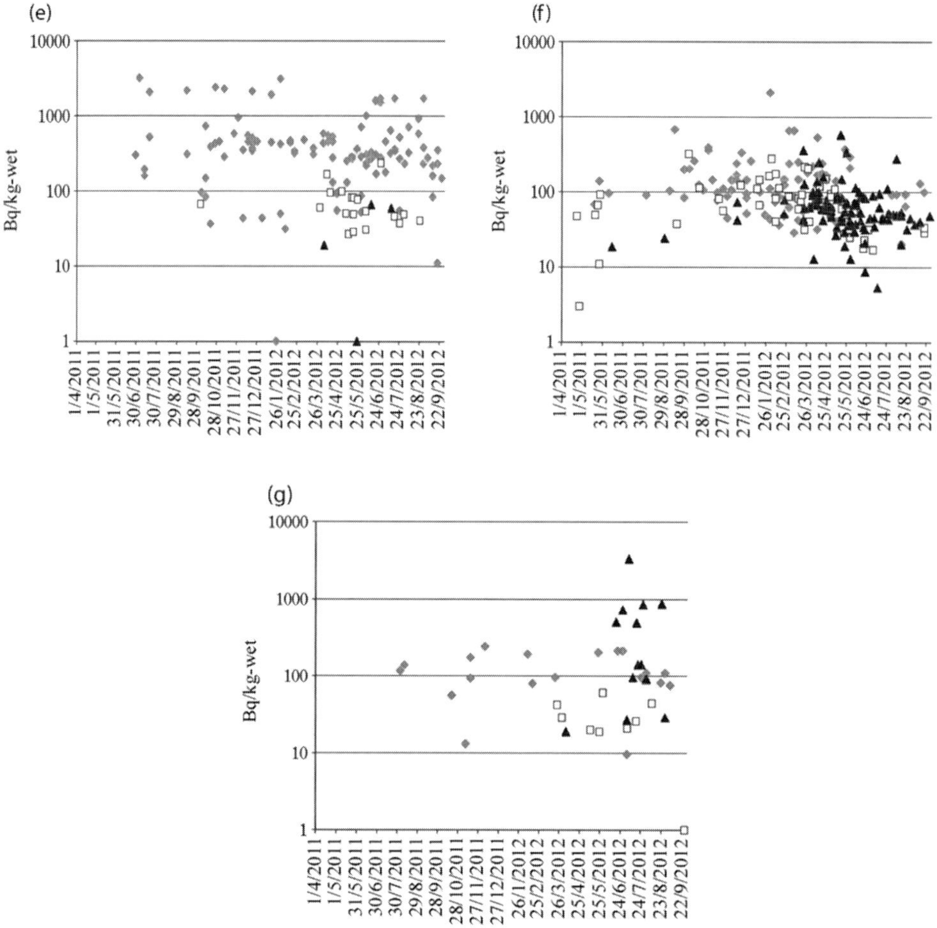

Figure 5.12 (*cont.*)

caesium radioactivity in some demersal fish, for example, the greenling, the olive flounder and the slime flounder, showed only a gradual decreasing trend. For the Japanese white rockfish commonly found in rocky reef areas in Fukushima, decreased radioactivity was not significant through the summer of 2012. Rockfish radioactivity levels started to decrease from 2012. Figure 5.13 shows the record of ^{137}Cs radioactivity in the greenling and the Japanese white rockfish through 2017. As we will discuss in Section 5.9, the decrease of radioactivity in these fish species was too slow, if the literature values of biological half-life for caesium are applicable.

A breakdown of caesium radioactivity data by prefecture (Figure 5.12) indicated that until a year after the accident, the radioactivity of the greenling, the olive flounder and the slime flounder was higher in Fukushima than in either Ibaraki or Miyagi. After April 2012, however, the levels in the Miyagi and Ibaraki samples

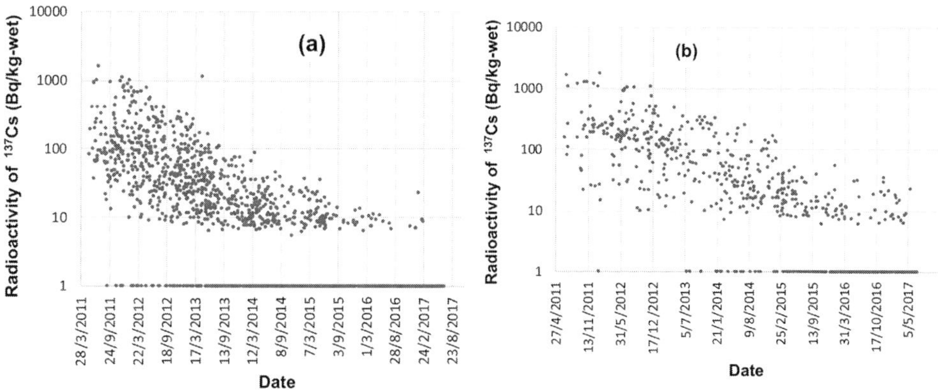

Figure 5.13 ^{137}Cs radioactivity data of (a) greenling and (b) Japanese white rockfish through summer 2017 (data compiled by the Fisheries Agency).

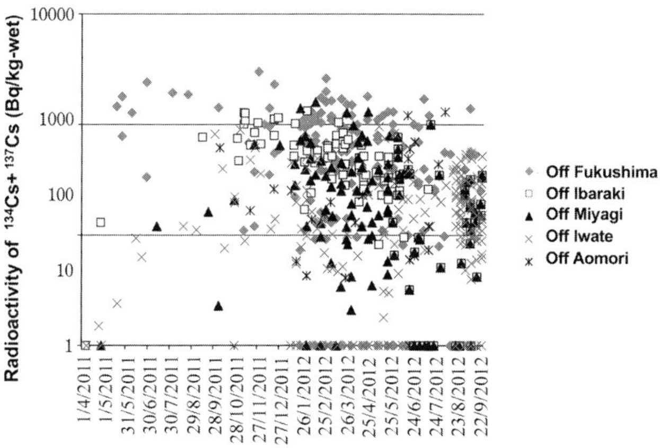

Figure 5.14 Caesium radioactivity (^{134}Cs + ^{137}Cs) data of Pacific cod during the period from April 2011 to September 2012 (data compiled by the Fisheries Agency).

became similar, and in the case of the greenling and the olive flounder, the radioactivity levels appeared to be higher in Miyagi than in Ibaraki. In the case of the blackish fish species, the caesium radioactivity of the Japanese sea bass in Fukushima showed a clear decreasing trend beginning in March 2012, but that in Miyagi showed an increasing trend. In the case of the black porgy, also a blackish species, the highest radioactivity in Fukushima was approximately 300 Bq/kg-wet, but in Miyagi several higher values were observed after June 2012, with the highest value being 3300 Bq/kg-wet. It is likely that the inflow of radioactive materials through rivers caused the higher values in Miyagi, and this possibility should be examined and monitored in the future.

Figure 5.14 shows the time evolution of the caesium radioactivity in the Pacific cod captured in several prefectures. Although the maximum value of the samples from Fukushima Prefecture was approximately 300 Bq/kg-wet, some samples from Aomori Prefecture exceeded the guideline level of 100 Bq/kg-wet for radioactive caesium. Based on these data, commercial shipping of the Pacific cod from the places of landing was halted for an extended area including Fukushima. Although the Pacific cod is a demersal species, it migrates and captures prey in a wide range of the sea, unlike the flatfish and rockfish species that tend to stay in a specific habitat. Therefore, the radioactivity of the Pacific cod may not become very high, but because of the wide migration area, individuals with radioactivity exceeding the guideline level might have been captured in an area very far from Fukushima. Because the distribution of contaminated benthic organisms was confined to the sea area close to Fukushima, radioactive contamination in food was likely not significant in remote areas. Therefore, the radioactivity of the Pacific cod may not be a persistent problem, given that the biological half-life of radioactive caesium is relatively short, as will be discussed later.

5.8.2 Radionuclides Other than Caesium Isotopes

In response to the global fallout of radioactive materials from nuclear testing in the atmosphere in the 1950s and 1960s, the Japanese government has implemented an environmental radioactivity survey of Japanese territory since 1957; radioactivity data of radionuclides such as ^{90}Sr and ^{137}Cs are reported. Since 1983, an annual survey of radionuclides in marine organisms, marine sediments and seawater has been conducted in major fishing grounds in coastal waters near nuclear power plants. Artificial radionuclides detected from the marine organisms in these studies have included ^{90}Sr, ^{137}Cs and $^{239+240}$Pu. The respective radioactivity levels of these radionuclides are generally low in marine organisms; the samples have to be incinerated, and the remaining ash is subjected to analysis. Therefore, volatile radionuclides including radioactive iodine isotopes are not measured in this survey. The Marine Environmental Radioactivity Comprehensive Evaluation Project, which was conducted from 2005, showed that the radioactivity levels of ^{90}Sr, ^{137}Cs and $^{239+240}$Pu in marine organisms in waters around the national nuclear fuel facilities (off the coast of Aomori Prefecture in the Pacific Ocean) were ND~0.01, ND~0.18 and ND~0.001 (Bq/kg-wet), respectively (MEXT, 2012).

In the national radioactivity survey of the marine environment in response to the Fukushima accident, a survey plan was established based on the Radioactivity Measurement Methods No. 24, ‘Pretreatment Methods of Gamma-Ray Spectrometer Samples in an Emergency Event’ (MEXT, 1992). Accordingly, fresh samples undergo γ-ray spectrometric analysis to determine the ^{131}I radioactivity. The observed radioactivity levels for kounago (juvenile fish of a sand lance species) collected in the coastal sea area off Ibaraki Prefecture and Fukushima Prefecture in early April 2011 were 1700–1900 Bq/kg-wet, with a maximum of 12 000 Bq/kg-wet. Because ^{131}I has a short half-life, the ^{131}I radioactivity in seawater decreased exponentially, and from around mid-June 2011, ^{131}I was only detected in some seaweed species (Fisheries Agency, 2012).

On 17 March 2011, the Ministry of Health, Labour and Welfare established provisional regulation guideline values based on a document issued by the Nuclear Safety Commission in 1980.[3] The ministry implemented the regulation that halted the supply of food containing radioactive material exceeding the guideline level (Department of Food Safety, Pharmaceutical and Food Safety Bureau, Ministry of Health, Labour and Welfare, 2011a). Because one may expect that radioactive caesium from the power plant accompanies radioactive strontium, the ratio of ^{90}Sr to radioactive caesium (sum of ^{134}Cs and ^{137}Cs) was assumed to be 0.1. The consumption limit for each food category was calculated based on this ratio because an actual analysis of ^{90}Sr would take a long time. Furthermore, because Sr behaves similarly to calcium (Ca), public concern regarding the accumulation of radioactive Sr in marine organisms was aroused. Therefore, an analysis of radioactive Sr in fish was officially introduced in the food monitoring of radioactivity from the Fukushima accident (Fisheries Agency, 2015).

The analysis detected 0.45 and 1.2 Bq/kg-wet of ^{89}Sr and ^{90}Sr, respectively, in an individual body of the Japanese rockfish that had a Cs radioactivity of 970 Bq/kg-wet. ^{90}Sr was also detected in samples of round-nose flounder, southern mackerel and icefish, although the radioactivity was below 0.5 Bq/kg-wet. The ratio of radioactive Sr to radioactive Cs ranged from 0.002 to 0.008. In general, the Sr radioactivity levels in fish and shellfish were often below the detection limit (ND; less than 0.04 Bq/kg-wet in this case), and with this level of radioactivity, it was uncertain whether the radioactive Sr in fish and shellfish came from the Fukushima accident or from global fallout. The $^{239+240}$Pu radioactivity levels in fish and shellfish obtained from the waters off Fukushima Prefecture (Department of Food Safety, Pharmaceutical and Food Safety Bureau, Ministry of Health, Labour and Welfare, 2011b) and off Aomori Prefecture (MEXT, 2012) were both ND~1 mBq/kg-wet, meaning it was difficult to conclude that the radioactivity was derived from the accident.

A radioactive isotope of silver (^{110m}Ag; half-life 250 days) was also detected in molluscs such as squid and octopus and in crustaceans such as shrimp and crab. Even before the Fukushima accident, ^{108m}Ag was reported for squid liver or midgut gland. The chemical nature of silver is similar to that of copper, which is contained in haemocyanin (a blood pigment) in these organisms, and thus silver is

[3] 'Indicators of Food and Drink Intake Restriction' in the 'Disaster Prevention Measures for the Nuclear Facilities' (Nuclear Safety Commission, June 1980).

concentrated along with copper in these organisms (Folsom *et al.*, 1970). In April 2011, following the Fukushima accident, ^{110m}Ag was detected in deposition materials from the atmosphere (Nuclear Regulation Authority, 2017a), and ^{110m}Ag was certainly released into the environment due to the accident. Within the 20 km zone of the FDNPS, ^{110m}Ag was detected at a level of 13–69 Bq/kg-wet from samples of squid and crab collected from April to June 2012 (Aono *et al.*, 2014).

The samples with detectable ^{110m}Ag showed relatively low Cs radioactivity; the Cs radioactivity typically ranged from below the detection limit to several factors lower than that of ^{110m}Ag (Department of Food Safety, Pharmaceutical and Food Safety Bureau, Ministry of Health, Labour and Welfare, 2011b; TEPCO, 2012). Furthermore, ^{110m}Ag of approximately 10 Bq/kg-wet was detected in benthic organisms (sea cucumber, worms, starfish, heart urchins, etc.) and plankton that could become the prey of fish and shellfish (Aono *et al.*, 2014). After the Chernobyl accident, ^{110m}Ag was also detected in seaweed in the Mediterranean Sea.

The provisional regulation guideline for food implemented in April 2012 was set for radioactive caesium (total of ^{134}Cs and ^{137}Cs), although the value accounted for the effects of other radionuclides (with half-lives of two years or more) accompanying radioactive Cs. These radionuclides include radioactive Sr and Pu.

5.9 Transfer Mechanisms of Radionuclides in the Marine Ecosystem

JOTA KANDA AND TAKASHI ISHIMARU

In the marine environment, the transfer pathways of radioactive materials into the bodies of marine organisms include assimilation of radioisotopes in food and absorption of dissolved radioactive substances in seawater through gills, body surfaces and gastrointestinal tracts. The relative contributions of these pathways may vary depending on the species of radionuclide and that of the organism. For some radionuclides, such as caesium isotopes, radionuclides incorporated into the body will eventually be discharged. The discharge of some elements is insignificant, and the radionuclides of these elements thus continuously accumulate in the body. The concentration ratio (CR) is used to indicate the degree of transfer of radionuclides in an organism in a marine environment. The CR is given by the following equation (Radioactive Waste Management Center, 1996; IAEA, 2004; Takata *et al.*, 2010b):

$$\text{CR} = (\text{Radioactivity concentration in the body } (\text{Bq/kg-wet}))/ \\ (\text{Radioactivity concentration in the seawater } (\text{Bq/L}) \text{ or } (\text{Bq/kg})).$$

The radioactivity concentration in an organism is usually expressed on the basis of the wet weight of the body. The radioactivity concentration in the seawater may also be expressed on a weight basis (Bq/kg). In this case, the CR value should be dimensionless, although the value is practically unchanged.

For marine fishery products from the coastal areas around the Fukushima Daiichi and Fukushima Daini nuclear power plants, ^{137}Cs radioactivity data were available before the accident (MEXT, 2010). The radioactivity of various fish, such as the Japanese sea bass and the Japanese rockfish, was 0.049–0.16 Bq/kg in 2009, and the concentration ratios calculated from the data were 29–114. For the various species of marine organisms and radionuclides, concentration ratio data have undergone review by the Radioactive Waste Management Center (1996) and IAEA (2004). The IAEA (2004) recommended concentration ratio values for caesium in marine organisms; the recommended values were 20 for phytoplankton, 40 for zooplankton and 100 for fish. The ratio varies substantially even among species within the same taxonomic group (Radioactive Waste Management Center, 1996; IAEA, 2004). As was the case for the distribution coefficients for marine sediments, the concentration ratio is an equilibrium value that is representative of sufficiently long periods under a given environmental condition for radioactivity.

When an individual marine organism is given a diet containing a radionuclide or is transferred to a water body containing a radionuclide, it takes time to accumulate the radionuclide in the body to the level that is predicted from the concentration ratio. In the case of transfer through food, the delays at the preceding steps of the food chain with different metabolic processes are summative. Furthermore, when radionuclides are taken up, some of the radionuclides will be concurrently discharged from the body. To quantify these dynamic processes, we need information about rate constants for both uptake and discharge (Kasamatsu and Ishikawa, 1997).

Because the rate constants must be experimentally determined in the laboratory, the available information remains limited (Radioactive Waste Management Center, 1996). For species with a significant discharge of radionuclides from the body, the body radioactivity will decrease if the incorporation of the radionuclide stops, although this may take time. The incorporation and discharge processes are also affected by environmental factors in the habitat. The rate of discharge, or the decrease of radioactivity, is expressed as biological (ecological) half-lives. Because caesium is not accumulated in a specific part of marine organisms, the discharge rate of caesium is relatively high. The biological half-life of caesium for fish is in the range of several days to several tens of days, depending on the fish species (Radioactive Waste Management Center, 1996). However, Matsumoto *et al.* (2015) suggested that the value for the Japanese white rockfish, especially for aged individuals, should be substantially longer. This possibility should be examined for other fish species with slower decreasing rates of radioactive caesium.

Because the seawater radioactivity changed significantly in a short period after the Fukushima accident, predicting the radioactivity in marine organisms was difficult based on the concentration ratio. For example, an investigation by the Japan Agency for Marine-Earth Science and Technology (JAMSTEC), which was conducted in the latter half of April 2011, showed that the apparent concentration ratios of ^{137}Cs for zooplankton were 200–840 (Honda *et al.*, 2012). The high concentration ratios likely reflected the rapid decline of caesium radioactivity in seawater and the time lag of accumulation and discharge of caesium in zooplankton; a relatively high concentration of ^{137}Cs remained in the body of zooplankton, although the seawater radioactivity had already decreased.

As mentioned above, caesium radioactivity in seawater decreased rapidly in the coastal waters of Fukushima Prefecture. Accordingly, the concentration of radioactive caesium in fish was expected to decrease rapidly following the reduced transfer from seawater. Among the caesium radioactivity of fish species shown in Figure 5.12, the radioactivity of the shirasu (larvae of the Japanese sardine) showed a decrease in a manner that followed the decrease in seawater radioactivity. However, the decrease of caesium radioactivity in several fish species was quite slow, as shown in Figures 5.12 and 5.13. Unless the longer biological half-life of caesium found in the Japanese white rockfish (Matsumoto *et al.*, 2015) were applicable to these fish species, the remaining possible cause would be the continuous input of radioactive caesium. Considering the rapidly decreased caesium radioactivity in seawater, transfer from seawater should be insignificant, and the only possible pathway of input is through food.

Plankton are a representative food for fish. Plankton have a shorter generation time, and the biological half-life of caesium is short in plankton; the caesium radioactivity in plankton should follow the decrease in seawater radioactivity. Although a survey by TUMSAT found that the caesium radioactivity was high in some plankton samples from the shallow waters near the coast of Fukushima in 2012 (Ishimaru *et al.*, 2012), at least some portion of this high radioactivity could be attributed to the highly radioactive caesium-bearing particles (Ishimaru *et al.*, 2016).

Benthic organisms also become food for fish. They live in the seabed sediments with higher caesium radioactivity levels than those of seawater. Therefore, the transfer of radioactive caesium to fish from benthos was also postulated. Marine sediments are a complex mixture of mineral particles, biological shells of siliceous and calcareous material, organic matter in biological remains (detritus) and in living organisms, and pore seawater retained among the particles. The radioactivity levels of benthic organisms decreased concurrently with the sediment radioactivity (Sohtome *et al.*, 2014), and the radioactivity level of benthic organisms was not high enough to explain the delayed decrease of radioactivity.

The degree of adsorption or binding of radioactive material to the sediment particles should vary depending on the type of particles. The transfer of radioactive material to the body of benthic organisms may also vary depending on the type of particle. Caesium is known to bind strongly to mineral particles, particularly to certain types of clay minerals. The biological transfer of radioactive caesium that was bound strongly to clay minerals is less likely, although radioactive caesium in detrital particles may be easily transferred to benthic organisms. The transfer of radioactive materials to benthic organisms should be closely related to the chemical form of the radioactive material. From the data collected in the investigations by TUMSAT as well as the data obtained by the Fukushima Prefectural Fisheries Experimental Station (Sohtome *et al.*, 2014), the radioactive caesium in benthic organisms showed a positive correlation with the radioactive caesium in the sediments. However, the caesium radioactivity in various species of benthic organisms showed substantial variations even at the same location, highlighting the complexity of the transfer process in marine sediments (Ishimaru *et al.*, 2012).

Marine organisms inherently include various species with a diverse range of taxa and lifestyles. In addition, the radioactivity of seawater substantially changed both temporally and spatially after the accident. Thus, predicting the radioactivity in marine organisms was very difficult. Furthermore, the data for marine life taken after the accident indicated that variations in radioactivity in different individual samples of the same species were large. The individuals living in different places and with different migration zones naturally showed different radioactivity values. Therefore, comprehensive monitoring of seawater, sediments and marine organisms will be necessary for some time to come.

5.10 Radioactive Caesium from the Fukushima Nuclear Power Plant in Migratory Marine Animals

ZOFIA BAUMANN, DANIEL J. MADIGAN AND NICHOLAS S. FISHER

The Kuroshio Extension system off Japan is one of the most productive oceanic regions worldwide. Diverse marine animals use these waters for part or all of their life cycle to forage and accumulate energetic reserves prior to reproduction or migration. Therefore, the massive release of radionuclides from the failed FDNPS in March and April 2011 provided an enormous source of radioactive contaminants to food webs that was of significance not just locally, but globally, as various species could become vectors of Fukushima contamination to other parts of the Pacific Ocean.

The spawning grounds of the Pacific bluefin tuna (*Thunnus orientalis*; hereafter PBFT) are located in the East China Sea and the Sea of Japan (Chen *et al.*, 2006;

Baumann *et al.*, 2015). At the end of the first year of life, some juvenile fish migrate from these coastal seas to the Kuroshio Current Extension in the Pacific, where they forage for several weeks to months (Itoh *et al.*, 2003). Some of these fish then make a trans-Pacific migration to the California Current Large Marine Ecosystem (Itoh *et al.*, 2003). In the spring of 2011, juvenile PBFT migrated from the shallow coastal seas to Pacific waters, which were contaminated by the radioactive plume originating in the damaged FDNPS (Madigan *et al.*, 2012). Like other fish from that region, the muscle tissue and internal organs of tunas became contaminated with the radioactive caesium isotopes ^{134}Cs and ^{137}Cs. As tunas are pelagic predators, the most likely exposure route to ^{137}Cs and ^{134}Cs escaping the damaged FDNPS was via ingestion of pelagic feeding forage fish such as sand lance (*Ammodytes* spp.), sardine (*Sardinops melanostictus*), jack mackerel (*Trachurus japonicus*), Pacific mackerel (*Scomber japonicus*) and others that were exposed to these released radionuclides. A small number (n = 8) of juvenile PBFT caught off Japan in August 2011 and October 2011 were found to have 6.7–20 Bq/kg-wet for ^{134}Cs and 8.5–21 Bq/kg-wet for ^{137}Cs (MEXT, 2012), well below the 100 Bq/kg-wet level of radiocaesium established by the Japanese government as safe for human consumption. In August 2011, small PBFT (~60–70 cm) estimated to be in the one-year-old age class (Madigan *et al.*, 2012) were collected off southern California. A subset of 15 fish was analysed for radioactive γ-emitting radioisotopes, and results clearly indicated that FDNPS was the source (Madigan *et al.*, 2012). Specifically, the presence of ^{134}Cs in muscle tissue marked the newly migrant PBFTs because any ^{134}Cs (half-life: two years) previously released into the Pacific by nuclear weapons testing in the 1950s and 1960s had decayed to undetectably low levels, whereas the longer-lived ^{137}Cs (half-life: 30 years) was still detectable at low levels in Pacific biota, water and sediments. Total radiocaesium levels in PBFT captured off California in August 2011 were about 10 Bq/kg-dry (~2.5 Bq/kg-wet) (Madigan *et al.*, 2012). By employing previously determined physiological parameters, including turnover rate of previously assimilated radiocaesium, somatic growth rate and ^{137}Cs and ^{134}Cs radioactivity levels in Pacific seawater (negligibly low or absent in the central and eastern Pacific, respectively), as well as radioactive decay rates of the two Cs isotopes, Madigan *et al.* (2012) estimated that total radiocaesium levels in PBFT were 15-fold higher when they first migrated from Japan than when captured off California. Although clearly detectable, the radioactivity levels in the PBFT caught in California waters and in Japanese waters were calculated to be far below levels that would likely be toxic to seafood consumers or to the fish themselves (Fisher *et al.*, 2013).

Analysis of the ^{134}Cs:^{137}Cs ratios in the fish indicated that the duration of PBFT trans-Pacific migration was estimated to range from one to four months (Madigan *et al.*, 2012). It was further demonstrated that the radionuclides released into the

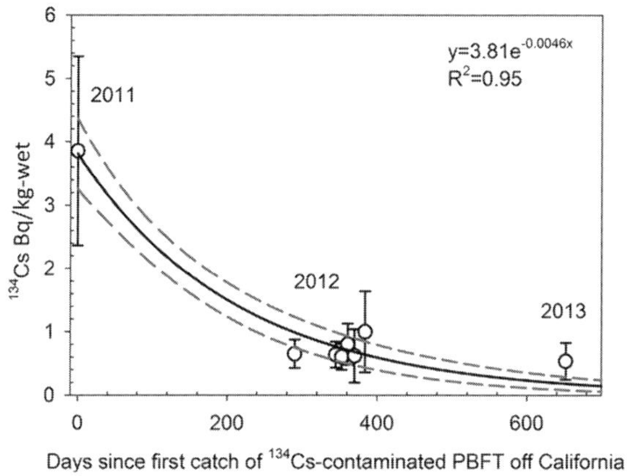

Figure 5.15 Decline [0.0046 ± 0.0004 (SE) d^{-1}] of ^{134}Cs in PBFT caught off California in summers of 2011, 2012 and 2013. The calculated effective half-life of ^{134}Cs in PBFT was estimated to be 151 days. One individual caught in 2012 with atypically high ^{134}Cs activity of 7.4 ± 0.6 Bq/kg-dry was treated as an outlier. Figure modified from Madigan *et al.* 2017b Copyright © 1980, CCC Republication.

Pacific from the damaged power plant could be used as tracers of tuna migration, providing new insights and tools to study the movements and ecology of fish species with depleted populations due to heavy commercial fishing (Madigan *et al.*, 2013; 2014; 2016; 2017a). Moreover, measurements of ^{134}Cs and ^{137}Cs in PBFT sampled off California in 2013 and 2014 enabled calculation of the effective biological half-life of ^{134}Cs, which was estimated to be 151 days (Figure 5.15). This effective half-life integrates decline due to biological turnover of ^{134}Cs acquired by PBFT while foraging in the western Pacific off Japan, as well as its radioactive decay (Madigan *et al.*, 2017a). This information is of significance because in the days and months following the accident, and especially following the detection of Fukushima radioactivity in PBFT caught off California, assessment of the risks and prediction of future radioactivity levels in these fish were difficult and caused considerable public anxiety (Fisher *et al.*, 2015). Subsequent measurements of radioactivity in PBFT caught off California in later years showed that by 2015, ^{134}Cs was not detectable in individuals that had recently migrated to eastern Pacific waters and ^{137}Cs was approaching levels that were prevalent prior to the accident (Madigan *et al.*, 2017b).

Muscle tissue samples from another tuna species, albacore (*Thunnus alalunga*), caught off the US Pacific Northwest in 2011 and 2012 were also found to contain Fukushima radiocaesium. Similar to California-caught PBFT, levels of radioactivity in albacore were very low (^{134}Cs: 0.02–0.36 Bq/kg-wet) (Neville *et al.*, 2014). In another migratory fish, chinook salmon (*Oncorhynchus tshawytscha*), caught off northern California in 2013, ^{134}Cs was not detected (Madigan *et al.*, 2017b). Madigan *et al.* (2017b) analysed other migratory marine species for γ-emitting

Table 5.1 *Summary of migratory species sampled in the North Pacific that were analysed for ^{134}C and ^{137}Cs.*

Species	Sampling area[a]	*n*	SL[b] ± SD (cm)	Year(s) acquired	^{137}Cs Bq/kg	Inferred origin[c]
Bigeye tuna	CPO	7	58 ± 3	2012	1.6 ± 0.6	CPO
Skipjack tuna	CPO	3	77 ± 28	2012	1.3 ± 0.1	CPO
Yellowfin tuna	CPO	4	48 ± 7	2012	1.5 ± 0.3	CPO
Yellowfin tuna	EPO	10	73 ± 12	2012/13	0.8 ± 0.3	EPO
Dolphin-fish	WPO	5	52 ± 2	2013	1.1 ± 0.3	WPO/CPO
Chinook salmon	EPO	34	80 ± 10	2013	1.0 ± 0.3	EPO/CPO
Common dolphin	EPO	4	205 ± 32	2015	0.9 ± 0.2	EPO
Bottlenose dolphin	EPO	2	180 ± 75	2015	0.7 ± 0.2	EPO
Fin whale	EPO	1	1714	2014	1.7 ± 0.1	WPO/CPO
[d]Olive ridley turtle	EPO	1	66	2015	1.0 ± 0.1	EPO
Swordfish	EPO	5	183 ± 11	2012/13	0.8 ± 0.2	EPO
Blue shark	EPO	6	118 ± 15	2012	1.3 ± 0.3	EPO/CPO
[e]Mako shark	EPO	1	373	2013	2.9 ± 0.2	EPO
Mako shark	EPO	3	127 ± 3	2012/13	1.3 ± 1.2	EPO
Opah	EPO	4	103 ± 5	2012	0.6 ± 0.1	EPO
Wahoo	EPO	1	122	2012	1.7 ± 0.1	EPO

Origin of each species was inferred based on signatures of C and N stable isotopes.
[a] EPO: eastern Pacific Ocean; CPO: central Pacific Ocean; WPO: western Pacific Ocean.
[b] SL = standard length; length for olive ridley turtle is carapace length.
[c] Origin inferred by comparing $\delta^{13}C$ and $\delta^{15}N$ to similar organisms from WPO, CPO and EPO.
[d] Only sample with measurable ^{134}Cs (0.1 Bq/kg).
[e] This mako shark was analysed individually due to its exceptional size.
Table modified from Madigan *et al.* 2017b Copyright © 1980, CCC Republication.

radioisotopes from various sources (Table 5.1). The migratory routes of these animals are depicted in Figure 5.16. Samples of opah (*Lampris guttatus*), swordfish (*Xiphias gladius*), blue shark (*Prionace glauca*), mako shark (*Isurus oxyrinchus*), yellowfin tuna (*Thunnus albacares*) and wahoo (*Acanthocybium solandri*) were collected during 2012 and 2013 within 200 km of the coast of California. Muscle tissue samples of three marine mammal species, common dolphin (*Delphinus capensis*), bottlenose dolphin (*Tursiops truncatus*) and fin whale (*Balaenoptera physalus*), as well as one reptile (olive ridley sea turtle, *Lepidochelys olivacea*) were also collected at coastal locations. These samples were taken from the carcasses of animals that had died of unknown causes, making them available for radioanalysis. In addition, yellowfin, skipjack (*Katsuwonus pelamis*) and bigeye (*Thunnus obesus*) tunas were sampled in Hawaiian waters (18°43′N, 158°170′W). Fukushima contamination was undetectable in all but one of the migratory animals examined in the Madigan *et al.* (2017b) study. The olive

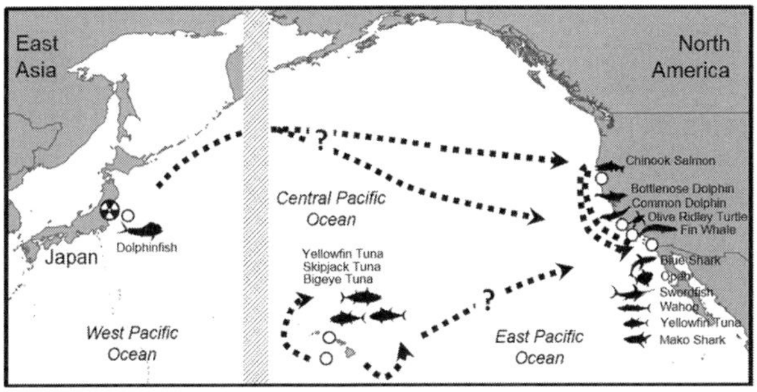

Figure 5.16 Sampling locations (open circles) in the western, central and eastern North Pacific Ocean. Species icons and common names are listed with the region in which they were sampled. Arrows show simplified patterns of animal migration patterns based on stable isotope analysis and published life history information. Question marks denote unknown, hypothesised migration routes. Figure modified from Madigan *et al.* 2017b Copyright © 1980, CCC Republication.

ridley sea turtle muscle tissue contained very low levels of ^{134}Cs (0.1 Bq/kg-dry), indicating FDNPS as a source. One individual wahoo caught north of O'ahu Island (Hawai'i) in October 2011 also contained detectable ^{134}Cs (0.39 Bq/kg-wet and 3.95 Bq/kg-wet of ^{134}Cs and ^{137}Cs, respectively), indicating Fukushima as the source (S. Charmasson, personal communication).

As speculation (with little evidence) about spreading contamination from Fukushima to distant ecoregions via animal movements grew following the disaster, it became important to assess these dire assertions. While sampling of the various species tended to be sporadic rather than systematic, the accumulated results showed that levels of Fukushima-derived radiocaesium, best indicated by the presence of ^{134}Cs, were below detection in most of these animals. These low levels refute the contention that marine migrators throughout the Pacific would become significantly contaminated for a sustained period and could be impacted by the radioactivity released from Fukushima into the Pacific.

The Fukushima disaster resulted in the largest accidental release of radioactivity into the ocean in history (Buesseler *et al.*, 2017), but strong ocean currents and the volume of the Pacific Ocean led to rapid dispersal and dilution of radioactivity. Trace levels of dissolved ^{134}Cs (2 Bq/m^3 of seawater) eventually reached the west coast of North America, approximately 35 months after the accident (Smith *et al.*, 2015). Currently, radioactivity levels of seawater measured anywhere in the Pacific, including in the plume that has reached the North American west coast, are very low and comparable to the ^{137}Cs levels that are a legacy of the nuclear weapons testing that occurred in the 1950s and 1960s (Buesseler *et al.*, 2017). Therefore, marine organisms inhabiting the North Pacific outside of the immediate vicinity of Fukushima contain trace levels of ^{137}Cs and the quickly decaying ^{134}Cs, which are at or below the limits of detection.

References

Aono, T., Y. Ito, T. Sohtome, *et al.* (2014). Observation of radionuclides in marine biota off the coast of Fukushima Prefecture after TEPCO's Fukushima Daiichi Nuclear Power Station accident. In S. Takahashi (ed.) *Radiation Monitoring and Dose Estimation of the Fukushima Nuclear Accident*. New York: Springer, pp. 115–23.

Aoyama, M. and K. Hirose (1995). The temporal and spatial variation of ^{137}Cs concentration in the Western North Pacific and its marginal seas during the period from 1979 to 1988. *J. Environ. Radioact.*, **29**, 57–74.

Aoyama, M., K. Hirose, T. Miyao and Y. Igarashi (2000). Low level Cs measurements in deep seawater samples. *Applied Radiation and Isotopes*, **53**(1), 159–62, doi:10.1016/S0969-8043(00)00128-7.

Aoyama, M., M. Fukasawa, K. Hirose, *et al.* (2011). Cross equator transport of ^{137}Cs from North Pacific Ocean to South Pacific Ocean (BEAGLE2003 cruises). *Prog. Oceanogr.,* **89**, 7–16.

Aoyama, M., D. Tsumune and Y. Hamajima (2012a). Distribution of ^{137}Cs and ^{134}Cs in the North Pacific Ocean: impacts of the TEPCO Fukushima-Daiichi NPP accident. *J. Radioanal. Nucl. Chem.*, **296**, 535–9.

Aoyama, M., D. Tsumune, M. Uematsu, F. Kondo and Y. Hamajima (2012b). Temporal variation of ^{134}Cs and ^{137}Cs activities in surface water at stations along coastal line near the Fukushima Dai-ichi Nuclear Power Plant accident site. *Japan. Geochem. J.*, **46**, 321–5.

Aoyama, M., M. Uematsu, D. Tsumune and Y. Hamajima (2013). Surface pathway of radioactive plume of TEPCO Fukushima NPP1 released ^{134}Cs and ^{137}Cs. *Biogeosciences,* **10**, 3067–78.

Aoyama, M., Y. Hamajima, M. Hult, *et al.* (2016a). ^{134}Cs and ^{137}Cs in the North Pacific Ocean derived from the March 2011 TEPCO Fukushima Dai-ichi Nuclear Power Plant accident, Japan. Part one: surface pathway and vertical distributions. *J. Oceanogr.*, **72**, 53–65.

Aoyama, M., M. Kajino, T. Y. Tanaka, *et al.* (2016b). ^{134}Cs and ^{137}Cs in the North Pacific Ocean derived from the March 2011 TEPCO Fukushima Dai-ichi Nuclear Power Plant accident, Japan. Part two: estimation of ^{134}Cs and ^{137}Cs inventories in the North Pacific Ocean. *J. Oceanogr.*, **72**, 67–76.

Aoyama, M., Y. Hamajima, Y. Inomata and E. Oka (2017). Recirculation of FNPP-1-derived radiocesium observed in winter 2015/2016 in coastal regions of Japan. *Appl. Radiat. Isotopes*, **126**, 83–7.

Aoyama, M., Y. Hamajima, Y. Inomata, *et al.* (2018). Radiocaesium derived from the TEPCO Fukushima accident in the North Pacific Ocean: surface transport processes until 2017. *J. Environ. Radioact.*, **189**, 93–102.

Baumann, H., R. D. J. Wells, J. R. Rooker, *et al.* (2015). Combining otolith microstructure and trace elemental analyses to infer the arrival of juvenile Pacific bluefin tuna in the California current ecosystem. *ICES Journal of Marine Science*, **72**(7), 2128–38.

Buesseler, K., M. Aoyama and M. Fukasawa (2011). Impacts of the Fukushima nuclear power plants on marine radioactivity. *Environ. Sci. Technol.*, **45**, 9931–5.

Buesseler, K., M. Dai, M. Aoyama, *et al.* (2017). Fukushima Daiichi-derived radionuclides in the ocean: transport, fate, and impacts. *Ann Rev Mar Sci*, **9**, 173–203.

Chen, K. S., P. Crone, and C. C. Hsu (2006). Reproductive biology of female Pacific bluefin tuna *Thunnus orientalis* from south-western North Pacific Ocean. *Fisheries Science*, **72**(5), 985–94.

Chino, M., H. Nakayama, H. Nagai, *et al.* (2011). Preliminary estimation of release amounts of ^{131}I and ^{137}Cs accidentally discharged from the Fukushima Daiichi nuclear power plant into the atmosphere. *J. Nucl. Sci. Technol.*, **48** (7), 1129–34, doi:10.1080/18811248.2011.9711799.

Department of Food Safety, Pharmaceutical and Food Safety Bureau, Ministry of Health, Labour and Welfare (2011a). Notice No. 0317 Article 3 of the Department of Food Safety, 17 March. http://bit.ly/2VsVECJ (accessed 19 September 2018).

Department of Food Safety, Pharmaceutical and Food Safety Bureau, Ministry of Health, Labour and Welfare (2011b). New standard limits for radionuclides in foods. www.mhlw.go.jp/english/topics/2011eq/dl/new_standard.pdf (accessed 19 September 2018).

Estournel, C., E. Bosc, M. Bocquet, *et al.* (2012) Assessment of the amount of Cesium-137 released into the Pacific Ocean after the Fukushima accident and analysis of its dispersion in Japanese coastal waters. *J. Geophys. Res. Oceans* **117**, C11.

Fisher, N. S., K. Beaugelin-Seiller, T. G. Hinton, *et al.* (2013). Evaluation of radiation doses and associated risk from the Fukushima nuclear accident to marine biota and human consumers of seafood. *Proc. Natl. Acad. Sci. USA*, **110**, (26), 10670–5.

Fisher, N. S., S. W. Fowler and D. J. Madigan (2015). Perspectives and reflections on the public reaction to recent Fukushima-related radionuclide studies and a call for enhanced training in environmental radioactivity. *Environ. Toxicol. Chem.* **34**(4), 707–709.

Fisheries Agency (2012). Results of the inspection on radioactive materials in fisheries products. Press releases, March 2011–March, 2012. www.jfa.maff.go.jp/e/inspection/pdf/e120330_2.pdf (accessed 19 September 2018).

Fisheries Agency (2015). Report on the monitoring of radionuclides in fishery products. http://bit.ly/2Vp7uhd (accessed 19 September 2018).

Folsom, T. R., R. Grismore and D. R. Young (1970). Long-lived gamma-ray emitting nuclide silver-108m found in Pacific marine organisms and used for dating. *Nature*, **227**, 941–3, doi:10.1038/227941a0

Honda, M. C., T. Aono, M. Aoyama, *et al.* (2012). Dispersion of artificial caesium-134 and -137 in the western North Pacific one month after the Fukushima accident. *Geochem. J.*, **46**(1), e1–e9.

IAEA (International Atomic Energy Agency) (2004). *Sediment Distribution Coefficients and Concentration Factors for Biota in the Marine Environment*. Vienna: IAEA.

Inomata, Y., M. Aoyama and K. Hirose (2009). Analysis of 50-y record of surface ^{137}Cs concentrations in the global ocean using the HAM-global database. *J. Environ. Monitor*, **11**, 116–25.

Inoue, M., H. Kofuji, Y. Hamajima, *et al.* (2012a). ^{134}Cs and ^{137}Cs activities in coastal seawater along northern Sanriku and Tsugaru Strait, northeastern Japan, after Fukushima Dai-ichi Nuclear Power Plant accident. *J. Environ. Radioactiv.*, **111**, 116–19, doi:10.1016/j-jenvrad. 2011.09.012.

Inoue, M., H. Kofuji, S. Nagao, *et al.* (2012b). Lateral variation of ^{134}Cs and ^{137}Cs concentrations in surface seawater in and around the Japan Sea after the Fukushima Dai-ichi Nuclear Power Plant accident. *J. Environ. Radioactiv.*, **109**, 45–51, doi:10.1016/j.jenvrad.2012.01.004.

Ishimaru, T., T. Aono, Y. W. Watanabe, *et al.* (2012). Distribution and ecological transfer of radiocesium in the coastal waters of Fukushima Prefecture. 2012 spring meeting of the Oceanographic Society of Japan, Tsukuba, 27 March.

Ishimaru, T., Y. Ito, M. Otsuki, *et al.* (2016). Cesium radioactivity of plankton samples collected in the coastal waters of Fukushima Prefecture and the highly radioactive cesium-bearing particles. 2016 autumn meeting of the Oceanographic Society of Japan, Kagoshima, 13 September.

Itoh, T., A. Nitta and S. Tsuji (2003). Migration patterns of young Pacific bluefin tuna (*Thunnus orientalis*) determined with archival tags. *Fishery Bulletin*, **101** (3), 514–34.

Kaeriyama, H., D. Ambe, Y. Shimizu, *et al.* (2013). Direct observation of ^{134}Cs and ^{137}Cs in surface seawater in the western and central North Pacific after the Fukushima Dai-ichi nuclear power plant accident. *Biogeosciences*, **10**, 4287–95, doi:10.5194/bg-10-4287-2013.

Kaeriyama, H., Y. Shimizu, T. Setou, *et al.* (2016). Intrusion of Fukushima-derived radiocaesium into subsurface water due to formation of mode waters in the North Pacific. *Sci. Rep.*, **6**, 22010.

Kanda, J. (2013). Continuing ^{137}Cs release to the sea from the Fukushima Dai-ichi Nuclear Power Plant through 2012. *Biogeosciences*, **10**, 6107–13.

Kasamatsu, F. and Y. Ishikawa (1997). Natural variation of radionuclide ^{137} Cs concentration in marine organisms with special reference to the effect of food habits and trophic level. *Mar. Ecol. Prog. Ser.*, **160**, 109–20.

Kawamura, H., T. Kobayashi, A. Furuno, *et al.* (2011). Preliminary numerical experiments on oceanic dispersion of ^{131}I and ^{137}Cs discharged into the ocean because of the Fukushima Daiichi nuclear power plant disaster. *J. Nucl. Sci. Tech.*, **48**, 1349–56.

Kumamoto, Y., M. Aoyama, Y. Hamajima, *et al.* (2014). Southward spreading of the Fukushima-derived radiocesium across the Kuroshio Extension in the North Pacific. *Sci. Rep.*, **4**, 4276.

Kusakabe, M., N. Inatomi, H. Takata, and T. Ikenoue (2017). Decline in radiocesium in seafloor sediments off Fukushima and nearby prefectures. *J. Oceanogr.*, **73**(5), 529–45.

Madigan, D. J., Z. Baumann and N. S. Fisher (2012). Pacific bluefin tuna transport Fukushima-derived radionuclides from Japan to California. *Proc. Natl. Acad. Sci. USA*, **109** (24), 9483–6.

Madigan, D. J., Z. Baumann, O. E. Snodgrass, *et al.* (2013). Radiocesium in Pacific bluefin tuna *Thunnus orientalis* in 2012 validates new tracer technique. *Environ. Sci. Technol.*, **47**, 2287–94.

Madigan, D. J., Z. Baumann, A. B. Carlisle, *et al.* (2014). Reconstructing transoceanic migration patterns of Pacific bluefin tuna using a chemical tracer toolbox. *Ecology*, **95**, 1674–83.

Madigan, D. J., W.-C. Chiang, N. J. Wallsgrove, *et al.* (2016). Intrinsic tracers reveal recent foraging ecology of giant Pacific bluefin tuna at their primary spawning grounds. *Mar. Ecol. Prog. Ser.*, **553**, 253–66.

Madigan, D. J., Z. Baumann, A. B. Carlisle, *et al.* (2017a). Isotopic insights into migration patterns of Pacific bluefin tuna in the eastern Pacific Ocean. *Can. J. Fish. Aquat. Sci.*, **75**, 260–70, doi: 10.1139/cjfas-2016-0504.

Madigan, D. J., Z. Baumann, O. E. Snodgrass, *et al.* (2017b). Assessing Fukushima-derived radiocesium in migratory Pacific predators. *Environ. Sci. Technol.*, **51**, 8962–71.

Matsumoto, A., Y. Shigeoka, H. Arakawa, *et al.* (2015). Biological half-life of radioactive cesium in Japanese rockfish contaminated by the Fukushima Daiichi nuclear power plant accident. *J. Environ. Radioact.*, **150**, 68–74, doi:10.1016/j.jenvrad.2015.08.003.

Masumoto, Y., Y. Miyazawa, D. Tsumune, *et al.* (2012). Oceanic dispersion simulation of cesium-137 from Fukushima Daiichi Nuclear Power Plant. *Elements*, **8**(3), 207–12.

MEXT (Ministry of Education, Culture, Sports, Science and Technology) (1992). *Radioactivity Measurement Methods No.24: 'Pretreatment Methods of Gamma-Ray Spectrometer Samples in an Emergency Event'*. Tokyo: MEXT (in Japanese).

MEXT (2010). *Annual Report of the Marine Environmental Radioactivity Comprehensive Evaluation Project, 2009*. Tokyo: MEXT (in Japanese).

MEXT (2012). Readings of marine soil monitoring in sea area. http://radioactivity.nsr.go.jp/en/list/260/list-1.html (accessed 19 September 2018).

Nagao, S. (2015) Transport behavior of radiocesium in river systems after the Fukushima Dai-ichi Nuclear Power Plant accident. *Chikyukagaku (Geochemistry)*, **49**, 217–26 (in Japanese).

Nagao, S., M. Kanamori, S. Ochiai, *et al.* (2013). Export of ^{134}Cs and ^{137}Cs in the Fukushima river systems at heavy rains by Typhoon Roke in September 2011. *Biogeosci.*, **10**, 6215–23.

Nagao, S., M. Kanamori, S. Ochiai, M. Inoue and M. Yamamoto (2015). Migration behavior of ^{134}Cs and ^{137}Cs in the Niida River water in Fukushima Prefecture, Japan during 2011–2012. *J. Radioanalytical Nuclear Chem.*, **303**, 1617–21.

Nakagawa, R., M. Ishida, D. Baba, *et al.* (2012). Spatiotemporal distribution of radioactive cesium released from Fukushima Daiichi Nuclear Power Station in the sediment of Tokyo Bay, Japan. In *Proceedings of the International Symposium on Environmental Monitoring and Dose Estimation of Residents After Accident of TEPCO's Fukushima Daiichi Nuclear Power Stations*. Osaka: Kyoto University Research Reactor Institute, pp. 133–6.

Neville, D. R., A. J. Phillips, R. D. Brodeur and K. A. Higley (2014). Trace levels of Fukushima disaster radionuclides in East Pacific albacore. *Environ. Sci. Technol.*, **48**(9), 4739–43.

Nuclear Regulation Authority (2017a). Readings of sea area monitoring. http://radioactivity.nsr.go.jp/en/list/205/list-1.html (accessed 19 September 2018).

Nuclear Regulation Authority, (2017b). Change of the radioactivity concentration of the sediment in off-shore sea area. http://bit.ly/2VkiGf0 (accessed 19 September 2018).

Ohara, T., Y. Morino and M. Nishizawa (2011). Where and how did radioactive materials emitted to the air from Fukushima nuclear power plant deposit?. *Kagaku*, **81**(12), 1254–8. (in Japanese).

Radioactive Waste Management Center (1996). *Concentration Factors of Radionuclides in the Marine Organisms*. Tokyo: Radioactive Waste Management Center (in Japanese).

Science Council of Japan (2014). *A Review of the Model Comparison of Transportation and Deposition of Radioactive Materials Released to the Environment as a Result of the Tokyo Electric Power Company's Fukushima Daiichi Nuclear Power Plant Accident*. Tokyo: Science Council of Japan. www.scj.go.jp/en/report/index.html.

Smith, J. N., R. M. Brown, W. J. Williams, *et al.* (2015). Arrival of the Fukushima radioactivity plume in North American continental waters. *Proc. Natl. Acad. Sci. USA*, **112**, 1310–15.

Sohtome, T., T. Wada, T. Mizuno, *et al.* (2014). Radiological impact of TEPCO's Fukushima Dai-ichi Nuclear Power Plant accident on invertebrates in the coastal benthic food web. *J. Environ. Radioact.*, **138**, 106–15, doi:10.1016/j.jenvrad.2014.08.008.

Takata, H., T. Aono, K. Tagami and S. Uchida (2010a). Sediment–water distribution coefficients of stable elements in four estuarine areas in Japan. *J. Nuclear Sci. Tech.*, **47**(1), 111–22, doi:10.3327/jnst.47.111.

Takata, H., T. Aono, K. Tagami and S. Uchida (2010b). Concentration ratios of stable elements for selected biota in Japanese estuarine areas. *Radiat. Environ. Biophys.*, **49** (4), 591–601, doi:10.1007/s00411-010-0317-x.

TEPCO (2011a). Past Measurement Result by Seawater Radiation Monitor near Fukushima Daiichi Nuclear Power Station, Archives, www.tepco.co.jp/en/nu/fukushima-np/f1/index2-e.html (accessed 19 September 2018)

TEPCO (2011b). Out flow of fluid containing radioactive materials to the ocean from areas near intake channel of Fukushima Daiichi Nuclear Power Station Unit 2 (continued report). Press release, 5 April. www.tepco.co.jp/en/press/corp-com/release/11040506-e.html (accessed 19 September 2018).

TEPCO (2011c). Submission of a report to Ministry of Economy, Trade and Industry, Nuclear and Industrial Safety Agency, on the tsunami investigation at Fukushima Daiichi and Daini Atomic Power Plants. Press release, 8 July. www.tepco.co.jp/cc/press/11070802-j.html (accessed 19 September 2018) (in Japanese).

TEPCO (2012). Results of radionuclide analysis of marine organisms within the 20-km zone of the Fukushima Daiichi Nuclear Power Station, September 28. http://bit.ly/2VqDMIu (accessed 19 September 2018).

Tsubono, T., K. Misumi, D. Tsumune, *et al.* (2016). Evaluation of radioactive cesium impact from atmospheric deposition and direct release fluxes into the North Pacific from the Fukushima Daiichi nuclear power plant. *Deep Sea Res. Part 1 Oceanogr. Res. Pap.*, **115**, 10–21.

Tsumune, D., T. Tsubono, M. Aoyama and K. Hirose (2012). Distribution of oceanic ^{137}Cs from the Fukushima Dai-ichi Nuclear Power Plant simulated numerically by a regional ocean model. *J. Environ. Radioact.*, **111**, 100–8.

Yamashiki, Y., Y. Onda, H. G. Smith, *et al.* (2014). Initial flux of sediment-associated radiocesium to the ocean from the largest river impacted by Fukushima Daiichi Nuclear Power Plant. *Sci. Rep.*, **4**, 3714.

Yoshida, N. and J. Kanda (2012). Tracking the Fukushima radionuclides. *Science*, **336**, 1115–16.

6

Diffusion and Deposition of Radioactive Materials in the Terrestrial Environment

Radioactive materials emitted from the Fukushima Daiichi Nuclear Power Station (FDNPS) were deposited on soils and trees in forested areas, agricultural land and urban areas. It is expected that the radioactively polluted soils and radioactive materials would spread through erosion of soils from mountains and rivers. In this chapter, we first examine the behaviour of radioactive materials deposited on the ground in upcountry districts. Second, we discuss the movement of radioactive materials from various types of flatlands and forests in mountainous areas based on their chemical forms in soils and on trees. Finally, we report on the accumulation and transfer of radioactive materials to vegetation such as trees.

6.1 Overview of the Large-Scale Measurement of Radioactive Materials Deposited on Ground Surfaces

ISAO TANIHATA, MAMORU FUJIWARA AND YUICHI ONDA

6.1.1 Measurements of Radioactivity in Soils

The radioactivity nuclide map project was performed in June and July 2011 over a wide area within a 100 km radius of the FDNPS site. This large-scale project was completed by a joint team consisting of many scientists from Japanese universities and the Japanese Ministry of Education, Culture, Sports, Science and Technology (MEXT). Three types of datas were collected in the project:

1. spatial dose rates 1 m above the ground surface (μSv/h);
2. amounts of radioactive materials within 5 cm of the ground surface (Bq/m^2); and
3. permeation depth distribution of radioactive materials, primarily ^{134}Cs and ^{137}Cs, to a depth of 20 cm.

The obtained results are briefly explained in this report. Detailed discussions are provided in a website report from the Nuclear Regulation Authority of Japan, titled 'Monitoring information of environmental radioactivity level' (http://radioactivity.nsr.go.jp/en/list/311/list-1.html), and a paper by Saito *et al.* (2015). Both the results from the radioactive nuclide map project by MEXT and those obtained by the Ministry of Agriculture, Forestry and Fisheries (MAFF) are shown on the same website. Detailed information on the fallout of radioactive materials in relation to the FDNPS accident can be found on the NRA website (http://radioactivity.nsr.go.jp/en).

6.1.2 Spatial Dose Rates at a Height of 1 m

The spatial dose rates based on measurements from 2200 locations were reported by MEXT on 2 August 2011. It was found that high dose rates were distributed across Fukushima, Miyagi and Ibaraki Prefectures, including the 20 km zone centred at the FDNPS. The high dose rate area extended to the northwest from the FDNPS and was connected to the high dose rate area in the lowlands called Naka-Dori in Fukushima Prefecture. Although dose rate distributions were previously estimated using radioactivity measurements from aeroplanes (airborne measurement), the more reliable and dense measurements of the spatial dose rates presented here provide a standard distribution for calibrating airborne data.

The map of the radioactivity spatial dose rates and the radioactivity deposition maps can be found at www.rcnp.osaka-u.ac.jp/dojo.

6.1.3 Radioactive Materials in Soils within 5 cm of the Surface

Based on the results of the pilot project performed in May 2011, it was concluded that almost all fallout of radioactive materials deposited after the FDNPS accident permeated into the soil and was chemically trapped in soils within 5 cm of the ground surface (Kato *et al.*, 2012a). Therefore, we decided to collect soil samples to a depth of 5 cm. The detailed sampling scheme is described by Onda *et al.* (2015).

In the soil sampling project in June 2011, approximately 11 000 samples were collected at 2200 locations. Nearly half of the collected soil samples were sent to the Japan Chemical Analysis Center (JCAC), and γ-rays emitted from the soil samples were analysed. The remaining soil samples were sent to the Center for Nuclear Study, University of Tokyo (CNS). After reviewing the descriptions and labels attached to the soil samples, the collected soil samples were redistributed to the research groups working on nuclear physics and earth sciences at 20 Japanese universities and institutes. Moreover, γ-ray analyses of the soil samples were also performed by these individual research groups. The results for the deposition density maps of ^{134}Cs and ^{137}Cs were reported on 30 August 2011 by MEXT. In addition, the results for the deposition density map of ^{131}I were released to the press on 21 September 2011. The final report was delivered in March 2012. The radioactivity ratio ^{134}Cs /^{137}Cs was nearly constant at 0.92, regardless of the soil sampling location.

In contrast it was found that $^{131}I/^{137}Cs$ depended on the sampling location, with relatively high values in the southern region of the FDNPS. This result could be attributed to the following:

1. There were various release events of radioactive materials from the FDNPS from 11 March to 16 March 2011. The $^{131}I/^{137}Cs$ dose ratios were different in each event.
2. Because the wind directions differed depending on the day, the released radioactive materials flowed in different directions.

The radioactivity deposition maps for ^{134}Cs, ^{137}Cs and ^{131}I can be found at www.rcnp.osaka-u.ac.jp/dojo. A detailed discussion is provided in the MEXT report.

6.1.4 Permeation Depth Distribution of Radioactive Materials

In the project, soil core samples were collected using a 30 cm iron pipe to study the permeation depth distribution of radioactive materials in soils. Without removing the soils from the pipe, we made a non-destructive inspection of γ-rays emitted from the soils within the pipe. All of the measurements were conducted at the RCNP, Osaka University.

The permeation depth distribution of the sampled radioactive materials was studied using three Canberra HPGe detectors with a relative γ-ray detection efficiency of 25%. The measuring method is illustrated in Figure 6.1. We set two lead blocks with thicknesses of 5 cm in front of the HPGe detector. The gap

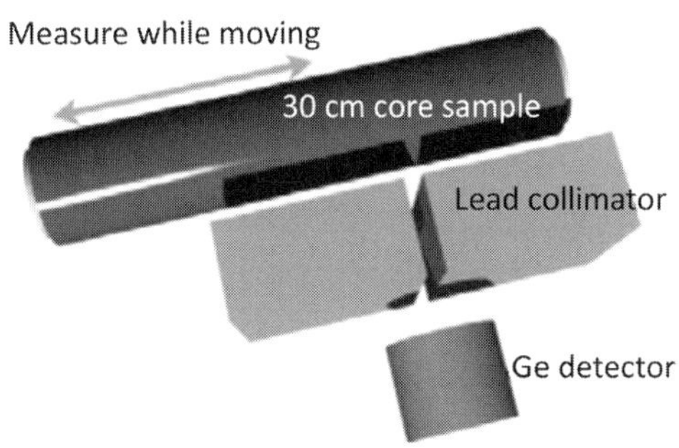

Figure 6.1 Radionuclide contained in soil core sample: measurement principle.

between the two blocks was 0.5 cm. This gap was used as a slit to define the direction from the core sample. Next, γ-rays passing through the gap from the 30 cm core sample to the HPGe detector were measured. The γ-ray counts due to 604 keV and 796 keV γ -rays from ^{134}Cs and 661 keV γ-rays from ^{137}Cs were analysed. Taking into account the γ-ray absorption by a piece of lead with a thickness of 5 cm, the intensities of 604 keV and 796 keV γ-rays attenuate to the levels of 8.6×10^{-4} and 6.7×10^{-3}, respectively. Therefore, it was expected that the directional definition of the measured γ-rays with a HPGe detector was good. In fact, a simulation confirmed that the resolution to identify the locations of γ-ray emission was approximately 5 mm with the full width at half-maximum (FWHM). The γ-ray absorption in the soil core sample inside the iron pipe was 10%. In addition, because the γ-ray absorptions were expected to be nearly identical along the soil core sample, the measured results for the radioactive materials did not affect the final shape of the permeation depth distribution. As shown in Figure 6.1, the actual γ-ray measurements were performed following a step-by-step procedure along the 30 cm soil core sample. The MEXT report provides more detailed explanations regarding the soil sampling method and measurements.

6.1.5 Results on the Permeation Depth Distribution of Radioactive Materials

Figure 6.2 shows some results for the permeation depth distribution of the sampled radioactive materials used as the pilot data in May 2011. In May 2011, radioactive ^{131}I isotopes released from the FDNPS did not completely decay, although approximately eight half-life intervals of ^{131}I had passed after the FDNPS accident.

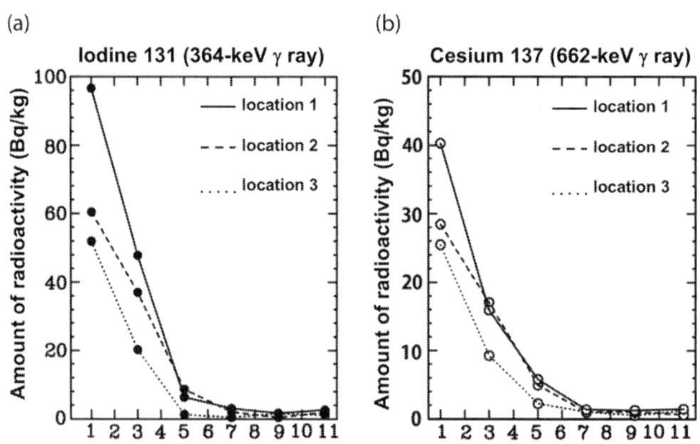

Figure 6.2 Example of depth distribution in pilot soil survey measured in May 2011. Results are shown at near the Iwaki City Health Center high-rise apartment, west of river (location 1), Iwaki City Health Center riverbed east (location 2) and the entrance to Miyama Elementary School (location 3).

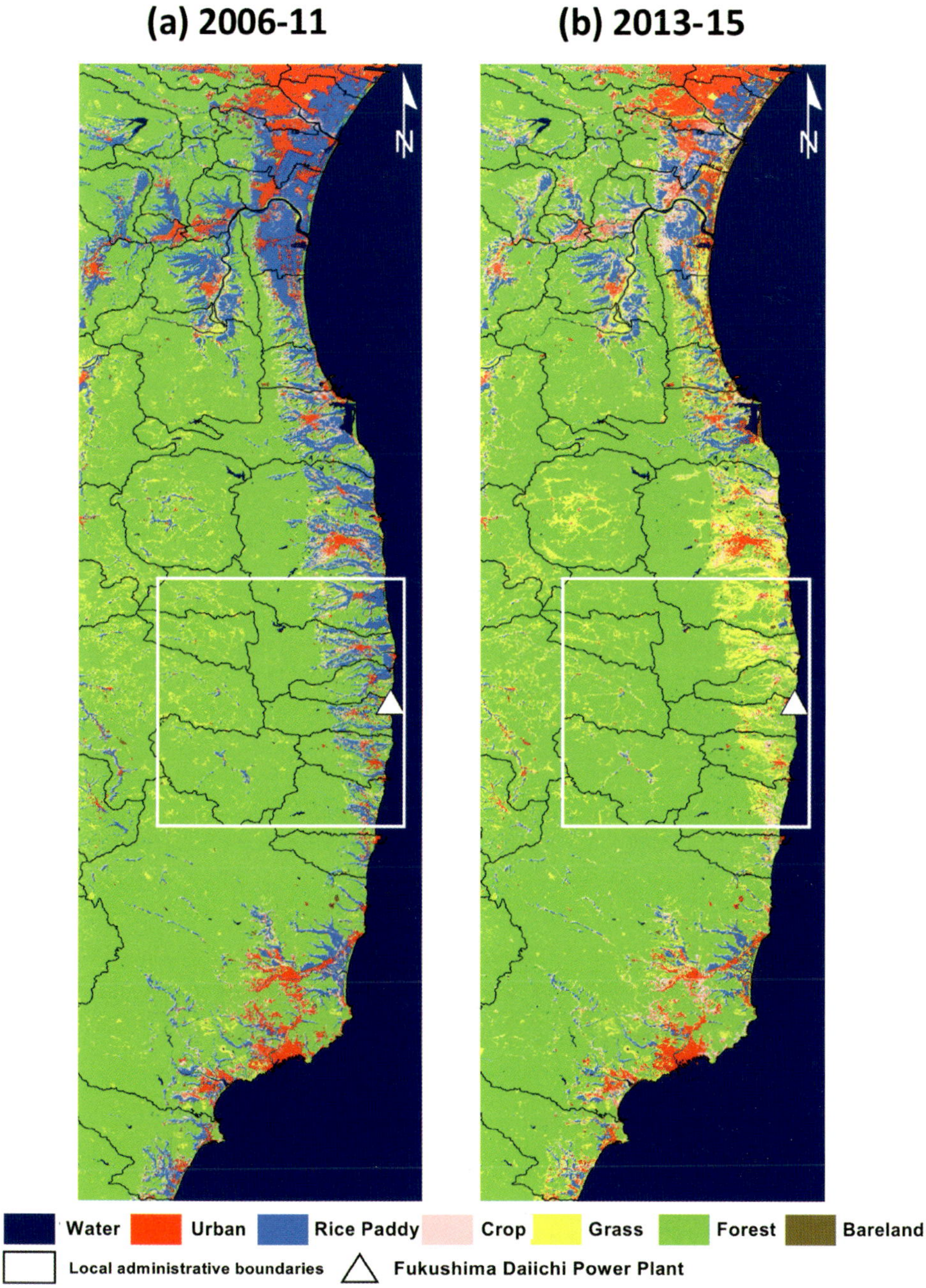

Figure 1.1 Land use changes before and after 2011 by (a) ALOS AVNIR-2 observation in the years 2006–11 and (b) Landsat-8 OLI in 2013–15 as found by Ishihara and Tadono (2017).
Provided by JAXA. A black and white version of this figure will appear in some formats.

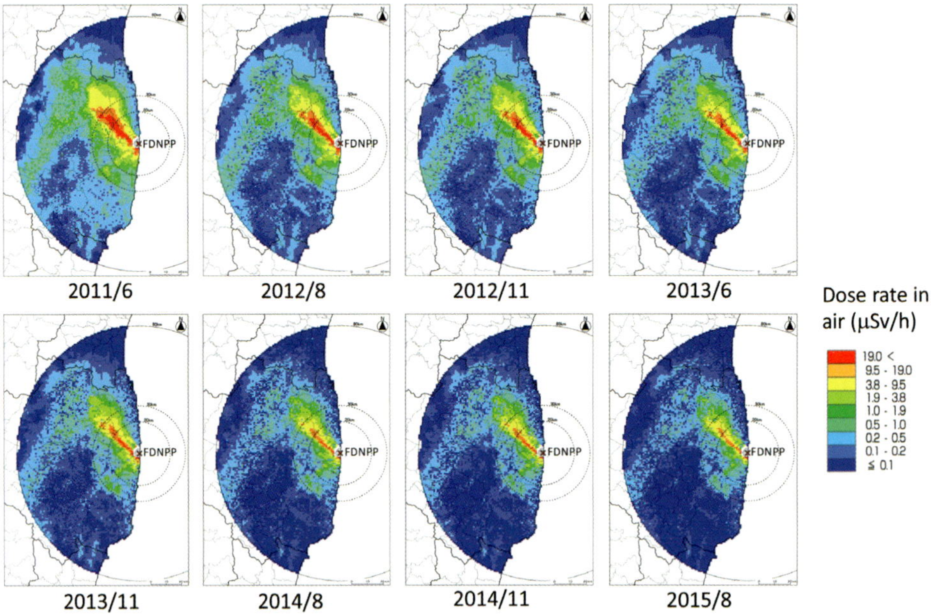

Figure 1.10 Air dose rate maps constructed by integrating data from undisturbed field measurements and aerial monitoring. A black and white version of this figure will appear in some formats.

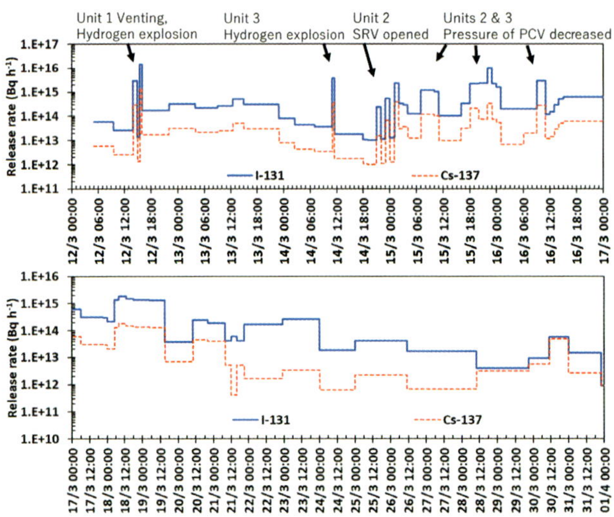

Figure 2.1 The time trend of the release rates of ^{131}I and ^{137}Cs discharged into the atmosphere from 12 March to 1 April 2011, estimated by the JAEA group (Katata *et al.*, 2015 which is partially revised by Chino *et al.*, 2016).

Data from 24 March to 1 April are by Terada *et al.* (2012). A black and white version of this figure will appear in some formats.

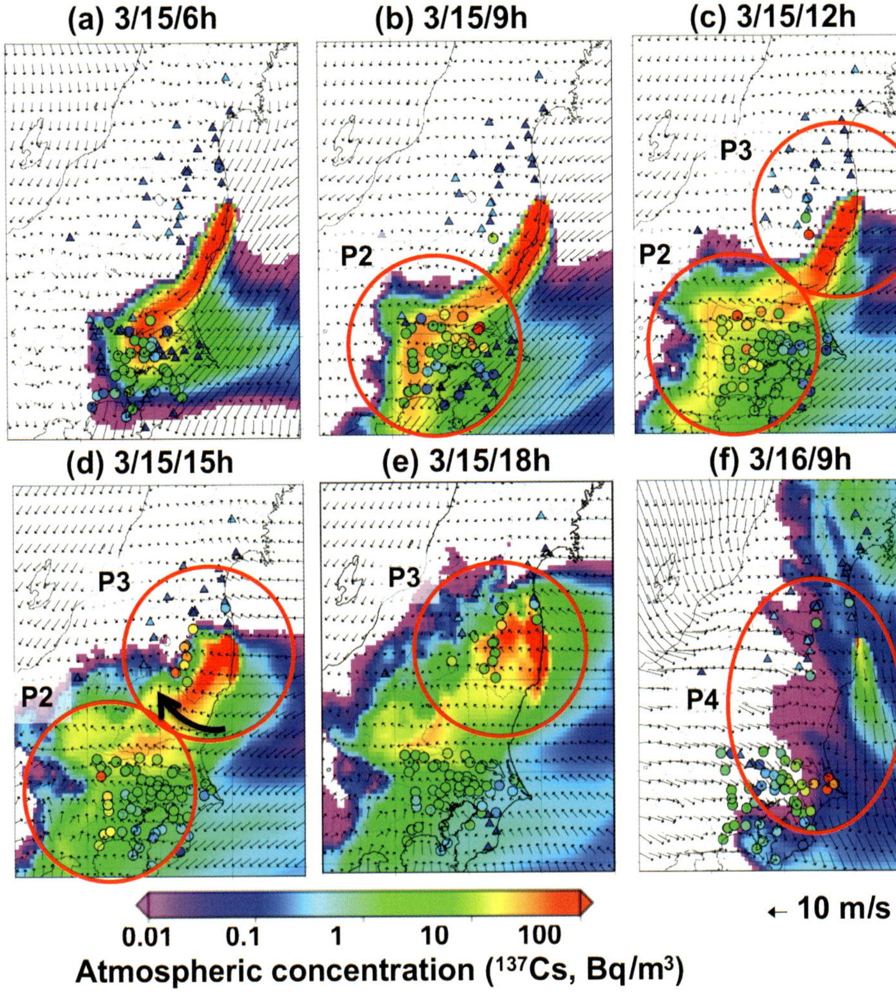

Figure 3.5 Maps of observations (coloured circles) and model ensemble means of atmospheric ^{137}Cs concentration for 15–16 March (Nakajima *et al.*, 2017). Areas identified as plumes with concentration larger than 10 Bq/m^3 are encircled and labelled with the name of the plume. JMA-MANAL wind vectors at 1000 hPa are superimposed. A black and white version of this figure will appear in some formats.

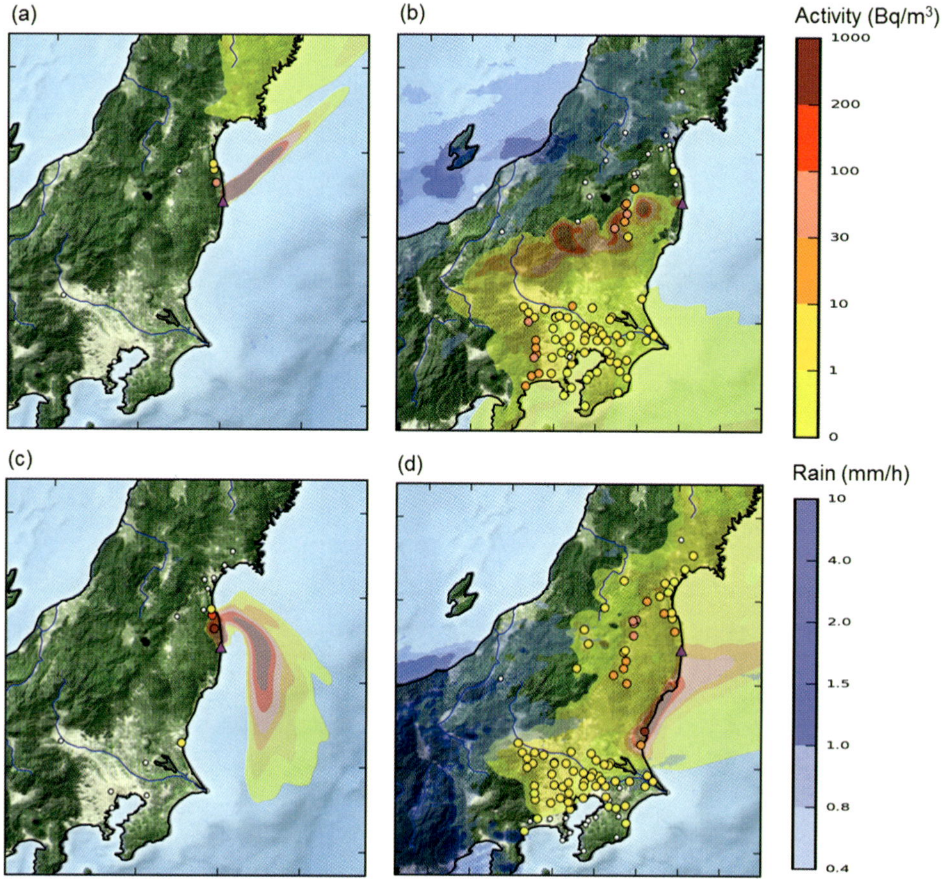

Figure 3.11 Simulated concentrations of ^{137}Cs compared with the measurements (coloured circles). Rainfall is represented by the blue palette. The triangle locates the FDNPS. Each map represents an instant of the main contamination episodes. (a) 12 March at 19:00 (UTC); (b) 15 March at 6:00 (UTC); (c) 18 March at 10:00 (UTC); and (d) 20 March at 20:00 (UTC). A black and white version of this figure will appear in some formats.

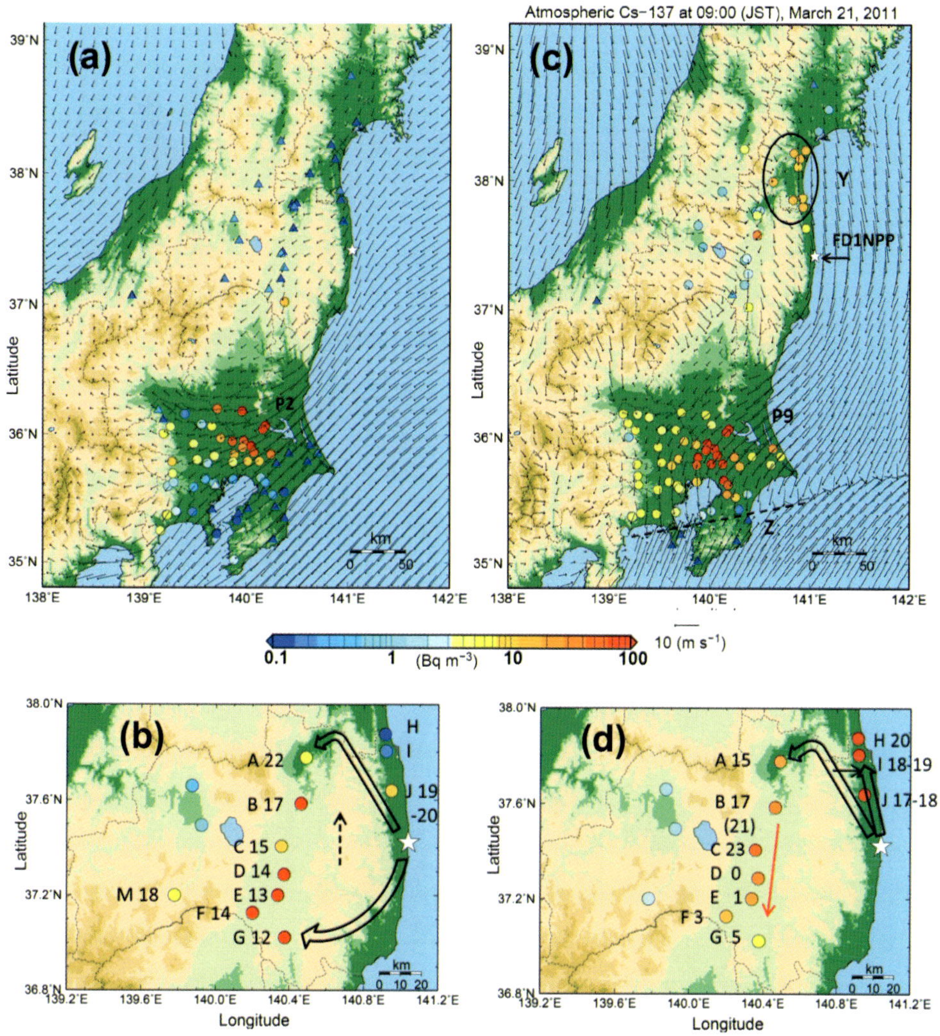

Figure 3.22 (a) Spatial distribution of atmospheric ^{137}Cs concentrations (coloured dot) for plume P2 at the SPM monitoring stations and wind vectors (black arrows) at 1000 hPa by JMA in Kantou and southern Tohoku areas, at 9:00 (JST), 15 March 2011. (b) The maximum ^{137}Cs concentrations with the coloured dot at the SPM stations on 15 March. (c) The same as (a), but at 9:00 (JST), 21 March 2011 for plume P9. (d) The same as (b) but during 20–21 March for plume P8. See the main text for meanings of wide arrows, red arrow, symbols Y and Z, and the numbers (Tsuruta *et al.*, 2014; 2017). A black and white version of this figure will appear in some formats.

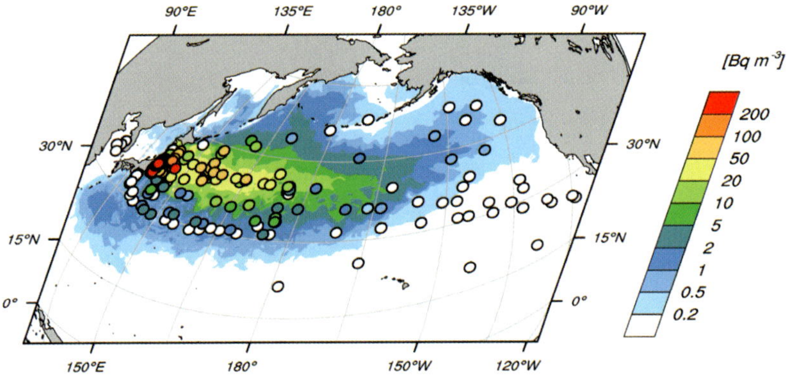

Figure 5.9 Horizontal distribution of ^{134}Cs activity concentrations in surface water based on the study of Tsubono *et al.* (2016) and observations of ^{134}Cs activity concentrations in October 2011, reproduced from figure 3.1 in Aoyama *et al.* (2018). Open circles with colour scale: observations of ^{134}Cs activity concentrations; colour shading: model results. A black and white version of this figure will appear in some formats.

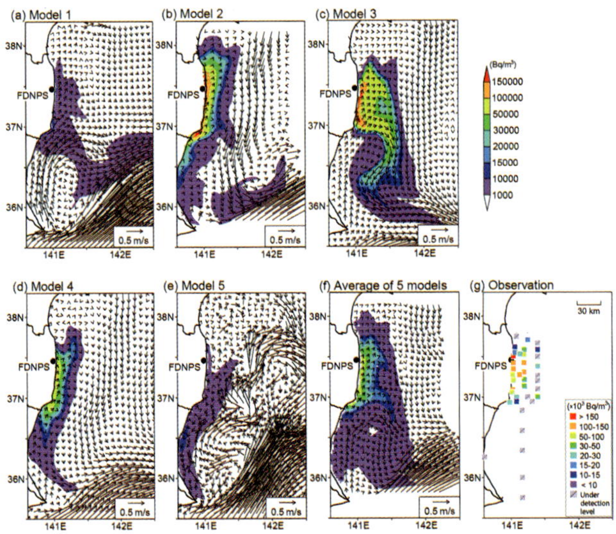

Figure 5.10 Horizontal distributions of the surface flow field and ^{137}Cs concentration in late April 2011. Panels (a) to (e) show results from five different models for the same period, while panel (f) indicates the average of the five model results. (g) Observed distribution of ^{137}Cs concentration during the period 21–30 April 2011 (Masumoto *et al.*, 2012). A black and white version of this figure will appear in some formats.

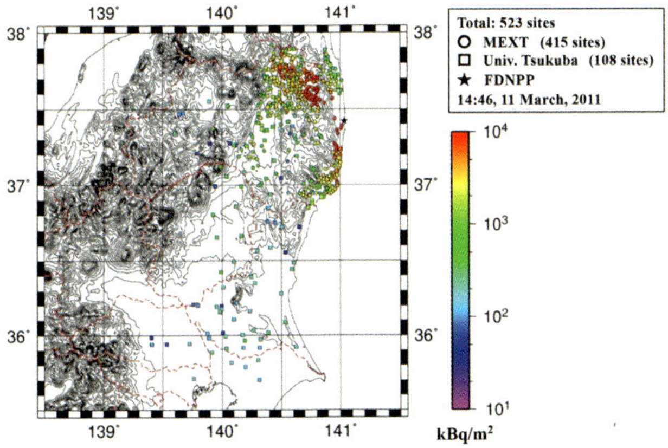

Figure 6.17 Map of ^{131}I deposition density in eastern Japan, integrating the ^{131}I data (108 sites, Kinoshita *et al.*, 2011) with the initial ^{131}I data (415 sites) by Sato *et al.* (2015). A black and white version of this figure will appear in some formats.

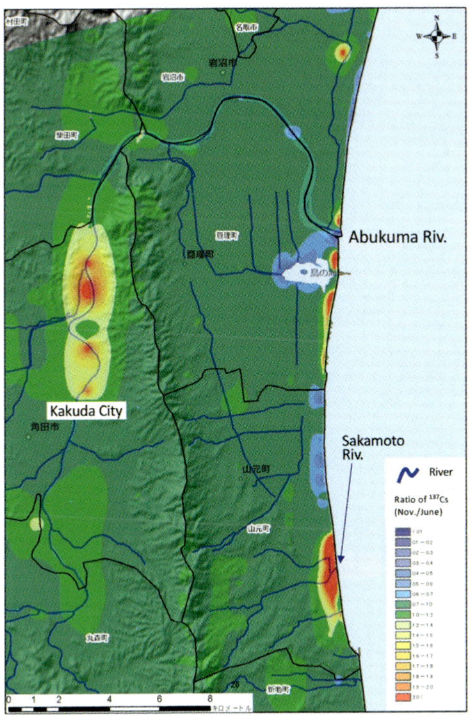

Figure 6.21 Ratio of air dose rate in November 2011 normalised to that in June, 2011 by airborne survey by MEXT (2011b), which suggests increased radiocaesium concentration in some areas. A black and white version of this figure will appear in some formats.

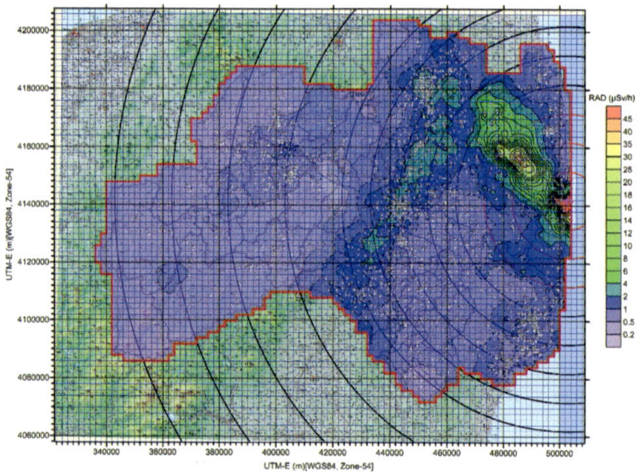

Figure 11.1 The air dose rate distribution at a height of 1 m at the end of March 2011 observed by the Fukushima University radiation measurement team is shown. This result was announced on 17 June 2011. The figure also includes the radiation monitoring results conducted at elementary and junior high schools over 12–16 April; the measurement of the dose rates within 20 km from the FDNPS was conducted by the Ministry of Education, Culture, Sports, Science and Technology. A black and white version of this figure will appear in some formats.

The intensity of the ^{131}I radioactivity decreased to approximately 1/256 of its original value. Thus, we determined the permeation depth distribution of radioactive ^{131}I isotopes (see Figure 6.2(a)). Because more than 90% of radioactive iodine and caesium were found to be trapped within the top 5 cm of the soil, we concluded that it was sufficient to sample soils within this region for measuring the deposition density map of the radioactive materials.

In June 2011, soil core samples were collected at approximately 300 locations. Among them, 100 reliable results were obtained for the permeation depth distribution of the radioactive materials. Figure 6.3 shows typical examples at various sampling locations.

The results indicated an exponential attenuation with increasing soil depth. At the location labelled 28N26, the attenuation was steep, with the γ-ray intensity decreasing by one-tenth for every soil depth increase of 18 mm. However, the one-tenth attenuation depth, $L_{1/10}$ was 48 mm at location 16N18. The permeation depth distributions for ^{134}Cs and ^{137}Cs were found to be nearly identical because the chemical properties of ^{134}Cs and ^{137}Cs are the same.

The permeation depth distributions measured using γ-ray measurements can be expressed as follows:

$$I(x) = I(0)\epsilon e^{-x/\lambda},$$

where $I(x)$ is the density of the radioactive materials at depth x, and ε is the detection efficiency of the γ-ray detector. Because the geometry of the detector arrangement is the same in the present measurements, the detection efficiency ε

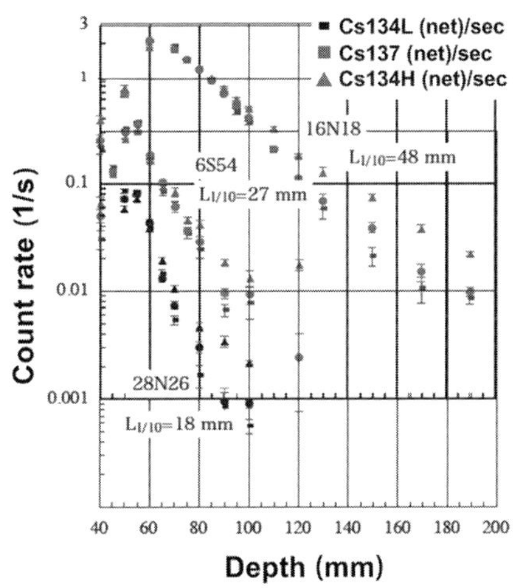

Figure 6.3 Example of depth distribution: depth distribution of radioactive material at three points shown in the figure.The distributions of ^{134}Cs were measured by 604 keV (indicated as Cs134L) and 796 keV (indicated as Cs134H) γ-rays and that of ^{137}Cs was measured by 661 keV γ-rays.

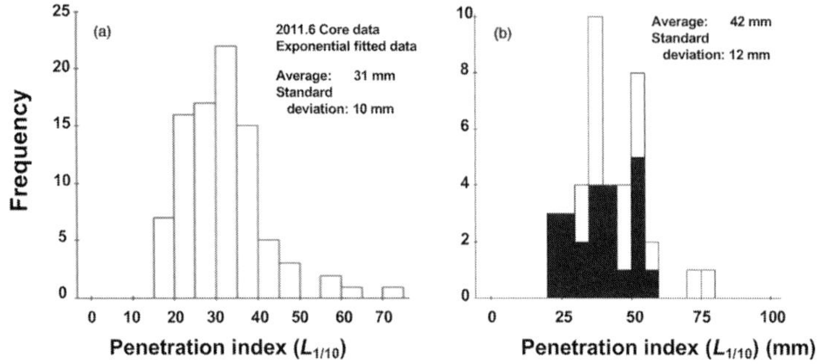

Figure 6.4 Depth distribution of radioactive Cs: results of soil samples collected in (a) June 2011 and (b) December 2011 and 2012. The sample data collected were in March, the part painted in black is the soil data within 20 km of the FDNPS.

does not depend on the depth x. Therefore, the attenuation coefficient λ can be determined without any corrections. The parameter λ corresponds to the depth at which the radioactive materials are reduced to $1/e = 1/2.72$. Because this parameter cannot be directly observed, we decided to use the one-tenth attenuation depth $L_{1/10}$ instead of the λ parameter. The relation between $L_{1/10}$ and λ is given as follows: $L_{1/10} = \lambda \ln 10 = 2.30\lambda$.

The second campaign was performed in December 2011 and March 2012. In March 2012, we collected soil samples in the inner region of the evacuation zone within 20 km of the FDNPS. Even after one year, the permeation depth distributions showed typical exponential attenuation.

Figure 6.4 shows the distribution of the index $L_{1/10}$. The average $L_{1/10}$ was 31 mm in June 2011. However, the average $L_{1/10}$ was 42 mm in December 2011 and March 2012. Although the permeation depth seemed to increase, we cannot conclude that there was a meaningful change for the following scientific reasons:

1. The standard deviation of $L_{1/10}$ exceeded 10 mm.
2. The sampling locations were not the same.

We must determine whether the permeation depth distributions will expand to deeper soil layers over time. At present, caesium radioactivity has been trapped in the shallow soils, and there is no indication that the radioactivity will permeate into deeper soil layers.

6.1.6 Measurement of the Fallout of Radioactive Materials

1. Although several serious releases of radioactive materials occurred from the FDNPS accident, radioactive materials emission from the FDNPS site during the soil sampling period was negligible.

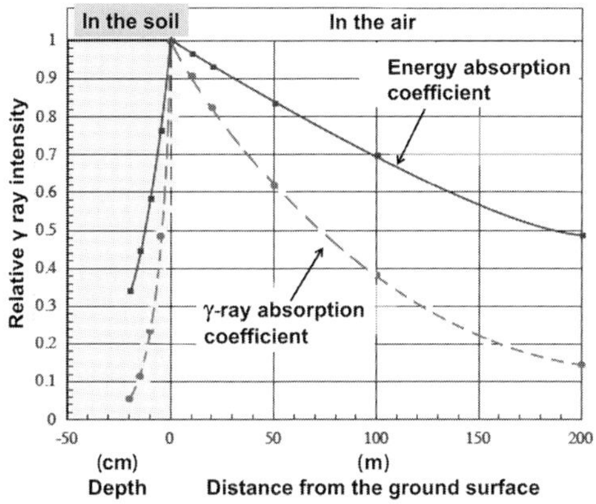

Figure 6.5 γ-ray attenuation curves in soil and air. The solid line shows the attenuation obtained from the energy absorption coefficient; the broken line is obtained from the γ-ray absorption coefficient.

2. γ-ray attenuation: the γ-ray attenuation effect is an important factor in understanding the spatial dose rates. Figure 6.5 shows the γ-ray attenuation curves in soil and air. The relative γ-ray attenuations in soil are shown as a function of the soil depth on the left of Figure 6.5. In Figure 6.5, the observed γ-ray intensity is normalised at the ground surface. The γ-ray intensity observed at the ground surface rapidly decreased with increasing soil depth for ^{137}Cs. For calculations using the γ-ray absorption coefficients, if γ-rays are scattered, relevant γ-rays are assumed to disappear. Therefore, the rescattering effect of γ-rays due to Compton scattering was neglected; calculations using the energy absorption coefficients are realistic for understanding the behaviour of the spatial dose rates. Based on the energy absorption coefficients, the ^{137}Cs γ-ray intensity was reduced by half at a depth of 12 cm, meaning that the ^{137}Cs γ-ray intensity is not attenuated much if the radioactivity is trapped within the top 5 cm of the soil.

   The relative γ-ray attenuations in air are shown as a function of the distance above the ground surface on the right side of Figure 6.5; the horizontal scale is given in metres and is 100 times larger than for the soil depth. The ^{137}Cs γ-ray intensity was found to be reduced by half after passing through 200 m of the atmosphere. Thus, γ-ray attenuation is very small in air.
3. Radioactive materials in soil and the decontamination effect:

   considering the aforementioned γ-ray attenuation and rescattering effect (sky shine effect), simulation results are shown in Figure 6.6. It was assumed that the spatial dose rate was measured at 1 m above the ground surface in a radioactively polluted flat plain within a 1 km circle for a 500 m air height. In Figure 6.6, we normalise the spatial dose rates to unity because the soil for decontamination

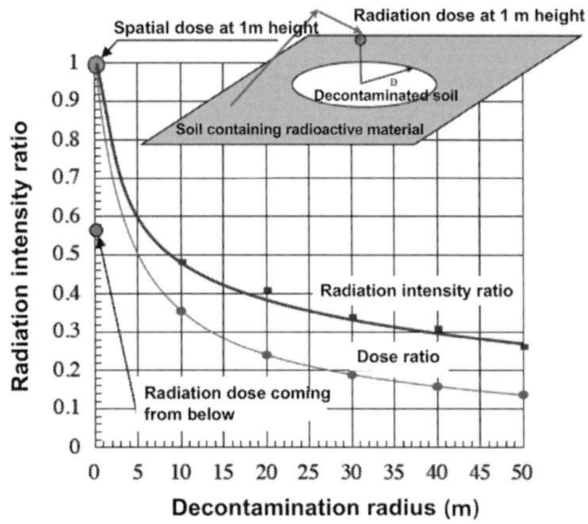

Figure 6.6 Soil radiation and decontaminating effect (R. Gurriaran, personal communication).

radius is 0. The γ-rays directly came from the ground surface and from the sky due to Compton scattering by the air (sky shine effect). In Figure 6.6, a shaded circle is drawn for a relative radiation dose of 0.57 for the case of non-decontamination. This radiation dose rate at a height of 1 m came directly from the ground surface. The remaining dose of 43% came from the sky due to the sky shine effect. For γ-rays coming from a far distance, the contribution was found to be higher than that of nearby sources because the scattering angle of γ-rays becomes relatively small and the source γ-ray intensity increases with increasing distance. Figure 6.6 shows that even if the soil decontamination area were to be expanded to a 50 m circle, approximately one-quarter of the γ-rays would still come from outside the circle.

6.1.7 Results and Their Interpretations

In the present project, we obtained the radioactivity nuclide map, spatial dose rate distribution and local maps for the radioactive caesium and iodine isotopes. As an example, Figure 6.7 shows the deposition density map for ^{137}Cs. The website www.rcnp.osaka-u.ac.jp/dojo shows the deposition density maps for other radioactive nuclides.

According to Figure 6.7, high dose rate areas existed to the northeast of the FDNPS. However, there was a low dose rate area near the FDNPS site, which is quite surprising, given the proximity to the accident site. The radioactive dose rates were relatively high in Fukushima City, Nihonmatsu City and Kohriyama City in the Nakadouri area. The constructed maps will be useful to ensure future security when residents return to their homes.

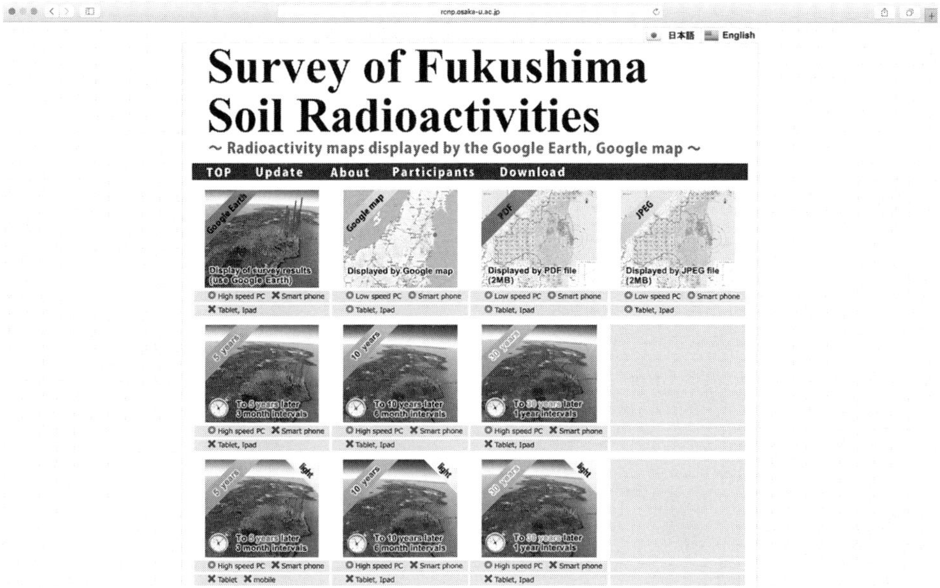

Figure 6.7 Soil contamination map of ^{137}Cs (www.rcnp.osaka-u.ac.jp/dojo).

6.1.8 Meaning of the Soil Radioactivity Measurements for Understanding the Initial Deposited Radioactive Nuclides over Broad Areas

Based on the results of the radioactivity nuclide map project, it is feasible to compare the actual measurements of the deposition density map for ^{137}Cs and the results from airborne measurements. Figure 6.8 shows a comparison between the two results obtained from the soil sampling and from aeroplane monitoring. The aeroplane monitoring data were calibrated based on the observations over a large area; these data reveal that the initial deposited radioactive materials originated from the FDNPS accident.

The measurement of soil radioactivity density for top soils in agricultural land has also been conducted (Takata *et al.*, 2014). However, the deposition density of radioactive materials was obtained in units of Bq/kg. This result was used to estimate the ^{137}Cs deposition density for the permeation of cultivated crops (>5000 Bq/kg). The depth of top soils in the agricultural land was assumed to be 15 cm. However, this depth is not universal and depends on the characteristic of the agricultural lands. The soil core sampling method was not standardised; sometimes, soil core samples were taken with the top soil layer deeper than 15 cm. In addition, the density information for the sampled top soils was often not attached. Therefore, we could not use the obtained data to assist in the creation of the deposition density map in units of Bq/m^2.

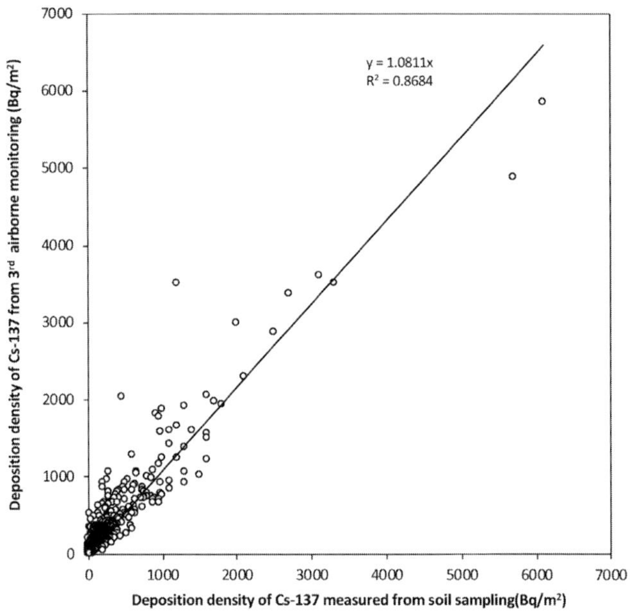

Figure 6.8 Result of nuclide analysis by soil collection and ^{137}Cs deposition measured by aircraft monitoring (http://radioactivity.nsr.go.jp/en/contents/6000/5235/view.html).

6.2 Radionuclide Transfer from Forest Environments

YUICHI ONDA

6.2.1 Accumulation and Transfer of Radioactive Materials in Forest Environments

Radioactive materials released accidentally from the FDNPS were widely deposited in forested areas, and we must investigate the accumulation and transfer of radioactive materials in forest environments. Thus, the project was launched to study the distribution and transfer of radiocaesium in forested areas.

Fallout radioactive materials in the forested area will first be trapped in the forest canopy, and some of the radioactive material will also be deposited on the forest floor. Thereafter radioactive materials will gradually migrate to the forest floor, and then some part will flow into and down rivers.

The very first forest interception of fallout radionuclides to forest canopies after the FDNPS accident (Kato *et al.*, 2012b) showed that 62–65% of ^{137}Cs was trapped in forest canopies, whereas 25–51% of ^{131}I was trapped in the forest canopy. Furthermore, Onda *et al.* (2015) reported that 40 days after the release, 69% of ^{137}Cs remained trapped in forest canopies, but only 18% of ^{131}I remained. The interception ratio can vary due to different tree density and leaf area index (LAI).

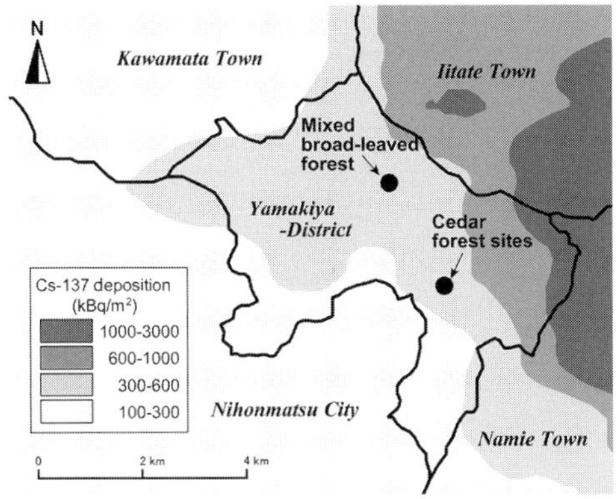

Figure 6.9 Locations of the study site and monitoring towers.

We measured the trend in the aerial dose rate at different heights, deposition of radiocaesium on the forest floor and concentrations of radiocaesium in throughfall, stemflow and fallen leaves.

We selected study sites in areas with stands of typical vegetation (cedar plantation): one young (18 years old) cedar stand and one mature (40 years old) cedar stand. One mixed broad-leaved stand composed of a mixture of various trees, such as Japanese oak and red pine, was also selected (Figure 6.9). Monitoring towers with heights of 8–12 m were installed in each forest stand to collect periodical measurements of the air dose rate for trees with different heights. Fresh leaves, dead leaves in the tree canopy and fallen leaves were also collected. The leaves were dried and fragmented to measure the radiocaesium concentrations per dry weight. To confirm the inventory of radiocaesium on the ground surface in each forest stand, soils were sampled at different depths. The soil samples were dried and measured for their radiocaesium concentration per dry weight. The monitoring started in July 2011. Detailed results are given by Kato *et al.* (2017; 2018).

The radiocaesium inventory at different depths in the cedar stands and mixed broad-leaved stand confirmed that approximately 50–90% of the deposited radiocaesium resided in the litter layer of the ground surface (Figure 6.10(a)). When the three forest sites were compared, the mature cedar stand showed greater amounts of radiocaesium than the young cedar stand and the mixed broad-leaved stand. The vertical distribution of the aerial dose rate was monitored both inside and outside the forest (Figure 6.10(b)). Outside the forest, the aerial dose rate exhibited the highest value at the ground surface and decreased with height. The dose rate became almost constant above 5 m. In the broad-leaved stand, the aerial dose rate increased near the ground surface and remained constant after a certain height. In contrast, in the cedar forest stands, both young and mature cedar stands displayed higher aerial dose rates towards the canopy. The higher dose rates in the cedar canopies were attributed to the significant accumulation of radiocaesium in the canopy.

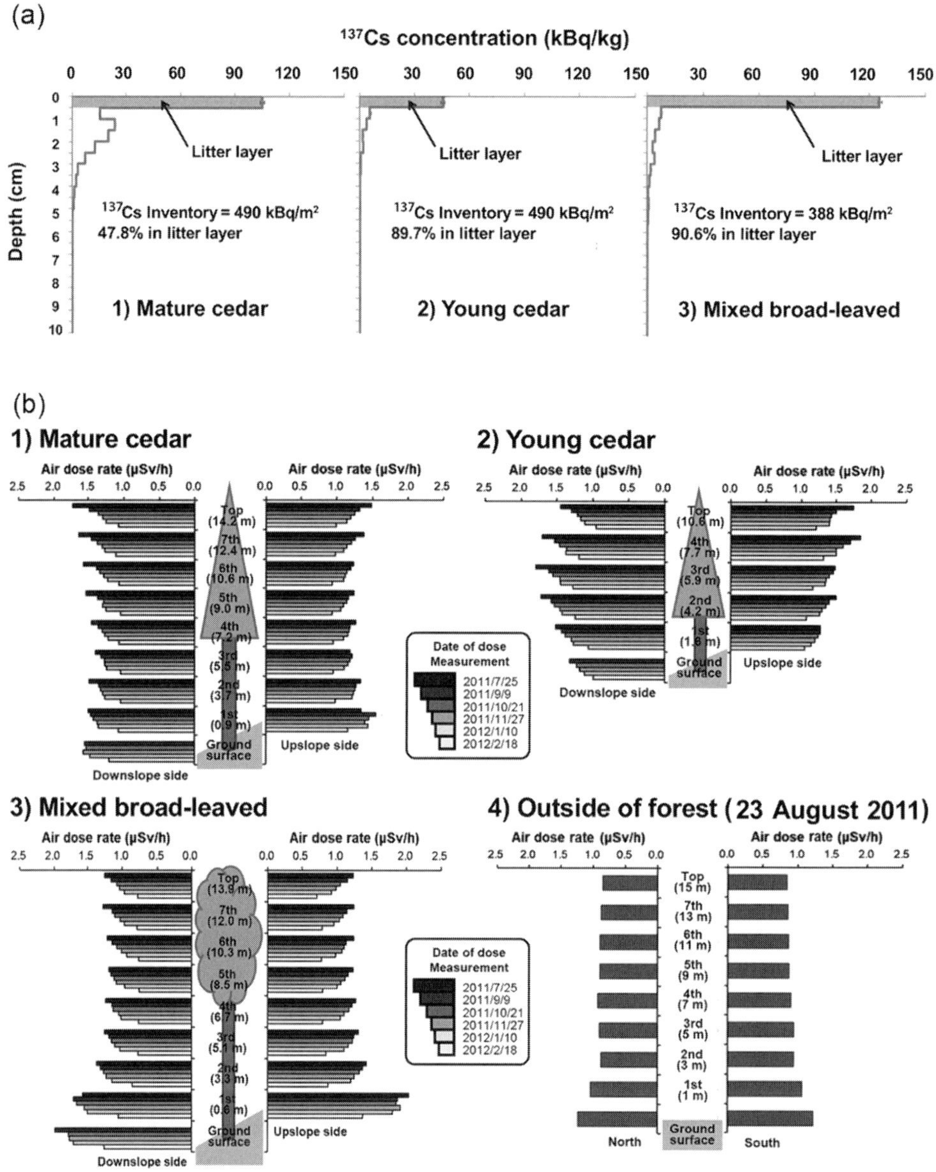

Figure 6.10 (a) Depth profile of ^{137}Cs concentrations in forest soils. (b) Vertical distribution of the measured air dose rate in the experimental forest stands. (c) ^{137}Cs concentrations in foliage samples collected from different heights of the canopy.

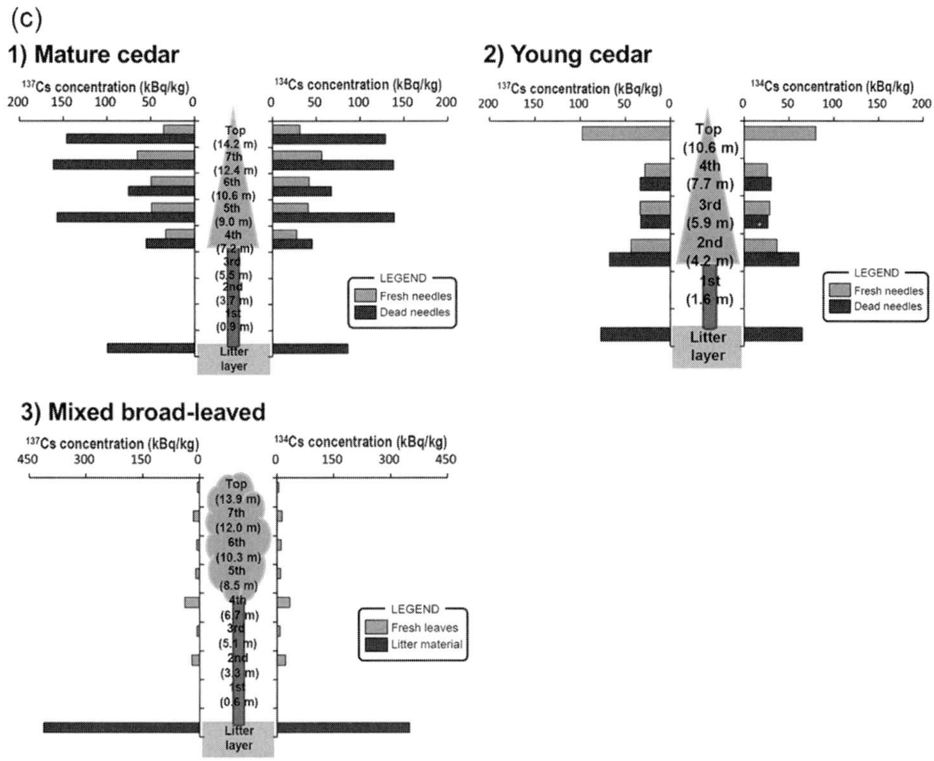

Figure 6.10 (*cont.*)

Radionuclides are easily absorbed into fresh leaves but not dead leaves. However, the radioactive concentrations in the fresh leaf samples were similar or even lower than those in the dead leaf samples (Figure 6.10(c)). Therefore, in July 2011, the amount of radiocaesium transferred from roots or leaves to tree bodies was smaller than the amount of radiocaesium directly deposited on the leaves. In contrast, massive fallout of radiocaesium occurred in mid-March, when no leaves were present in the mixed broad-leaved stand, which suggests that a large amount of radiocaesium was deposited on the dead leaves on the ground in the mixed broad-leaved stand. Therefore, the amount of radiocaesium deposited on the litter layer was larger in the mixed broad-leaved stand than in the cedar stands, which resulted in higher dose rates near the ground surface.

The radiocaesium inventories in the forest soil profile indicated a gradual increase with the transfer of radiocaesium from fallen leaves or leaves in the canopy to the forest floor through rainfall, and then a gradual increase of the inventory and downward migration to the deeper part of the soil (Takahashi *et al.*, 2015).

These data have become the basis for the subsequent decontamination in forest environments. In the mixed broad-leaved stand, the removal of the litter layer on the ground surface is effective because the litter layer that contains fallen leaves has a large amount of radiocaesium. Meanwhile, in the cedar forest stands, cutting trees is recommended for decontamination because fresh leaves have to be removed as well due to the high radiocaesium concentrations in fresh and dead leaves near the canopy.

The total inventory of radionuclides should be decreased by conducting urgent decontamination of the affected forest areas in the first year because the radionuclides are expected to move to the tree bodies in the future. This is especially true for the cedar stands, where cutting down the trees is recommended where deposition density is high, the area can be replanted and the new-generation trees subsequently used. The reuse of wood is also possible if the trees are cut and the trunks are less contaminated. However, actual forest decontamination work has been limited to collecting litter within 20 metres of houses.

6.2.2 Processes of Transfer, Diffusion and Enrichment of Fallout Radionuclides in Terrestrial Environments

Regarding the fine sediments discharged into rivers, we monitored the radiocaesium runoff associated with sediment erosion from various types of land use. We established a standardised plot for soil erosion (22.1 m) from farmland, bare land, grassland, pasture and cedar plantation forest areas to monitor the soil erosion amount and radiocaesium runoff. The result demonstrated that sediments including approximately 0.03% radiocaesium were transferred into rivers during the monitoring period of 1.5 months in the bare land plot. In contrast, in the pasture plot and the forest plot, the sediment yield due to rainfall can be prevented by vegetation, which resulted in extremely low amounts of radiocaesium runoff into the rivers (Yoshimura *et al.*, 2015).

In the case of cropland, not all sediment reached the streams because some portion of the sediment was deposited during transport. However, in the case of paddy fields, when sediment is discharged 100% flows into and is transported by rivers. For paddy fields, the initial preferential discharge of highly contaminated fine particulate matter (Wakahara *et al.*, 2014) by puddling was found to be transferred to streams. In addition, the physical mixing with deeper uncontaminated soil explains the faster rate of decline in particulate ^{137}Cs activity concentration in the initial three years after the nuclear release (Yoshimura *et al.*, 2015).

We focused on the forms of radiocaesium transported in rivers and compared the concentrations of the radionuclides dissolved in the river water with those in suspended sediments. The results showed that more than 90% of the radiocaesium flowed down the river in the form of suspended sediment at every monitoring

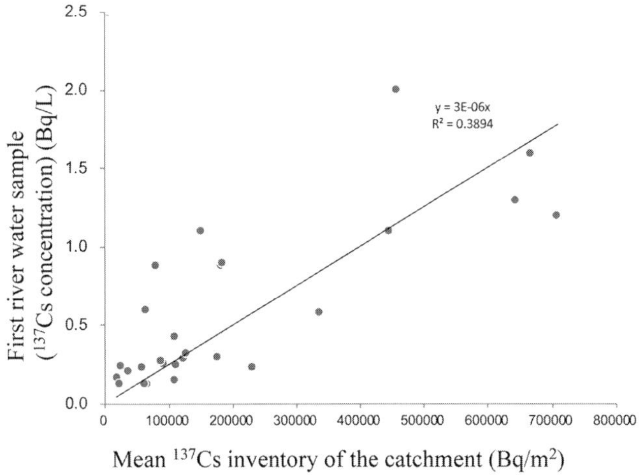

Figure 6.11 Relationship between the mean inventory of the soil collected within the upper catchment of the river and the ^{137}Cs concentrations in the river water samples.

station. The maximum total ^{134}Cs and ^{137}Cs concentration observed in the main channel of the Abukuma River was 126 000 Bq/kg, which was more than 10 times greater than the standard value for the sediment (Yamashiki *et al.*, 2014). Moreover, sediments with a similarly high concentration were deposited in a dam lake located on the main stream of the Abukuma River.

We also examined the relationships between the radiocaesium concentrations in the river water, riverbed sediment, suspended sediment and soil samples collected at 2200 locations.

Figure 6.11 displays the relationship between the ^{137}Cs concentrations in the river water samples and the ^{137}Cs mean inventory of the soil collected within the upper catchment of the river. The data were provided by the Japan Food Research Laboratories commissioned by MEXT. According to the data, higher concentrations in the upper catchment resulted in higher concentrations in the river water samples. When radiocaesium concentrations in the suspended sediment samples and those in the soil collected within the upper catchment were compared, a weak (r^2 = 0.39) correlation was found.

Regarding the riverbed sediments, radionuclides are likely to be adsorbed onto fine-grained materials; the radiocaesium concentrations have been reported to be proportional to the power of 0.65 power of the specific surface area (He and Walling, 1996; Figure 6.12). Therefore, if the radionuclide concentration in the riverbed sediments at a certain location is used as an indicator of the contamination level, the particle size of the sediments at different locations should be considered.

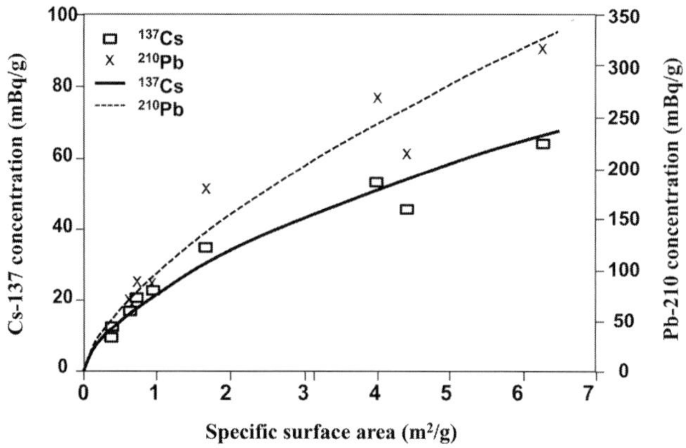

Figure 6.12 Relationship between specific surface area and radioisotope. Reprinted from J. Environ. Radioact., 30. He and Walling (1996). Interpreting particle size effects in the adsorption of ^{137}Cs and unsupported ^{210}Pb by mineral soils and sediments. pp. 117–137. Copyright 1996, with permission from Elsevier.

6.3 Sediment and Radionuclide Transfer from the Land to the Ocean: International Research Perspectives

OLIVIER EVRARD AND J. PATRICK LACEBY

Only a few nuclear accidents have resulted in a significant fallout of radionuclides. In particular, two accidents produced widespread radiocaesium deposition on soils in continental environments: Chernobyl in 1986 and Fukushima in 2011 (Steinhauser *et al.*, 2014). In Japan, the main contamination plume of radioactive material from the FDNPS accident was deposited on soils that predominantly drain to the Pacific Ocean through a network of coastal rivers (Figure 6.13). Although highly regrettable, this accident provided a unique opportunity to improve our understanding of the processes controlling the transfer of sediment and particle-bound contaminants (e.g. radiocaesium) in catchments subject to typhoon events.

An extensive programme was developed and implemented shortly after the accident, under the authority of MEXT in conjunction with the Japan Atomic Energy Agency (JAEA) to produce reliable maps of the initial soil contamination from radionuclides and its evolution over time (Saito and Onda, 2015). This considerable and important fieldwork campaign involved the contribution of hundreds of specialists from dozens of Japanese institutions. From an international perspective, experts from the French Institute of Radioprotection and Nuclear Safety (IRSN) contributed to this *in-situ* γ-spectrometry monitoring of soils contaminated by the FDNPS fallout (Mikami *et al.*, 2015).

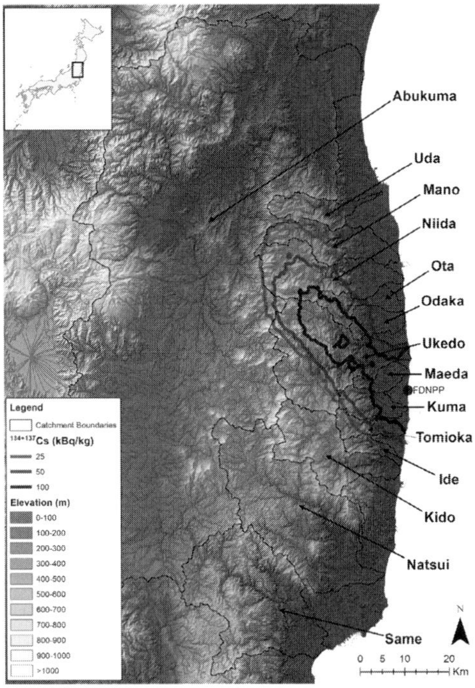

Figure 6.13 Coastal rivers draining the main radioactive plume with radiocaesium activities in kBq/kg (Chartin *et al.*, 2013, based on data from MEXT).

After the accident, the Japan Science and Technology Agency (JST) opened a specific call-for-projects in April 2011 entitled 'J-RAPID' in order to 'support collaborative research activities between Japanese and foreign researchers' to conduct urgent investigations 'following natural or anthropogenic disasters and similar unanticipated events' (JST, 2012). The first J-RAPID call was mainly open to collaborations between Japanese and American or French researchers. Although the call for research proposals mainly focused on the investigation of the Great East Japan Earthquake, a number of projects examined the impact of the tsunami and the FDNPS accident. Among these projects, 'TOFU' (November 2011–March 2013) investigated the early propagation of contaminated sediment in coastal rivers draining the main radioactive plume in Fukushima Prefecture (Figure 6.13). Another project entitled 'FREEBIRD' investigated the impact of radiation exposure on bird populations (Garnier-Laplace *et al.*, 2011).

Although these projects were selected rapidly after the accident, the first river-monitoring fieldwork campaigns involving international researchers missed the first typhoons (Songda in May 2011 and Ma-on in July 2011) that initiated the redistribution of sediment contaminated with radionuclides. However, preliminary investigations of radionuclide migration with depth in soils (Kato *et al.*, 2012a) and crucial flood monitoring were conducted by Japanese researchers early in 2011. For instance, they demonstrated that typhoon Roke, which occurred in September 2011, contributed 30–50% of the radiocaesium annual flux to the Pacific Ocean in two upstream tributaries of the Niida River catchment (Nagao *et al.*, 2013).

In the largest river draining the contaminated region (the Abukuma River, Figure 6.13), monitoring stations were available to quantify the radiocaesium fluxes exported to the ocean (Yamashiki *et al.*, 2014). For most of the smaller coastal catchments devoid of perennial measurement stations, sediment and radiocaesium fluxes were estimated using spatially distributed soil erosion models (Kitamura *et al.*, 2014). Another technique to compensate for the unavailability of river-monitoring stations during the early post-accident phase consisted of investigating the evolution of radionuclide concentrations in exposed flood deposits collected every six months in rivers draining the plume. These results demonstrated the very strong reactivity of the region's river systems to weather events. For example, after a significant export of radioactive contamination during a succession of typhoons in 2011, the dispersion of sediment was amplified by the spring floods and the subsequent tropical storms and typhoons that occurred in 2012 (Chartin *et al.*, 2013).

In parallel with this urgent post-accident research, the French president, Nicolas Sarkozy, decided in June 2011 to dedicate a significant budget (€50 million) to strengthen the country's research capacity in radioprotection and nuclear safety (ANR, 2012). With 19 operational nuclear power plants, France is one of the most exposed countries to nuclear risks worldwide (Christoudias *et al.*, 2014). Although most projects focused on the improvement of nuclear reactor safety or post-accident remediation (Chagvardieff, 2014), the specific goal of the AMORAD project (2013–19) is to improve modelling approaches to simulate the dispersion of radionuclides in the environment, with a major focus on post-accident radiocaesium dynamics in the Fukushima region. The preliminary results from this project demonstrate that during low-flow periods, radiocaesium is mainly transported with the organic fraction of sediment, whereas the mineral fraction is dominant during high-flow events such as those that occur during major typhoons (Naulier *et al.*, 2017), confirming previous results of Japanese researchers (Ueda *et al.*, 2013). Furthermore, researchers demonstrated that the main sources of particle-bound radiocaesium in Fukushima rivers were subsoils supplied by landslides or channel bank erosion (45 ± 26%), agricultural areas such as paddy fields (38 ± 19%) and, to a much lesser extent, forests (17 ± 10%), despite their extensive coverage of more than two-thirds of the highly contaminated area in the region (Laceby *et al.*, 2016b). Transfers of radionuclides in forests and their potential supply of contaminated material to river networks will remain a major topic of interest in the near future (Loffredo *et al.*, 2014), as woodlands constitute the main reservoir of radiocaesium in the fallout-impacted region (Coppin *et al.*, 2016).

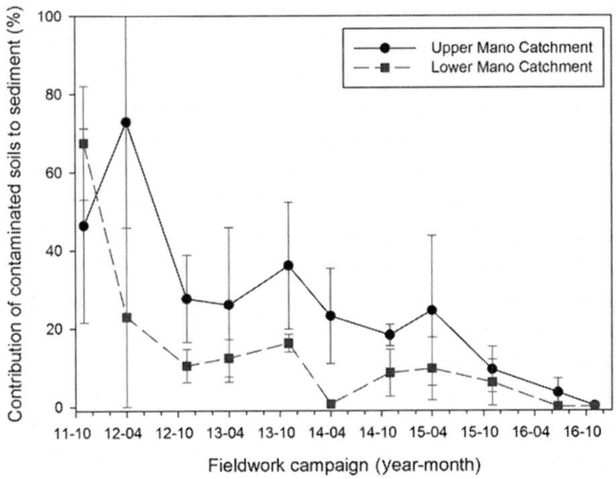

Figure 6.14 Mean contribution of contaminated soils to sediment transiting the Mano catchment in Fukushima Prefecture (for methods, see Evrard *et al.*, 2016).

The investigation of typhoon impacts on sediment and sediment-bound contaminants in the Fukushima region represents an original contribution of this post-accident research to the international literature (Evrard *et al.*, 2015), as these events are known to provide a significant contribution to the annual sediment fluxes in Southeast Asia (Syvitski *et al.*, 2005). Although typhoons only contribute 22% of the annual rainfall in the area, their occurrence generates 40% of the annual rainfall erosivity (Laceby *et al.*, 2016a). Typhoons Roke (September 2011) and Etau (September 2014) were the most erosive events, generating a maximum connectivity between hillslopes and rivers (Chartin *et al.*, 2017) and exporting a significant quantity of radionuclides to the Pacific Ocean (Yamashiki *et al.*, 2014; Kurikami *et al.*, 2016). This flush of sediment contaminated with radionuclides during typhoons, together with the significant contribution of subsoil sources and the rapid progress of decontamination works in the region, facilitated a rapid and substantial decrease in radiocaesium concentrations measured in flood sediment deposits collected in the coastal catchments draining the main radioactive plume of Fukushima Prefecture (Evrard *et al.*, 2016). This decrease reached more than 90% on average between November 2011 and November 2016 (Figure 6.14), further exemplifying the reactivity of the river systems in this region to the climate and rainfall regimes.

Post-accident research in Fukushima Prefecture confirmed the expected occurrence of a rapid leaching phase during the first two years following the catastrophe (Garcia-Sanchez and Konoplev, 2009; Delmas *et al.*, 2017). The turnover of sediment in these coastal river systems was very rapid, although it was largely dependent on the frequency and intensity of typhoon events. Currently, a large component of the long-term research on the fate of radionuclides is being undertaken by an international group of scientists of the Institute of Environmental Radioactivity (IER), University of Fukushima, Japan.

This research centre was created in July 2013 to provide a significant contribution to the international research effort carried out to improve our knowledge of the impact and the fate of the radionuclides released into the environment in post-accident conditions. The IER involves several senior scientists who developed expertise on radionuclide transfers following the Chernobyl accident. They will undoubtedly place the important findings learned after the Fukushima accident into a broader post-accident international research perspective (Konoplev *et al.*, 2016).

In the future, research should focus on the contribution of forests to radiocaesium transport in rivers, as they now represent the largest significant reservoir of contamination in the region, having overtaken paddy fields because of the success of remediation efforts targeting those fields. Although most research has focused on radiocaesium isotopes, there is great potential to research other radionuclides such as plutonium isotopes (e.g. ^{241}Pu) to further improve our knowledge of spatial and temporal dynamics of the fallout radionuclide contamination (Evrard *et al.*, 2014; Steinhauser, 2014; Cao *et al.*, 2016). The analysis of long-lived radioisotopes (e.g. ^{135}Cs; half-life 2×10^6 years) could also provide novel insights into the events that occurred at FDNPS in March 2011 and the longer-term mobility of the contamination in the region (Zheng *et al.*, 2014).

Although the international scientific community should strive to maximise its understanding of radionuclide behaviour in this exceptional post-accident circumstance, it is important to emphasise that each potential fallout situation will be unique. If another accident occurs in the future, the fate of radioisotopes will mainly depend on the initial pattern of fallout deposition and the environmental conditions prevailing during and after the event. In addition to the considerable effort undertaken to map the initial distribution of radionuclides in the soils, we believe that a similar investment should be afforded to establish a network of continuous river-monitoring stations that could ideally be included in the network of critical zone observatories (Banwart *et al.*, 2013). This network would undoubtedly stimulate multidisciplinary research and help design appropriate management solutions in a region confronted by major environmental and societal challenges.

6.4 Distribution and Migration of Radioiodine in Terrestrial Environment

TETSUYA MATSUNAKA AND KIMIKAZU SASA

Radioiodine is one of the most important radionuclides released during the FDNPS accident. The total amounts of radionuclides discharged into the atmosphere were estimated to be 8.1 GBq for ^{129}I (half-life: 1.57×10^7 years) (Hou *et al.*, 2013) and 120–200 PBq for ^{131}I (half-life: 8.02 days) (Chino *et al.*, 2011; Katata *et al.*, 2012;

Terada *et al.*, 2012; Kobayashi *et al.*, 2013). Understanding the distribution and behaviour of ^{131}I in the contaminated area is important for assessing not only the radiological hazard posed by this radioisotope's high fission yield and γ-ray energy, but also the radioactive contamination of the terrestrial biosphere. Short-lived ^{131}I was implicated in the elevated occurrence of thyroid cancer observed in children after the 1986 Chernobyl accident (Kazakov *et al.*, 1992; Jacob *et al.*, 1998). Radiation monitoring inside the highly contaminated restricted area at Fukushima showed ^{131}I deposition of 2.8–55 kBq/m^2 (decay corrected to 14 June 2011) (MEXT, 2011a). Although the accident-derived ^{131}I in soil disappeared in a few months, the long-lived ^{129}I can be used as a tracer to retrospectively infer the level of ^{131}I released, as there was a strong correlation between ^{131}I and ^{129}I levels in contaminated surface soils in the Chernobyl area (Mironov *et al.*, 2002). In soil, I^- and IO_3^- are the major inorganic species of iodine (Zhang and Hou, 2013), while humic substances such as humic acid, fulvic acid and humin associated with iodine are the major organic forms (Hansen *et al.*, 2011). In the contaminated soil in Chernobyl, 70% of the ^{129}I was bound to organic matter and Fe–Mn oxide phases, while most of the ^{137}Cs remained in the soil after the sequential chemical extraction (Hou *et al.*, 2003). The ^{129}I, which is less radiologically harmful than ^{131}I, provides information about internal exposure by ^{131}I and long-term migration of ^{129}I different from the behaviour of ^{137}Cs in the terrestrial environment.

6.4.1 Pre- and Post-Accident ^{129}I and ^{137}Cs Levels, and $^{129}I/^{137}Cs$ Ratios in Soil

Information on the pre-accident levels of ^{129}I and ^{137}Cs and the $^{129}I/^{137}Cs$ ratio in the Fukushima area is essential to assess the extent of deposition and behavioural response of accident-derived ^{129}I and ^{137}Cs. This section aims to establish the deposition density and extent of subsurface infiltration of accident-derived ^{129}I and ^{137}Cs in the top 0–30 cm of the soil layer in the Fukushima restricted area. We compared cumulative inventories of ^{129}I and ^{137}Cs, concentrations of ^{129}I and ^{137}Cs and the $^{129}I/^{137}Cs$ ratio among 30 cm soil cores collected 3–4 years before and 20 months after the accident at Site S-1 (4 km west of the accident site) and Site S-2 (8 km west of the accident site). Iodine for the measurement of ^{129}I concentration and the $^{129}I/^{127}I$ ratio from the soil samples was purified using the sample preparation scheme including the pyrohydrolysis method developed by Muramatsu *et al.* (2008). The level of ^{129}I was determined with an accelerator mass spectrometer (AMS) configured using the method of Matsuzaki *et al.* (2007). The detailed procedure for the measurement of ^{129}I was described by Matsunaka *et al.* (2015). The $^{129}I/^{127}I$ ratio for the prepared AgI sample including an iodine carrier with an $^{129}I/^{127}I$ ratio of 1.8×10^{-13} (Matsuzaki *et al.*, 2007) was measured using the AMS system at the Micro Analysis Laboratory Tandem Accelerator (MALT) at the University of Tokyo.

The original $^{129}I/^{127}I$ ratios and ^{129}I concentrations in the soil samples were calculated using the ^{127}I concentration obtained from inductively coupled plasma mass spectrometry and the $^{129}I/^{127}I$ ratio obtained from the AMS, and by subtracting the ^{127}I and ^{129}I contents of the background during the sample preparation. Pre-accident levels of ^{129}I in surface soils from a depth of 0 to 30 cm at the study sites were estimated to have been 59.3–63.3 mBq/m^2 for deposition density and 24.8–422 μBq/kg for concentration. Pre-accident levels of ^{137}Cs in surface soils within 30 cm depth were estimated to have been 2.51–7.88 kBq/m^2 for deposition density and 0.56–48.4 Bq/kg for concentration. The Fukushima nuclear accident resulted in the deposition of approximately 0.90–2.33 Bq/m^2 of ^{129}I and 0.80–4.04 MBq/m^2 of ^{137}Cs within 8 km west of the accident site, which were 14–39 and 320–510 times higher than local pre-accident levels, respectively. The deposition densities of accident-derived ^{129}I and ^{137}Cs at S-1, 4 km west of the accident site, were three and five times higher than at S-2, located 8 km west of the accident site. Approximately 90% of accident-derived ^{129}I and ^{137}Cs measured in the upper 30 cm of soils at both sites was absorbed in the surface layer from 0 to 44–95 kg/m^2 (0 to 4.3–6.2 cm depth) and from 0 to 16–25 kg/m^2 (0 to 1.0–3.1 cm depth), respectively (Figures 6.15 (a)–(b), (d)–(e)). Moreover, at both S-1 and S-2 the relaxation mass depths (h_0) of ^{129}I in the soil (10.8–11.2 kg/m^2) were larger than those of ^{137}Cs (8.1–10.6 kg/m^2). The ^{129}I penetrated to a slightly larger depth in the surface layer than did ^{137}Cs, probably as a result of the extent of permeability of the soil. This finding is in accordance with the logarithmic increase in the $^{129}I/^{137}Cs$ ratio with a depth found in the surface layer after the accident from 5.0×10^{-7} to 3.1×10^{-6} at S-1 and from 8.6×10^{-7} to 1.7×10^{-5} at S-2 (Figures 6.15(c), (f)). The ratios in the layer of 0–5 kg/m^2 at S-1 and 0–16 kg/m^2 in the topmost layer at S-2 are coincident with the activity ratio of ~3×10^{-7} in the unit 1, unit 2 and unit 3 reactors of the FDNPS (Nishihara *et al.*, 2012), and 1–3 orders of magnitude lower than the pre-accident levels, which ranged from 8.7×10^{-6} to 1.3×10^{-4}. These results suggest that the deposited ^{129}I and ^{137}Cs on our study sites after the accident had not significantly fractionated during the transport process from the accident site, and the contribution ratio of ^{129}I and ^{137}Cs in soil is different from that of the global fallout. These profiles of the $^{129}I/^{137}Cs$ ratio in the surface layer after the accident illuminate the different behaviour of the two radionuclides, namely, a downward migration of ^{129}I relative to ^{137}Cs in the surface layer. Approximately 7–9% of the accident-derived ^{129}I was present in the lower layer, from 44 to 100 kg/m^2 (4.3–8.6 cm depth) at S-1 and from 95 to 160 kg/m^2 (6.2–10.2 cm) at S-2 (Figures 6.15(a), (d)).

6.4.2 Distribution of ^{131}I and Determination of $^{129}I/^{131}I$ Ratio and in Surface Soil

The extent of ^{131}I soil deposition is critical to understand because of the role of ^{131}I in adversely affecting human health, yet research is limited, with Kinoshita

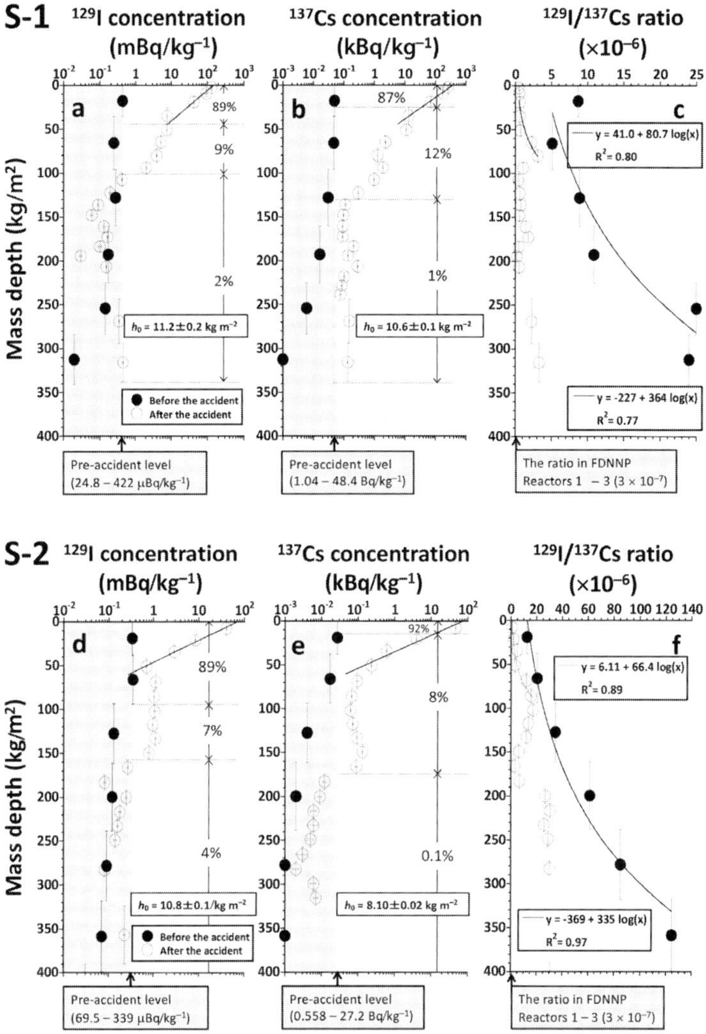

Figure 6.15 Depth profiles of ^{129}I concentration (a, d), ^{137}Cs concentration (b, e) and simulated fittings of an exponential depth distribution (black line) for the 30 cm soil cores collected before and after the Fukushima nuclear accident at S-1 and S-2 (Matsunaka *et al.*, 2016). Depth profiles of ^{129}I/^{137}Cs ratios in the soil cores before and after the FDNPS accident at the two sampling sites are shown in (c, f) (Matsunaka *et al.*, 2016). The black circles and white circles denote pre-accident and post-accident levels, respectively. The regression line and coefficient of determination (R^2) between mass depth and the ^{129}I/^{137}Cs ratio are shown in (c, f). Pre-accident levels of ^{129}I and ^{137}Cs, and the relaxation mass depths (h_0) of the radionuclides are shown in (a, d) and (b, e), respectively. The data for ^{129}I concentration in the surface layer of 0–5 cm depth and h_0 of ^{129}I are from Matsunaka *et al.* (2015). Reprinted from J. Environ. Radioact., 30. He and Wallling (1996). Interpreting particle size effects in the adsorption of ^{137}Cs and unsupported ^{210}Pb by mineral soils and sediments. pp. 117–137. Copyright 1996, with permission from Elsevier.

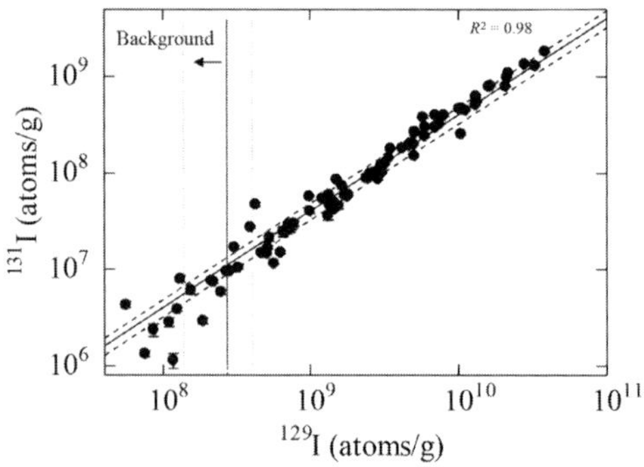

Figure 6.16 Correlation between concentrations of ^{129}I and ^{131}I in surface soils in and around Fukushima. The average ratio of ^{129}I/^{131}I is estimated to have been 26.0 ± 5.7 as of 11 March 2011.

et al. (2011) and MEXT (2011a) among the few available studies. ^{131}I has the tendency to accumulate in the human thyroid gland. Knowledge of the deposition of ^{131}I during the initial stage after the accident in different locations at Fukushima is essential for estimating the radiation dosimetry for ^{131}I, yet the period of investigation is limited to within a few months after the accident because of the short half-life of ^{131}I. In this section, the aim is to determine the ^{129}I/^{131}I atomic ratio in the surface soil of the Fukushima area in order to reconstruct the ^{131}I deposition using the long-lived ^{129}I, which has the same behavioural response as ^{131}I in the environment. For this purpose, we investigated the concentrations of ^{129}I among 5 cm surface soil samples at 108 sites in and around the Fukushima area, which were measured for ^{131}I immediately after the accident by γ-ray spectrometry (Kinoshita *et al.*, 2011). The ^{129}I concentrations in the soil were analysed using AMS at the University of Tokyo. Pre-accident levels of ^{129}I concentration were estimated to have been $(2.7 \pm 1.4) \times 10^8$ atoms/g in the Fukushima area. Figure 6.16 shows the correlation between post-accident concentrations of ^{129}I and ^{131}I in surface soils in the Fukushima area with a correlation coefficient (R^2) of 0.98. The average ratio of ^{129}I/^{131}I was estimated to be 26.0 ± 5.7 as of 11 March 2011. Previous studies had reported ^{129}I/^{131}I ratios of 31.6 ± 8.9 (Miyake *et al.*, 2012), 26.1 ± 5.8 (Miyake *et al.*, 2015) and 21[1] (Muramatsu *et al.*, 2015) in the surface soils of Fukushima. The calculations for ^{129}I/^{131}I ratios by the ORIGEN2 code resulted in values of 31.4 for the unit 1 reactor, 21.9 for the unit 2 reactor and 20.8 for the unit 3 reactor (Nishihara *et al.*, 2012). Our result of 26.0 ± 5.7 for the ^{129}I/^{131}I ratio was consistent with the estimates by the ORIGEN2 code. Figure 6.17 shows the distribution of the ^{131}I deposition density in eastern Japan; these data were gleaned by integrating ^{131}I data (108 sites; Kinoshita *et al.*, 2011) with the initial ^{131}I data (415 sites) from

[1] There is no standard deviation for the ^{129}I/^{131}I ratio estimated by Muramatsu *et al.* (2015).

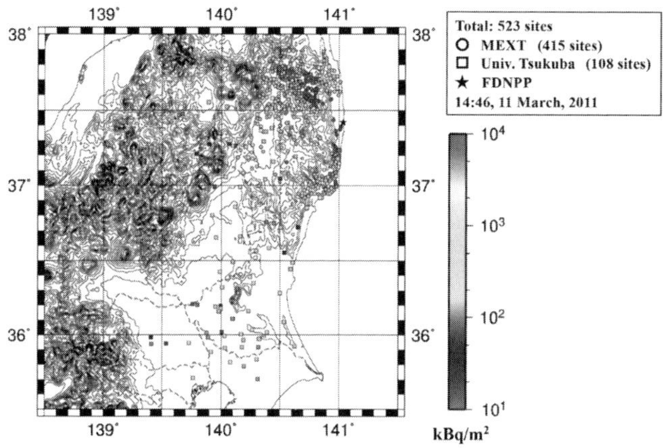

Figure 6.17 Map of ^{131}I deposition density in eastern Japan, integrating the ^{131}I data (108 sites, Kinoshita *et al.*, 2011) with the initial ^{131}I data (415 sites) by MEXT (2011a). A black and white version of this figure will appear in some formats. For the colour version, refer to the plate section.

MEXT (2011a). It is necessary to reveal the spatial distribution of ^{131}I released from the FDNPS accident. The result showed that there is a strong correlation between ^{131}I and ^{129}I concentrations. In addition, the average ratio of $^{129}I/^{131}I$ is equivalent to the calculated ratio obtained by ORIGEN2 code. It means that we will be able to reconstruct a reliable and precise ^{131}I deposition map by analysing ^{129}I concentrations in surface soils in Fukushima.

6.5 Understanding the Migration Behaviour of Radiocaesium at the Molecular Scale

YOSHIO TAKAHASHI, KAZUYA TANAKA AND AYA SAKAGUCHI

The adsorption of radiocaesium on soil particles or particulate matter in water has been discussed in the previous sections. However, the adsorption is not an irreversible reaction; equilibrium can be reached, which suggests that radiocaesium can be leached from these particles. Approximately 50–90% of the radiocaesium emitted from the FDNPS and collected on aerosol filters was leached out by water (Tanaka *et al.*, 2013). In the following, several studies conducted on the migration of radiocaesium after its deposition on the soil surface are discussed.

To analyse the distribution of radiocaesium after its deposition, a thin section of weathered granite from the surface soil in Yamakiya, Kawamata Town, Fukushima Prefecture, was prepared and examined by autoradiography (Figure 6.18; Tanaka *et al.*, 2013). The distribution of radiocaesium, indicated by black spots, was not correlated with that of potassium, which normally exhibits a geochemical behaviour similar to that of radiocaesium, as observed by scanning electron microscopy

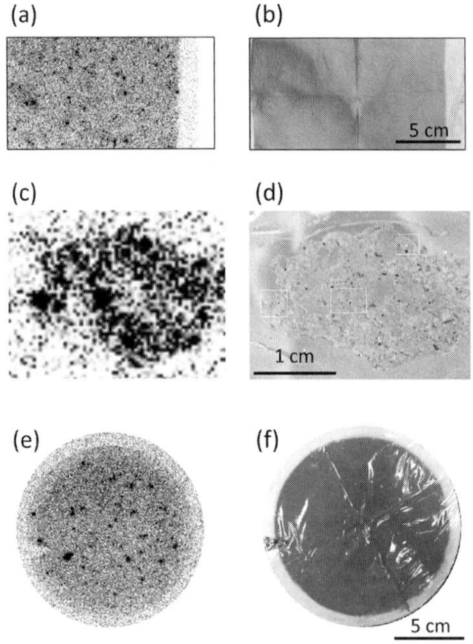

Figure 6.18 Autoradiographic images (a, c, e) and optical images (b, d, f) for aerosol filters (a,b; collected in Kawasaki City in Japan on 20 March 2011); a section of soil surface (c, d; weathered granite in Kawamata Town) which had not been disturbed after the FDNPS accident; and membrane filter used for the filtration of river water (e, f; downstream of Kuchibuto River collected on 31 July 2011; pore size: 3 μm) in Tanaka *et al.* (2013). Reprinted by permission from Springer Nature: Heterogeneous distribution of radiocesium in aerosols, soil and particulate matters emitted by the Fukushima Daiichi Nuclear Power Plant accident: Retention of micro-scale heterogeneity during the migration of radiocesium from the air into ground and river systems, Tanaka et al © 2013.

with energy dispersive X-ray spectroscopy (SEM-EDX). This result indicated that the distribution of radiocaesium was not completely controlled by the chemical and mineralogical processes under equilibrium in the soil–water system, which influence the distribution of potassium. This distribution pattern can be explained either by (1) the deposition of water-insoluble radiocaesium-enriched particles (Adachi *et al.*, 2013) and/or (2) the adsorption of water-dissolved radiocaesium onto the granite surface or its secondary minerals at the micrometre scale. The latter process can occur via adsorption on phyllosilicates, which are widely distributed in weathered granitic rocks. Phyllosilicates such as vermiculite have a strong ability to adsorb radiocaesium.

The fixation of radiocaesium in the soil surface through its transformation into its insoluble form has been suggested based on the depth profile in the soil layers (Kato *et al.*, 2012a; Tanaka *et al.*, 2012). An example depth profile of radiocaesium was examined for a soil core sample collected in Koriyama City on 13 April 2011

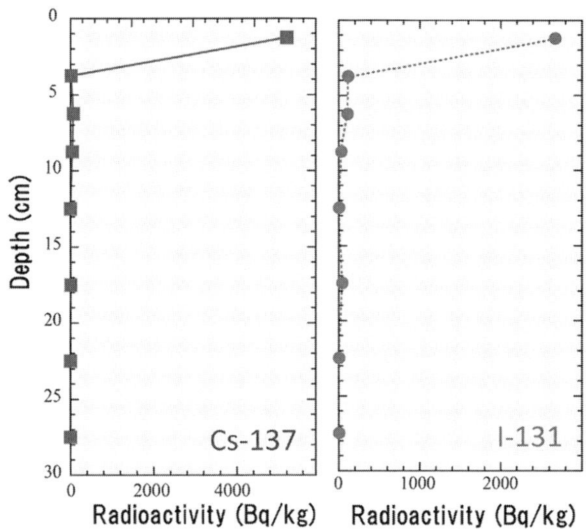

Figure 6.19 Depth profiles of ^{131}I and ^{137}Cs in the soil core samples collected in Koriyama City on 13 April 2011 (decay was corrected as to the sampling date; Tanaka *et al.*, 2012).

(Figure 6.19). The results showed that more than 90% of the radiocaesium was fixed within the top 5 cm of the profile (Tanaka *et al.*, 2012), suggesting that the radiocaesium was in its insoluble chemical form after its deposition on the soil surface. It was also reported that the depth profile remained essentially unchanged, even after the rainy season in Japan (Matsunaga *et al.*, 2013).

The radiocaesium fixed at the soil surface is generally eroded, enters rivers and is finally transported to the ocean. In the Abukuma River and its tributary the Kuchibuto River that runs through the Yamakiya District (Kawamata Town), more than 70% of the radiocaesium was within particulate matter fractions collected by filtration (>0.45 μm; Figure 6.20; Sakaguchi *et al.*, 2015), indicating that the radiocaesium was transported as particulate matter, such as clay minerals, in the river water. The distribution of radiocaesium collected on the filters and measured by autoradiography was also found to be heterogeneous (Figure 6.18; Tanaka *et al.*, 2013). As was already suggested for the similar distribution in the surface soil, this heterogeneous distribution can be explained either by (1) the presence of radiocaesium-enriched particles, as found in aerosol filters, and/or (2) a limited amount of particulate matter enriched in radiocaesium that may be transported from areas with high radiocaesium deposition. The latter process is plausible because the desorption of radiocaesium from particulate matter is very limited, which prevents a homogeneous distribution of radiocaesium in the river water. However, it has been suggested that radiocaesium is desorbed to a larger degree in estuarine areas due to the increased salinity in these areas (Sakaguchi *et al.*, 2015; Kakehi *et al.*, 2016).

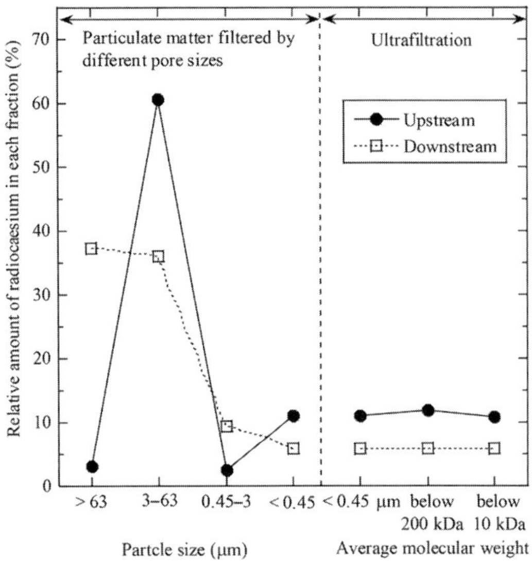

Figure 6.20 Relative ratios of radiocaesium in various particle size fractions of particulate matters in river water from Kuchibuto River, collected on 12 December 2011.

The distribution of radiocaesium in Fukushima Prefecture and its neighbouring prefectures has been measured periodically by airborne monitoring surveys by MEXT. At present, the measurements are conducted by the Nuclear Regulation Authority. A comparison of the monitoring results in June and November 2011 showed a secondary enrichment of radiocaesium in the midstream basin of the Abukuma River (Kakuda City, Miyagi Prefecture) and in various estuarine areas in Fukushima and Miyagi Prefectures. The increase of the radiocaesium concentration in the areas may have been related to the sedimentation of the particulate matter adsorbing radiocaesium transported through the rivers (Figure 6.21; MEXT, 2011b).

These facts showed that the heterogeneous deposition of radiocaesium on soil surfaces caused its heterogeneity in clay mineral particles, which in turn may be related to the heterogeneous distribution of radiocaesium, even in the particulate matter in river water. As indicated above, this heterogeneity can be caused either by (1) the presence of radiocaesium-enriched particles that may be produced within the reactors or immediately following the emission from the reactors and/or (2) the strong affinity of radiocaesium for soil particles at soil surfaces. The former process has been studied using X-ray spectroscopy and electron microscopy (Adachi *et al.*, 2013; Abe *et al.*, 2014). The latter process, in particular the interaction between caesium and clay minerals, has been extensively studied using extended X-ray absorption fine structure (EXAFS) spectroscopy.

One example of an EXAFS spectrum for caesium in soil and river sediment samples collected in the Fukushima area is given in Figure 6.22 (Qin *et al.*, 2012).

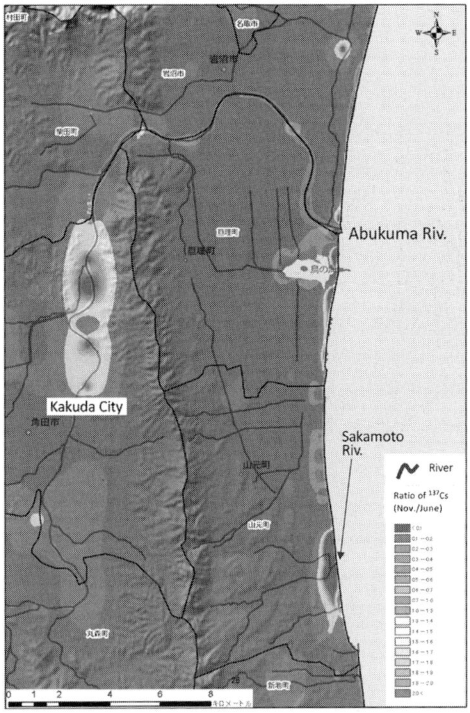

Figure 6.21 Ratio of air dose rate in November 2011 normalised to that in June, 2011 by airborne survey by MEXT (2011b), which suggests increased radiocaesium concentration in some areas. A black and white version of this figure will appear in some formats. For the colour version, refer to the plate section.

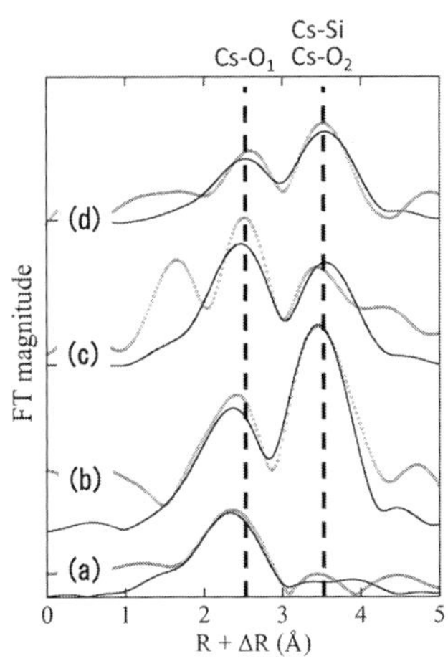

Figure 6.22 Radial structural functions (RSFs) of caesium L3-edge EXAFS for hydrated caesium ions (a) and caesium ions adsorbed on vermiculite (b), soil in Kawamata Town (c) and river sediment in Kuchibuto River (d) in Qin *et al.* (2012). Dashed and solid lines indicate experimental and fitted data, respectively.

This spectrum reflects the local structural environment around Cs atoms adsorbed onto clay minerals at the atomic scale, which can be obtained in synchrotron radiation facilities, such as the Photon Factory in High Energy Accelerator Research Organization (KEK; Tsukuba, Japan) and SPring-8 (Hyogo, Japan). The spectrum reveals direct bonding of caesium in the interlayer of clay minerals in soils and sediments.

Caesium is an alkali metal element and is present as a monovalent cation in the environment, where it can be readily dissolved in water. In the presence of 2:1 phyllosilicates, such as mica and clay minerals, caesium cations can be stably bound to the siloxane ring in the phyllosilicate structures. These phyllosilicates consist of a two-dimensionally spread sheet of octahedral aluminium oxide and tetrahedral silicon oxide layers, and the 2:1 phyllosilicate is formed by sandwiching the aluminium oxide layer by two silica layers. Each 2:1 layer piles up to form the layered structure. The silica layer has a six-membered ring structure of SiO_2 (siloxane ring), which is exposed to the interlayer space between the two 2:1 layer structures. The association of caesium ions and 2:1 phyllosilicate is related to the large ion size of the caesium ion. The size of the siloxane ring created by the three silicon and three oxygen atoms is fitted to the size of the caesium ion (= 1.78 Å in six coordination number), which leads to the formation of inner-sphere complexes of caesium ions with the siloxane ring. Similar stable complexes may be formed for potassium and ammonium ions.

These structural characteristics can be elucidated by EXAFS spectroscopy. Hydrated caesium cations have eight hydrated water molecules within the first coordination sphere of caesium, which is reflected by the peak at $R + \Delta R = 2.2$ Å in the RSF of caesium in L3-edge EXAFS (Figure 6.22(a)). Adsorption of caesium ions onto vermiculite results in the appearance of an intense second peak at $R + \Delta R = 3.4$ Å. This peak is caused by the interaction of caesium ions with silicon and oxygen in the siloxane layer and reveals the formation of inner-sphere complexes of caesium ions and vermiculite. When we added stable isotopes of caesium to a suspension of soil and river sediment samples from Fukushima, a similar peak at $R + \Delta R = 3.4$ Å was found, indicating the formation of inner-sphere complexes of caesium ions in the samples. It is true that the spectra were obtained for stable caesium ions added to the system, but radiocaesium in the environment with much lower concentrations is likely to form the same stable complexes with vermiculite once dissolved in water. Thus, the second peak of the RSF indicates that the soils and sediments examined here contained such phyllosilicate minerals that can form the inner-sphere complexes. The high stability of the inner-sphere complex of caesium ions and phyllosilicate is to some extent responsible for the heterogeneous distribution of radiocaesium in the environment.

To better understand the mobility of radiocaesium, various factors that can affect the adsorption of radiocaesium to phyllosilicates should be fully studied.

For example, natural polyorganic matter, such as humic substances, can inhibit the intrusion of caesium ions into the interlayer of the phyllosilicate and thus impede the formation of the inner-sphere complex (Fan *et al.*, 2014). It has also been suggested that radiocaesium adsorbed on the phyllosilicates can be desorbed in the presence of high concentrations of potassium and ammonium ions. Hence, radiocaesium can be more soluble in an environment with higher organic content and in estuarine and marine waters, which have higher salinity than fresh water.

Despite the much shorter half-life of radioiodine (^{131}I), its health effects can be more serious than those of radiocaesium. Depth profiles of iodine examined in the soil layers sampled from Fukushima, which were previously used to analyse the radiocaesium distribution, indicated that radioiodine was also retained at the soil surface. When iodine is reduced to iodide, it becomes soluble and readily migrates in the soil (Shimamoto *et al.*, 2010), which is inconsistent with the observations in the Fukushima area (Figure 6.19). The retention of iodine is likely to be caused by the incorporation of radioiodine into humic substances via electrophilic substitution of iodine for the aromatic ring in humic substances – a process that was introduced by Shimamoto *et al.* (2011) using synchrotron radiation techniques. The low water solubility of iodine in soil from the Fukushima area, due to the formation of organoiodine species, was also suggested by Tanaka *et al.* (2012) based on sequential extraction experiments for radioiodine deposited after the FDNPS accident. Less than 10% and approximately 30% of ^{131}I were leached by water and an NaOH solution (pH 10.5), respectively. Subsequently, the NaOH solution with ^{131}I was acidified to pH 2, which caused more than 60% of the ^{131}I in the solution to precipitate, possibly with humic substances, suggesting that a portion of the iodine was associated with organics in the soil, which explains the low rate of leaching of ^{131}I from the soil by water based on the Fukushima observations.

As shown above, the chemical state of radionuclides is closely related to their fates in the environment. The molecular-scale information that can be obtained by synchrotron radiation methods, electron microscopy and indirect methods, such as sequential extraction, is essential to identify the chemical species, which in turn can be used to model the transfer of radionuclides, including chemical reactions, in the environment via reactive transport models.

6.6 Effects on Agricultural Products and Wild Plants

CHISATO TAKENAKA

6.6.1 Effects on Agricultural Products and Edible Forest Products

Radioactive concentration in foods is a major concern for people because of the risk of internal exposure to radioactivity. Beginning on 19 March 2011, immediately following the FDNPS accident, the measurement results of radioactive concentrations in various foods (products from agricultural farms, livestock, marine areas and forests) obtained from the Kanto and Tohoku districts were published on the homepage of the Ministry of Health, Labour, and Welfare (www.maff.go.jp/noutiku_eikyo/mhlw3.html). According to these data, the farm products contaminated with radioactive material changed after the accident. From late March through April, spinach exhibited the highest frequency of high ^{131}I concentrations. The maximum ^{131}I concentration was recorded at 54.1 kBq/kg (in fresh weight) in spinach obtained on 18 March. The tentative standard value for the radioactive concentration in food was defined as 500 Bq/kg fresh weight. After the spinach, ^{131}I activity of 3.5 kBq/kg was reported in shiitake mushrooms (*Cortinellus shiitake*) from bed log cultivation on 14 April. Because the half-life of ^{131}I is eight days, food having high ^{131}I concentrations was not detected after late April 2011.

Table 6.1 summarises the farm and forest products containing radiocaesium exceeding the tentative standard value (500 Bq/kg) after 17 March 2011. Because 100 Bq/kg was established as the new, stricter standard value for food products by the Ministry of Health, Labour and Welfare in April 2012, products containing more than this amount and between 100 and 500 Bq/kg after April 2012 are shown separately in Table 6.2. Green vegetables (*Chenopodiaceae* spinach, *Brassicaceae* Japanese mustard spinach cole) were listed as highly radiocaesium-contaminated foods, with high frequencies from March to April 2011.

From May to June 2011, highly concentrated radiocaesium was detected in bamboo shoots, tea, shiitake mushrooms from bed log cultivation, ostrich ferns and citron. After September, radiocaesium was detected at high concentrations in mushrooms, and rice was listed after November 2011. After April 2012, many farm and forest products were reported as foods exceeding the standard value because of the new, stricter standard. It is characteristic that high concentrations of radiocaesium were detected in many edible wild plants (sprouts of ostrich fern, *Acanthopanax sciadophylloides*, *Aralia elata*, *Osmunda japonica*), whereas green vegetables that were reported in 2011 were not reported in 2012.

Although many green vegetables were measured for their radioactive content in 2012 in the same way as in 2011, no green vegetable samples exceeded the standard value for radiocaesium. This means that the radioactive contamination of the farm products immediately following the accident was caused by the direct deposition of radioactive fallout. At the time of the accident, most green vegetables had already developed their leaves; therefore, the radioactive material deposition from the atmosphere to the green vegetables occurred directly. Because highly concentrated radiocaesium was found in vegetables from greenhouse cultivation (e.g. parsley), the radiocaesium contamination was caused not

Table 6.1 *Farm and forest products with levels of radiocaesium above the tentative standard value (500Bq/kg).*

2011					2012	
March–April	May–June	July–August	September–October	November–December	January–February	March
Spinach (62), broccoli (22), log-grown shiitake (14), komatsuna (6), aburana (5), shinobuna (5), kukitachina (5), cabbage (5), kousaitai (4), mizuna (4), parsley (3), vitamin-na (2), *Wasabia japonica* (2), *Oenanthe javanica* (2), kabu (2), santouna (1), chijirena (1), kakina (1)	Bamboo shoots (62), tea (28), log-grown shiitake (34), ostrich ferns (3), kabu (1), parsley (1), *Prunus mume* (11)	*Citrus junos* (4), log-grown shiitake (3), tea (1), *Eriobotrya japonica* (1), *Ficus carica* L. (1), wheat (1), aburana (1), *Pholiota microspora* (1), *Lactarius volemus* (1)	Log-grown shiitake (8), *Lactarius volemus* (8), *Lactarius hatsudake* (8), *Suillus bovinus* (2), *Grifola frondosa* (2), *Pholiota microspora* (2), *Citrus junos* (2), *Punica granatum* (1), *Tricholoma matsutake* (1), *Sarcodon aspratus* (1), *Hypholoma sublateritium* (1), *Lyophyllum decastes* (1), *Pholiota lubrica* (1), *Castanea crenata* (1)	Rice (17), log-grown shiitake (13), *Hypholoma sublateritium* (7), *Pholiota microspora* (1), *Panellus serotinus* (1)	Rice (4), log-grown shiitake (4), *Citrus junos* (1), *Wasabia japonica* (2)	Log-grown shiitake (4), bamboo shoot (1)

The number of samples of each product is shown in parentheses.
Retrieved from www.maff.go.jp/noutiku_eikyo/mhlw3.html.

Table 6.2 *Farm and forest products with levels of radiocaesium above the standard value (100 Bq/kg) from April to June in 2012.*

April		May		June	
More than 500 Bq/kg	500~100 Bq/kg	More than 500 Bq/kg	500~100 Bq/kg	More than 500 Bq/kg	500~100 Bq/kg
Log-grown shiitake (20), ostrich ferns (2), *Chengiopanax sciadophylloides* (2), bamboo shoots (1), *Wasabia japonica* (1), *Aralia elata* (1), *Osmunda japonica* (1)	Log-grown shiitake (62), bamboo shoots (25), ostrich ferns(11), *Aralia elata* (7), *Petasites japonicus* (4), *Chengiopanax sciadophylloides* (1), *Osmunda japonica* (1), *Zanthoxylum piperitum* (1)	*Chengiopanax sciadophylloides* (18), log-grown shiitake (11), bamboo shoots (1), *Pteridium aquilinum* (1), *Aralia elata* (1), *Osmunda japonica* (1)	Log-grown shiitake (81), *Chengiopanax sciadophylloides* (30), *Aralia elata* (10), ostrich ferns (9), *Osmunda japonica* (9), bamboo shoots (8), *Pteridium aquilinum* (7), *Zanthoxylum piperitum* (2), *Oenanthe javanica* (2), *Elatostema umbellatum* (1)	Not found	Bamboo shoots (8), log-grown shiitake (3), *Prunus mume* (2), *Petasites japonicus* (1), *Angelica keiskei* (1), *Parasenecio delphiniifolius* (1), *Phyllostachys nigra* (1)

The number of samples of each product is shown in parentheses.
Retrieved from www.maff.go.jp/noutiku_eikyo/mhlw3.html.

only through wet deposition but also through dry deposition. On the other hand, one reason why few highly contaminated edible wild plants were detected in 2011 might be related to the smaller number of samples. Bamboo shoots were reported as a highly contaminated food for two consecutive years; this might have occurred because of transport of the radiocaesium through the underground bamboo stem.

For rice, to determine whether to plant in the spring of 2011, the radiocaesium transfer index from the soil to unpolished rice was defined by the government as 0.1 on 8 April 2011 (www.maff.go.jp/j/kanbo/joho/ saigai/ine_sakutuke.html). Therefore, a value of 5 kBq/kg in soil was the standard for rice planting to produce rice containing radiocaesium at a level below 500 Bq/kg, based on the government standard. The standard transfer index of 0.1 appeared to be quite high based on previous data from the Japanese Society of Soil Science and Plant Nutrition. However, at the time of the rice-planting season in 2011, the Ministry of Agriculture, Forestry and Fisheries defined the product limit area for rice based on three zones: restricted area, deliberate evacuation area and evacuation-prepared area in case of emergency. This effort may have resulted in the production of the highly radiocaesium-contaminated rice listed in Table 6.1 because rice could be planted in the hotspot zone, where the radiocaesium activity was high, even within the residential permitted area. In addition, it is well known that the availability of radiocaesium in plants depends on the soil properties. Therefore, the rice planted in sandy soil or low-clay soil could absorb more radiocaesium than that planted in clay soil, even if the total radiocaesium concentration in the soil was less than 5 kBq/kg.

As shown in Table 6.1, contaminated shiitake mushrooms from bed log cultivation were reported frequently, and after September 2011 high-radioactivity concentrations were detected in various other mushrooms, such as *Lactarius volemus*, *Lactarius hatsudake* and *Grifola frondosa*. Fukushima, Ibaraki, Tochigi and Gunma Prefectures are well known for being the main production areas of shiitake mushrooms from bed log cultivation.[2] It is suspected that the logs for shiitake cultivation were contaminated with radiocaesium by direct deposition while lying in the open air after the accident. In contrast, *L. volemus* and *L. hatsudake*, which grow on the forest floor, should be affected by the direct deposition of radiocaesium on the forest floor because they lacked leaves at the time of the accident. It has been well known that mushrooms absorb radiocaesium derived from global fallout at high concentrations through spawns (Yoshida *et al.*, 1994). Moreover, the mushrooms expanded in the shallow layer with the accumulation of radiocaesium at high concentrations because a large amount of radiocaesium was distributed in the surface soil layer (Kato *et al.*, 2012a).

[2] Special forestry basics document: http://bit.ly/2G2LVOs.

6.6.2 Effects on Wild Plants

Because caesium is an alkaline element similar to potassium, it is absorbed by the roots of plants. It is well known that the absorption ability varies depending on plant species. Therefore, plants with a high ability to absorb caesium are expected to be good candidate plants for the decontamination of radiocaesium. After the Chernobyl disaster, numerous studies using various plants were conducted in Russia and Europe, and it was reported that various plants such as amaranthus, redroot pigweed and quinoas accumulated radiocaesium (Lasat *et al.* 1998; Broadley *et al.* 1999; Fuhrmann *et al.* 2003). To locate the radiocaesium-accumulating plants in Fukushima, Japan, we started a field project in May 2011, collected every plant, both herbaceous and woody plants, at the soil sampling site and then measured the radiocaesium concentrations in the aboveground portions.

The radiocaesium level in the surface soils in 2011 ranged from 0.045 to 82.8 kBq/kg, whereas the concentration of radiocaesium in plant leaves (138 samples) that were newly sprouted after the accident varied from an undetectable level to 41 kBq/kg. The plants with the undetectable level of radiocaesium were sampled from both high- and low-contamination areas. This result clearly indicated the variability in the concentration of radiocaesium in leaves among the sampled plant species.

The characteristics of the radiocaesium concentration in plants can be evaluated using a concentration ratio (CR value):

$$\mathrm{CR} = {}^{137}\mathrm{Cs}\ \text{concentration in a plant}\ (\mathrm{Bq/kg}) / {}^{137}\mathrm{Cs}\ \text{concentration in the soil}\ (\mathrm{Bq/kg})$$

In general, the CR value is related to the radiocaesium uptake ability of a plant from the soil through its roots. However, the plant samples collected in 2011 contained radiocaesium that was absorbed not only from roots but also from the plant surface, such as from leaves. Therefore, the CR values for plants taken in 2011 indicated the capture of radiocaesium via both the aboveground and underground portions of the plants.

Figure 6.23 shows the data for 20 high CR values from the analysed plant individual samples collected in 2011. The highest CR value was 220 in the leaves of *Pieris japonica*; 13 of these high values were observed in the leaves of tree species. The reason for the high CR values in the leaves of evergreen tree species such as *P. japonica*, *Eurya japonica* and *Camellia sasanqua* is that the radiocaesium was deposited on the leaves that existed on the tree crown before the accident; it was absorbed at the leaf surface, transported to the buds, and subsequently included in the newly developed leaves. However, the transport of radiocaesium taken up by the roots should be possible during the first stage of contamination.

In herbaceous plants, relatively high CR values were found in *Houttuynia cordata* (CR: 17) and *Chenopodium ficifolium Smith* (CR: 5.2), which differed from previous reports that found high CR values in *Amaranthus* (CR: 2.2–3.2) and

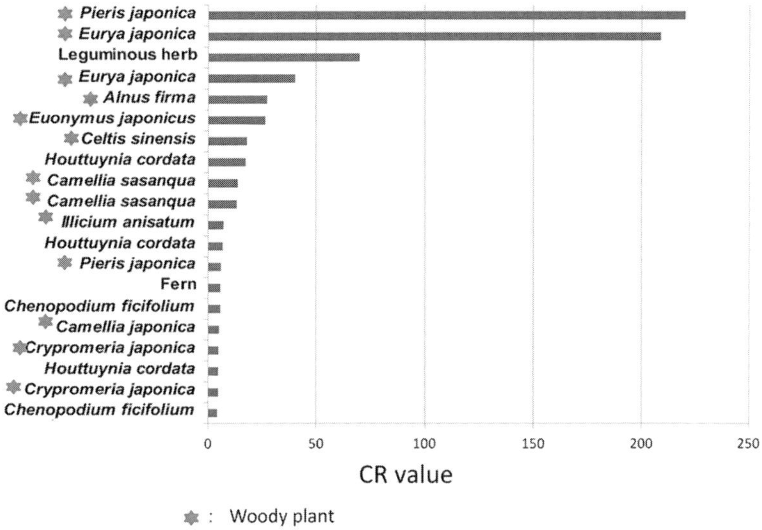

Figure 6.23 The top 20 samples in CR values among all plant samples collected in 2011.

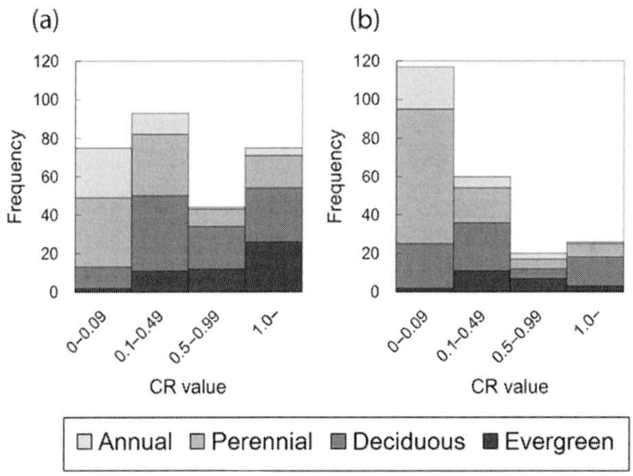

Figure 6.24 Histograms of CR values in various wild plants collected in 2011 (a) and 2012(b) in Fukushima.

ryegrass (CR: 0.92–2.82) (Vandenhove *et al.*, 1996; Lasat *et al.*, 1998). These high values in herbaceous plants, which had not sprouted at the time of the accident, were likely caused by the absorption of the initial input of radiocaesium in its bioavailable chemical form, which gradually changed to its more stable chemical form, i.e. the adsorbed form on clay minerals.

In 2012, we conducted a survey on the radiocaesium concentrations of wild plants in the same way as was done in 2011. A comparison of the histograms of CR values of wild plants between 2011 and 2012 (Figure 6.24) showed that the CR values declined from 2011 to 2012, especially in evergreen tree species. This change in evergreen tree species implied that the surface absorption of radiocaesium contributed to the high CR values in 2011. In addition, it is considered that the change in the chemical form of radiocaesium in soils was related to the CR change from 2011 to 2012 (Sugiura *et al.*, 2016).

During the summer of 2011, sunflowers were planted at many places in Fukushima Prefecture. This represented an effort to remove radiocaesium from the soil because sunflowers were considered to be accumulators of radiocaesium. However, the Ministry of Agriculture, Forestry and Fisheries published a press release showing that sunflower accumulated only 52 Bq/kg (in fresh weight) of radiocaesium in the aboveground tissue; moreover, the decontamination efficiency of radiocaesium by sunflowers was low. To confirm whether the value (52 Bq/kg) was representative, we collected and analysed sunflower samples from Date City, Fukushima. We found a high concentration of 27 kBq/kg (in dry weight) and a high concentration ratio value of 0.28. The sampling site where the sunflower plants with high concentrations were harvested was characterised by a concrete slope and sandy soils. Radiocaesium deposited on the concrete slope would flow into the soils under the slope, and the sandy soil exhibits less adsorption capacity for the caesium ion. This result indicated that sunflowers can accumulate radiocaesium under suitable conditions for uptake. Therefore, the press release regarding the ability of sunflower to absorb radiocaesium was not accurate.

6.6.3 Effects on Trees

Figure 6.25 shows photographs and the results of autoradiography for green and dead needles collected from the crowns of Japanese cedars from the Yamakiya area

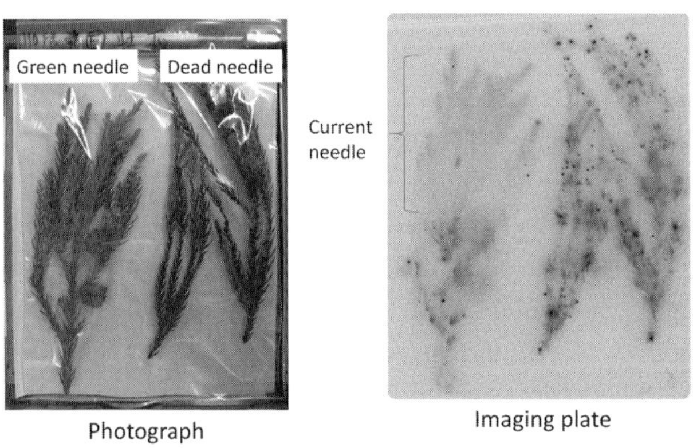

Figure 6.25 Distribution of radioactivity in green and dead needles of *Crypromeria japonica* collected in September 2011 from Fukushima.

in Fukushima in September 2011. Many black dots and images were found in the autoradiograph. In the autoradiograph of the dead needles, many black dots were distributed. In contrast, the autoradiograph of the green needles, which included both old needles that had fallen at the time of the accident and new needles that emerged after the accident, showed a heterogeneous distribution of radioactivity. Although the image of the old needles appeared to be similar to that of the dead needles, the image of the new needles was slightly different in terms of the distribution. These autoradiographs suggested that radiocaesium deposited on the surface of old needles could be absorbed into the needle tissue and transported to the needles that emerged after the accident. Surface absorption of radiocaesium occurred on not only the leaf surface but also the bark surface of trees. Sato *et al.* (2015) reported the possibility of inward migration via the bark from the analysis of radiocaesium concentrations in deciduous fruit tree species in 2011. Wang *et al.* (2016) conducted an application experiment using stable Cs solution on the surface of bark of Japanese cedar and reported Cs absorption through the bark. Therefore, it is suggested that the surface absorption of radiocaesium was an important part of the incorporation process into trees in 2011.

After radiocaesium is incorporated into plant tissue, it can be transported through the same physiological mechanism as potassium because both elements are alkaline metals. Potassium is an essential element for plants and is contained in various important parts including the pollen. Therefore, in the case of the male flower and pollen of the Japanese cedar, radiocaesium, which was once deposited on the cedar forest and absorbed into plant tissue, was expected to be dispersed via the scattering of pollen in the early spring season of 2012. Male flowers of the Japanese cedar were collected in November 2011, and the radiocaesium contained in the male flower and pollen was analysed.

Figure 6.26 shows photographs of the male flowers and their cross-sections (Takenaka and Kiyono, 2012). The male flowers had matured in November. The pollen was separated from the male flower and observed using autoradiography (Figure 6.27). The pollen separated from the male flower from Fukushima showed a distinct radioactive image based on its autoradiograph, indicating that the radiocaesium absorbed from the plant surface, probably at the needle surface, had been transported to the male flower and to the pollen.

Just after the accident, since the vegetation had started new growth, radiocaesium could be incorporated into various plants through roots, bark and leaf surfaces. The effects of initial incorporation into plants might continue for a few years, depending on the plant species, but the cycle of radiocaesium in vegetation ecosystems will reach a steady state sooner or later. It is important to watch the radiocaesium dynamics in vegetation ecosystems and then take measures against radiocaesium contamination.

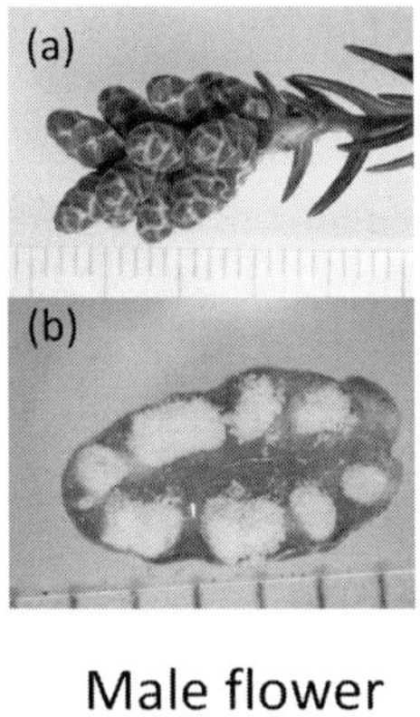

Figure 6.26 Male flower of *Crypromeria japonica* collected in November 2011. (a) appearance; (b) cross-section

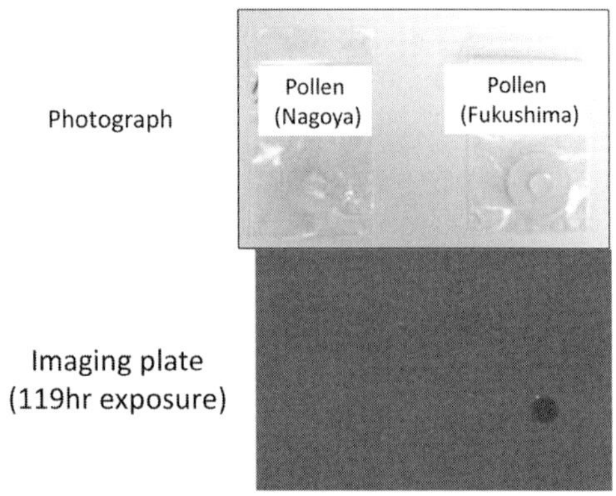

Figure 6.27 Radioactivity in pollen of *Crypromeria japonica* collected in November 2011.

References

Abe, Y., Y. Iizawa, Y. Terada, *et al.* (2014). Detection of uranium and chemical state analysis of individual radioactive microparticles emitted from the Fukushima nuclear accident using multiple synchrotron radiation X-ray analyses. *Anal. Chem.*, **17**, 8521–5.

Adachi, K., M. Kajino, Y. Zaizen and Y. Igarashi (2013). Emission of spherical cesium-bearing particles from an early stage of the Fukushima nuclear accident. *Sci. Rep.*, **3**, 2554.

ANR (2012). Appel à projets 'Recherche en matière de sûreté nucléaire et de radioprotection'. http://bit.ly/2BXZhZV (accessed 19 September 2018) (in French).

Banwart, S., J. Chorover, J. Gaillardet, *et al.* (2013). *Sustaining Earth's Critical Zone: Basic Science and Interdisciplinary Solutions for Global Challenges*. Sheffield: University of Sheffield.

Broadley M. R., N. J. Willey and A. Mead (1999). A method to assess taxonomic variation in shoot caesium concentration among flowering plants. *Environ. Pollut.*, **106**, 341–9.

Cao, L., J. Zheng, H. Tsukada, *et al.* (2016). Simultaneous determination of radiocesium (^{135}Cs, ^{137}Cs) and plutonium (^{239}Pu, ^{240}Pu) isotopes in river suspended particles by ICP-MS/MS and SF-ICP-MS. *Talanta*, **159**, 55–63.

Chagvardieff, P. (2014). La phytoremédiation des sols contaminés par des éléments chimiques radioactifs. *Revue Générale Nucléaire*, **5**, 49–51.

Chartin, C., O. Evrard, Y. Onda, *et al.* (2013). Tracking the early dispersion of contaminated sediment along rivers draining the Fukushima radioactive pollution plume. *Anthropocene*, **1**, 23–34.

Chartin, C., O. Evrard, J. P. Laceby, *et al.* (2017). The impact of typhoons on sediment connectivity: lessons learnt from contaminated coastal catchments of the Fukushima Prefecture (Japan). *Earth Surf. Process. Landf.*, **42**(2), 306–17.

Chino, M., H. Nakayama, H. Nagai, *et al.* (2011). Preliminary estimation of release amounts of ^{131}I and ^{137}Cs accidentally discharged from the Fukushima Daiichi Nuclear Power Plant into the atmosphere. *J. Nucl. Sci. Technol.*, **48**, 1129–34.

Christoudias, T., Y. Proestos and J. Lelieveld (2014). Global risk from the atmospheric dispersion of radionuclides by nuclear power plant accidents in the coming decades. *Atmos. Chem. Phys.*, **14**(9), 4607–16.

Coppin, F., P. Hurtevent, N. Loffredo, *et al.* (2016). Radiocaesium partitioning in Japanese cedar forests following the 'early' phase of Fukushima fallout redistribution. *Sci. Rep.*, **6**, 37618.

Delmas, M., L. Garcia-Sanchez, V. Nicoulaud-Gouin and Y. Onda (2017). Improving transfer functions to describe radiocesium wash-off fluxes for the Niida River by a Bayesian approach. *J. Environ. Radioact.*, **167**, 100–9.

Evrard, O., F. Pointurier, Y. Onda, *et al.* (2014). Novel insights into Fukushima nuclear accident from isotopic evidence of plutonium spread along coastal rivers. *Environ. Sci. Technol.*, **48**(16), 9334–40.

Evrard, O., J. P. Laceby, H. Lepage, *et al.* (2015). Radiocesium transfer from hillslopes to the Pacific Ocean after the Fukushima Nuclear Power Plant accident: a review. *J. Environ. Radioact.*, **148**, 92–110.

Evrard, O., J. P. Laceby, Y. Onda, *et al.* (2016). Quantifying the dilution of the radiocesium contamination in Fukushima coastal river sediment (2011–2015). *Sci. Rep.*, **6**, 34828.

Fan, Q. H., M. Tanaka, K. Tanaka, A., Sakaguchi and Y. Takahashi (2014). An EXAFS study on the effect of natural organic matter and the expansibility of clay mineral on cesium adsorption and mobility. *Geochim. Cosmochim. Acta*, **135**, 49–65.

Fuhrmann, M., M. Lasat, S. Ebbs, J. Cornish and L. Kochian (2003). Uptake and release of cesium-137 by five plant species as influenced by soil amendments in field experiments. *J. Environ. Qual.*, **32**, 2272–9.

Garcia-Sanchez, L. and A. V. Konoplev (2009). Watershed wash-off of atmospherically deposited radionuclides: a review of normalized entrainment coefficients. *J. Environ. Radioact.*, **100**(9), 774–8.

Garnier-Laplace, J., K. Beaugelin-Seiller and T. G. Hinton. (2011). Fukushima wildlife dose reconstruction signals ecological consequences. *Environ. Sci. Technol.*, **45**(12), 5077–8.

Hansen, V., P. Roos, A. Aldahan, X. L. Hou and G. Possnert (2011). Partition of iodine (^{129}I and ^{127}I) isotopes in soils and marine sediments. *J. Environ. Radioact.*, **102**, 1096–104.

He, Q. and D. E. Walling (1996). Interpreting particle size effects in the adsorption of ^{137}Cs and unsupported ^{210}Pb by mineral soils and sediments. *J. Environ. Radioact.,* **30**, 117–37.

Hou, X. L., C. L. Fogh, J. Kucera, *et al.* (2003). Iodine-129 and cesium-137 in Chernobyl contamination soil and their chemical fractionation. *Sci. Total Environ.*, **308**, 97–109.

Hou, X. L., P. P. Povinec, L. Y. Zhang, *et al.* (2013). Iodine-129 in seawater offshore Fukushima: distribution, inorganic speciation, sources, and budget. *Environ. Sci. Technol.*, **47**, 3091–8.

Jacob, P., G. Goulko, W. F. Heidenreich, *et al.* (1998). Thyroid cancer risk to children calculated. *Nature*, **392**, 31–2.

JST (2012). J-RAPID Strategic International Research Cooperative Program. Japan Science and Technology Agency. www.jst.go.jp/inter/english/sicp/country/j-rapid.html (accessed 19 September 2018).

Kakehi, S., H. Kaeriyama, D. Ambe, *et al.* (2016). Radioactive cesium dynamics derived from hydrographic observations in the Abukuma River Estuary, Japan. *J. Environ. Radioact.*, **153**, 1–9.

Katata, G., M. Ota, H. Terada, M. Chino and H. Nagai (2012). Atmospheric discharge and dispersion of radionuclides during the Fukushima Dai-ichi Nuclear Power Plant accident. Part I: source term estimation and local-scale atmospheric dispersion in early phase of the accident. *J. Environ. Radioact.*, **109**, 103–13.

Kato, H., Y. Onda and T. Tesfaye (2012a). Depth distribution of ^{137}Cs, ^{134}Cs, and ^{131}I in soil profile after Fukushima Daiichi Nuclear Power Plant accident. *J. Environ. Radioact.*, **111**, 59–64.

Kato, H., Y. Onda and T. Gomi (2012b). Interception of the Fukushima reactor accident-derived ^{137}Cs, ^{134}Cs and ^{131}I by coniferous forest canopies. *Geophys. Res. Lett.*, **39** (20), L20403, doi:10.1029/2012GL052928.

Kato, H., Y. Onda, K. Hisadome, N. Loffredo and A. Kawamori (2017). Temporal changes in radiocesium deposition in various forest stands following the Fukushima Dai-ichi Nuclear Power Plant accident. *J. Environ. Radioact.*, **166**, 449–57.

Kato, H., Y. Onda, T. Wakahara and A. Kawamori (2018). Spatial pattern of atmospherically deposited radiocesium on the forest floor in the early phase of the Fukushima Daiichi Nuclear Power Plant accident. *Sci. Total Environ.*, **615**, 187–96.

Kazakov, V. S., E. P. Demidchik and L. N. Astakhova (1992). Thyroid cancer after Chernobyl. *Nature,* **359**, 21–2.

Kinoshita, N., K. Sueki, K. Sasa, *et al.* (2011). Assessment of individual radionuclide distributions from the Fukushima nuclear accident covering central-east Japan. *Proc. Natl. Acad. Sci. USA*, **108**, 19526–9.

Kitamura, A., M. Yamaguchi, H. Kurikami, M. Yui and Y. Onishi (2014). Predicting sediment and cesium-137 discharge from catchments in eastern Fukushima. *Anthropocene*, **5**, 22–31.

Kobayashi, T., H. Nagai, M. Chino and H. Kawamura (2013). Source term estimation of atmospheric release due to the Fukushima Dai-ichi Nuclear Power Plant accident by atmospheric and oceanic dispersion simulations. *J. Nucl. Sci. Technol.*, **50**, 255–64.

Konoplev, A., V. Golosov, G. Laptev, *et al.* (2016). Behavior of accidentally released radiocesium in soil–water environment: looking at Fukushima from a Chernobyl perspective. *J. Environ. Radioact.*, **151**(3), 568–78.

Kurikami, H., H. Funaki, A. Malins, A. Kitamura and Y. Onishi (2016). Numerical study of sediment and ^{137}Cs discharge out of reservoirs during various scale rainfall events. *J. Environ. Radioact.*, **164**, 73–83.

Laceby, J. P., C. Chartin, O. Evrard, *et al.* (2016a). Rainfall erosivity in catchments contaminated with fallout from the Fukushima Daiichi nuclear power plant accident. *Hydrol. Earth Syst. Sci.*, **20**(6), 2467–82.

Laceby, J. P., S. Huon, Y. Onda, V. Vaury and O. Evrard (2016b). Do forests represent a long-term source of contaminated particulate matter in the Fukushima Prefecture? *J. Environ. Manage.*, **183**(3), 742–53.

Lasat, M. M., M. Fuhrmann, S. D. Ebbs, J. E. Cornish and L. V. Kochian (1998). Phytoremediation of a radiocesium-contaminated soil: evaluation of cesium-137 bioaccumulation in the shoots of three plant species. *J. Environ. Qual.*, **27**, 165–9.

Loffredo, N., Y. Onda, A. Kawamori and H. Kato (2014). Modeling of leachable ^{137}Cs in throughfall and stemflow for Japanese forest canopies after Fukushima Daiichi Nuclear Power Plant accident. *Sci. Total Environ*, **493**, 701–7.

Matsunaga, T., J. Koarashi, M. A. Atarashi, *et al.* (2013). Comparison of the vertical distributions of Fukushima nuclear accident radiocesium in soil before and after the first rainy season, with physicochemical and mineralogical interpretations. *Sci. Total Environ.*, **447**, 301–14.

Matsunaka, T., K. Sasa, K. Sueki, *et al.* (2015). Post-accident response of near-surface ^{129}I levels and ^{129}I/^{127}I ratios in areas close to the Fukushima Dai-ichi Nuclear Power Plant. *Japan. Nucl. Instrum. Methods Phys. Res. Sect. B*, **361**, 569–73.

Matsunaka, T., K. Sasa, K. Sueki, *et al.* (2016). Pre- and post-accident ^{129}I and ^{137}Cs levels, and ^{129}I/^{137}Cs ratios in soil near the Fukushima Dai-ichi Nuclear Power Plant. *Japan. J. Environ. Radioact.*, **151**, 209–17.

Matsuzaki, H., Y. Muramatsu, K. Kato, M. Yasumoto and C. Nakano (2007). Development of ^{129}I-AMS system at MALT and measurements of ^{129}I concentrations in several Japanese soils. *Nucl. Instrum. Methods Phys. Res. B*, **259**(1), 721–6.

MEXT (Ministry of Education, Culture, Sports, Science and Technology) (2011a). Extension site of distribution map of radiation dose, etc. http://ramap.jmc.or.jp/map/eng (accessed 19 September 2018).

MEXT (2011b). 4th Airborne monitoring, http://bit.ly/2VrjJtA (accessed 19 September 2018).

Mikami, S., T. Maeyama, Y. Hoshide, *et al.* (2015). Spatial distributions of radionuclides deposited onto ground soil around the Fukushima Dai-ichi Nuclear Power Plant and their temporal change until December 2012. *J. Environ. Radioact.*, **139**, 320–43.

Mironov V., V. Kudrjashova, F. Yioub and G. M. Raisbeck (2002). Use of ^{129}I and ^{137}Cs in soils for the estimation of ^{131}I deposition in Belarus as a result of the Chernobyl accident. *J. Environ. Radioact.*, **59**, 293–307.

Miyake, Y., H. Matsuzaki, T. Fujiwara, *et al.* (2012). Isotopic ratio of radioactive iodine (^{129}I/^{131}I) released from Fukushima Daiichi NPP accident. *Geochem. J.*, **46**, 327–33.

Miyake, Y., H. Matsuzaki, K. Sasa and T. Takahashi (2015). Measurement of long-lived radionuclides in surface soil around F1NPP accident site by accelerator mass spectrometry. *Nucl. Instr. Meth. B*, **361**, 627–31.

Muramatsu, Y., Y. Takada, H. Matsuzaki and S. Yoshida (2008). AMS analysis of ^{129}I in Japanese soil samples collected from background areas far from nuclear facilities. *Quat. Geochro.*, **3**, 291–7.

Muramatsu, Y., H. Matsuzaki, C. Toyama and T. Ohno (2015). Analysis of ^{129}I in the soils of Fukushima Prefecture: preliminary reconstruction of ^{131}I deposition related to the accident at Fukushima Daiichi Nuclear Power Plant (FDNPP). *J. Environ. Radioact.*, **139**, 59–64.

Nagao, S., M. Kanamori, S. Ochiai, *et al.* (2013). Export of ^{134}Cs and ^{137}Cs in the Fukushima river systems at heavy rains by Typhoon Roke in September 2011. *Biogeosciences*, **10**(10), 6215–23.

Naulier, M., F. Eyrolle-Boyer, P. Boyer, J. M. Metivier and Y. Onda (2017). Particulate organic matter in rivers of Fukushima: an unexpected carrier phase for radiocesiums. *Sci. Total Environ.*, **579**, 1560–71.

Nishihara, K., H. Iwamoto and K. Suyama (2012). Estimation of Fuel compositions in Fukushima-Daiichi Nuclear Power Plant. In *JAEA-Data/Code 012-018*. Ibaraki: Japan Atomic Energy Agency, pp. 1–190.

Onda, Y., H. Kato, M. Hoshi, *et al.* (2015). Soil sampling and analytical strategies for mapping fallout in nuclear emergencies. *J. Environ. Radioact.*, **139**, 300–7.

Qin, H., Y. Yokoyama, Q. Fan, *et al.* (2012). Investigation of cesium adsorption on soil and sediment samples from Fukushima Prefecture by sequential extraction and EXAFS technique. *Geochem. J.*, **46**, 355–60.

Saito, K. and Y. Onda (2015). Outline of the national mapping projects implemented after the Fukushima accident. *J. Environ. Radioact.*, **139**, 240–9.

Saito, K., I. Tanihata, M. Fujiwara, *et al.* (2015). Detailed deposition density maps constructed by large-scale soil sampling for gamma-ray emitting radioactive nuclides from the Fukushima Daiichi Nuclear Power Plant accident. *J. Environ. Radioact.*, **139**, 308–19.

Sakaguchi, A., K. Tanaka, H. Iwatani, *et al.* (2015). Size distribution studies of ^{137}Cs in river water in the Abukuma Riverine system following the Fukushima Dai-ichi Nuclear Power Plant accident. *J. Environ. Radioact.*, **139**, 379–89.

Sato, M., D. Takata, K. Tanoi, T. Ohtsuki and Y. Muramatsu (2015). Radiocesium transfer into the fruit of deciduous fruit trees contaminated during dormancy. *Soil Sci. Plant Nutr.*, **61**(1), 156–64, doi:10.1080/00380768.2014.975103.

Shimamoto, Y. S., T. Itai and Y. Takahashi (2010). Soil column experiments for iodate and iodide using K-edge XANES and HPLC-ICP-MS. *J. Geochem. Exploration*, **107**, 117–23.

Shimamoto, Y. S., Y. Takahashi and Y. Terada (2011). Formation of organic iodine supplied as iodide in a soil–water system in Chiba. *Japan. Environ. Sci. Technol.*, **45**, 2086–91.

Steinhauser, G. (2014). Fukushima's forgotten radionuclides: a review of the understudied radioactive emissions. *Environ. Sci. Technol.*, **48**(9), 4649–63.

Steinhauser, G., A. Brandl and T. E. Johnson (2014). Comparison of the Chernobyl and Fukushima nuclear accidents: a review of the environmental impacts. *Sci. Total Environ.*, **470–471**, 800–17.

Sugiura, Y., M. Shibata, Y. Ogata, *et al.* (2016). Evaluation of radiocesium concentrations in new leaves of wild plants two years after the Fukushima Dai-ichi Nuclear Power Plant accident. *J. Environ. Radioact.*, **160**, 8–24.

Syvitski, J. P. M., A. J. Kettner, S. D. Peckham and S.-J. Kao (2005). Predicting the flux of sediment to the coastal zone: application to the Lanyang Watershed, Northern Taiwan. *J. Coast. Res.*, **213**, 580–7.

Takahashi, J., K. Tamura, T. Suda, R. Matsumura and Y. Onda (2015). Vertical distribution and temporal changes of ^{137}Cs in soil profiles under various land uses after Fukushima Dai-ichi Nuclear Power Plant Accident. *J. Environ. Radioact.*, **139**, 351–61.

Takata, Y., K. Kohyama, H. Obara, *et al.* (2014). Spatial prediction of radioactive Cs concentration in agricultural soil in eastern Japan. *Soil Sci. Plant Nutr.*, **60**, 393–403.

Takenaka, C. and Y. Kiyono (2012). Redistribution of radionuclides through pollen dispersion. *Shinringijutu*, **840**,189–23 (in Japanese).

Tanaka, K., Y. Takahashi, A. Sakaguchi, *et al.* (2012). Vertical profiles of iodine-131 and cesium-137 in soils in Fukushima prefecture related to the Fukushima Daiichi Nuclear Power Station accident. *Geochem. J.*, **46**, 73–6.

Tanaka, K., A. Sakaguchi, Y. Kanai, H., *et al.* (2013). Heterogeneous distribution of radiocesium in aerosols, soil and particulate matters emitted by the Fukushima Daiichi Nuclear Power Plant accident: retention of micro-scale heterogeneity during the migration of radiocesium from the air into ground and river systems. *J. Radioanal. Nucl. Chem.*, **295**, 1927–37.

Terada, H., G. Katata, M. Chino and H. Nagai (2012). Atmospheric discharge and dispersion of radionuclides during the Fukushima Dai-ichi Nuclear Power Plant accident. Part II: verification of the source term and analysis of regional-scale atmospheric dispersion. *J. Environ. Radioact.*, **112**, 141–54.

Ueda, S., Hasegawa, H., Kakiuchi, H., *et al.* (2013). Fluvial discharges of radiocaesium from watersheds contaminated by the Fukushima Dai-ichi Nuclear Power Plant accident, Japan. *J. Environ. Radioact.*, **118**, 96–104.

Vandenhove H., M. VanHees, S. DeBrouwer and C. M. Vandecasteele (1996). Transfer of radiocaesium from podzol to ryegrass as affected by AFCF concentration. *Sci. Total Environ.*, **187**, 237–45.

Wakahara, T., Y. Onda, H. Kato, A. Sakaguchi and K. Yoshimura (2014). Radiocesium discharge from paddy fields with different initial scrapings for decontamination after the Fukushima Dai-ichi Nuclear Power Plant accident. *Environ. Sci. Process. Impact*, **16** (11), 2580–91.

Wang W., Y. Hanai, C. Takenaka, *et al.* (2016). Cesium adsorption and absorption through bark of Japanese cedar (*Cryptomeria japonica*). *J. Forest Res.*, **21**, 251–8.

Yamashiki, Y., Y. Onda, H. Smith, *et al.* (2014). Initial flux of sediment-associated radiocaesium to the ocean from the largest river impacted by Fukushima Daiichi Nuclear Power Plant. *Sci. Rep.*, **4**, 3714, doi:10.1038/srep03714.

Yoshida, S., Y. Muramatsu and M. Ogawa (1994). Radiocesium concentrations in mushrooms collected in Japan. *J. Environ. Radioactiv.*, **22**,141–54.

Yoshimura, K., Y. Onda and H. Kato (2015). Evaluation of radiocaesium wash-off by soil erosion from various land uses using USLE plots. *J. Environ. Radioact.*, **139**, 362–9.

Zhang, L. Y. and X. L. Hou (2013). Speciation analysis of ^{129}I and its applications in environmental research. *Radiochim. Acta*, **101**, 525–40.

Zheng, J., K. Tagami, W. Bu, *et al.* (2014). ^{135}Cs/^{137}Cs isotopic ratio as a new tracer of radiocesium released from the Fukushima nuclear accident. *Environ. Sci. Technol.*, **48**(10), 5433–8.

Part II

Development and Future Issues for the Infrastructure of Disaster Prevention

Preliminary remarks

TOKUSHI SHIBATA

The information needed for countermeasures in the case of disasters like the Fukushima Daiichi Nuclear Power Station (FDNPS) accident, are related to the nuclear power plant and environmental contamination:

1. Information related to the nuclear power plant, including information on:
   - the impact of an earthquake around the nuclear power plant;
   - the impact of a tsunami around the nuclear power plant;
   - the operating conditions of the nuclear power plant;
   - progress towards a cold shutdown of the nuclear power plant;
   - the radiation levels inside the nuclear power plant and around the plant; and
   - the release of radioactive material.
2. Information related to environmental contamination, including information on:
   - the nationwide impact of the earthquake;
   - the nationwide impact of the tsunami;
   - the release of radioactive material;
   - nationwide radiation levels;
   - nationwide contamination due to radioactive material; and
   - the principle of radiation protection in the case of an emergency.

The countermeasures taken by the Nuclear Emergency Response Headquarters and other governmental organisations were as follows:

11 March 2011

19:03 The declaration of a nuclear emergency was made.

21:23 The Nuclear Emergency Response Headquarters declared an area within a 3 km radius of the no. 1 nuclear power station as the evacuation area and within a 3–10 km radius as a zone in which people must take shelter indoors.

12 March 2011

03:06 Vent handling for the no. 1 nuclear power station was announced.

05:44 The evacuation order was issued for the area within 10 km of the FDNPS.

15:36 The housing of the no. 1 nuclear power station exploded.

18:25 A 'stay-inside' order was issued for the area within 20 km.

14 March 2011

11:01 The housing of the no. 3 nuclear power station exploded.

15 March 2011

The area within 20–30 km of the FDNPS was declared a shelter-in-place zone.

21 April 2011

The area within 20 km of the FDNPS was declared as a restricted area.

22 April 2011

An evacuation order for the area within 20–30 km of the FDNPS was lifted and was set as the Evacuation-Prepared Area and Deliberate Evacuation Area.

In the case of a nuclear power plant accident, if we assume the release of a massive amount of radioactive material, the most important measure is to protect people from radiation exposure. The measures adopted may depend on the amount of radioactive material released.

When the impact of the accident covers a broad area, as in the case of the FDNPS accident, the measures include not only evacuation and taking shelter indoors but also actions to prevent internal exposure. A monitoring system to predict the behaviour of the released radioactive material and atmospheric dispersion model results is required to enact various measures, such as evacuation and taking shelter indoors. The decontamination work is also important for reducing radiation exposure.

In the FDNPS accident, some important information was not actively used, such as results obtained from the airborne survey carried out by the US Department of Energy (DOE) and preventive stable iodine administration. Details of these issues are given below.

The DOE conducted airborne monitoring over 17–19 March following the FDNPS accident, covering an area up to 45 km from the FDNPS using two US jet fighters with air survey systems that can measure the distribution of measured air doses and display the information on maps. From the results, the region including Namie and Iitate Villages, which stretched northwest to approximately 30 km from the FDNPS, showed radiation levels of more than 125 μSv/h. These results were sent twice to the Ministry of Foreign Affairs of Japan through the US Embassy by email on 18 March and 20 March. The Ministry of Foreign Affairs of Japan, immediately after receiving the emails, transferred them to the Nuclear and Industrial Safety Agency (NISA) and the Ministry of Education, Culture, Sports, Science and Technology (MEXT), which are in charge of measuring radiation dose amounts. Neither organisation released these data nor transferred them to the Office of the Prime Minister or the Nuclear Safety Commission. The DOE announced the measured dose distribution in the USA on 23 March 2011. If these data were used to set the Evacuation-Prepared Area, for example, it would have been possible to set it one month earlier.

Because the measured data obtained by airborne monitoring were actual measured values, they showed the dose rate map for the surveyed area with some ambiguities, unlike simulated values using model calculations. Thus, the fact that the data were not used to protect people was a serious mistake. We should also consider the case of simulated data obtained by model calculations. The results might show the wrong radiation dose map based on inappropriate assumptions; in such a situation, the decision-making group without experts in radiation protection may be reluctant to adopt the results for the evacuation plan. If the details of the measurements, such as the method, measuring instrument, measuring system and calibration method, are not given, the decision-making group may also be reluctant to adopt the results for the evacuation plan, placing a heavy burden on the evacuees. Therefore, decision-making groups without experts in radiation protection tend to be late in actually making decisions. The data provided by the DOE might have been properly used for making a plan for evacuation if experts in radiation protection were included in the decision-making group.

For preventive stable iodine administration in the case of nuclear disasters, the Nuclear Safety Commission has set procedures. According to the regional disaster prevention plans of Fukushima Prefecture, the Fukushima Prefecture headquarters for disaster control should order the distribution and administration of stable iodine preparation based on the order by the Nuclear Emergency Response Headquarters or the judgement of the governor of Fukushima Prefecture.

The Nuclear Safety Commission was told that the Nuclear Safety Commission and the medical team of the Nuclear Emergency Response Headquarters had discussed the screening level for the administration of stable iodine preparation and confirmed it for a screening level exceeding 10 000 cpm. The Nuclear Safety Commission sent a message with additional comments on the screening level for the administration of stable iodine by fax to the medical team of the Nuclear

Emergency Response Headquarters on 13 March. However, this message was not delivered to the Headquarters Response Office; the order without this message was sent to Fukushima Prefecture, and the municipalities became concerned.

Fukushima Prefecture was waiting on the order for the administration of stable iodine from the Nuclear Emergency Response Headquarters. Despite the fact that the prefecture governor can order the administration without instructions from the Nuclear Emergency Response Headquarters, Fukushima Prefecture did not consider the distribution and the administration of stable iodine. The reactions of the local governments who stocked stable iodine were different. Futaba-cho, Tomioka-cho and Ohkuma-cho distributed stable iodine and ordered that it be administered to people in the towns. The local governments that distributed the stable iodine to people but did not handle the administration included Iwaki Town and Naraha Town. Namie Town arranged for stable iodine to be administered at evacuation centres but did not distribute it to people directly because there were no instructions from the Nuclear Emergency Response Headquarters.

Based on these two examples, because there were no experts at the Nuclear Emergency Headquarters, on the task forces of the different ministries and agencies and on the task forces of the local governments, important information was not utilised.

For preparation and disaster prevention, it is vital to secure experts at the headquarters of the government and the local governments, in addition to preparing monitoring and simulation systems.

7

Monitoring System

7.1 Introduction

HIROMI YAMAZAWA

The environmental consequences of the atmospheric release of radioactive materials from the accident at the Fukushima Daiichi Nuclear Power Station (FDNPS) were not sufficiently determined in the early stages of the accident, causing serious problems related to off-site countermeasures. One of the key questions faced not only by inhabitants in the affected areas but also by the public, including experts of relevant fields, is whether the confusion and problems in the emergency responses could have been avoided if the spatial extent and temporal evolution of the radioactive plume had been captured by monitoring. In this chapter, we will review the situation of the emergency preparedness related to the monitoring infrastructure at the time of the accident by examining whether the monitoring infrastructure was capable of coping with a large-scale nuclear disaster to determine an appropriate state of preparedness.

To make the following discussion specific and practical, it is essential to understand the exposure pathways between inhabitants and environmental radioactive materials. The pathways to be avoided in a nuclear accident are as follows (in order of their speed):

1. external dose from the plume (cloud shine);
2. internal dose due to the inhalation of radioactive materials in the plume;
3. external dose from radioactive material deposited on the ground surface (ground shine); and
4. internal dose from ingested radioactive material in food and drinking water.

Because the first two pathways are very quick, it is crucial that the necessary information is available through a routinely operated monitoring system.

The latter two pathways are relatively slow, which allows time to formulate and implement countermeasures. Actually, in the present case, car-borne and airborne dose rate monitoring activities effectively functioned to depict dose rate maps. Examinations of agricultural products were also conducted to regulate contaminated foods out of the market. According to the official dose evaluation jointly performed by the national and Fukushima prefectural governments, as of the evaluation in May 2012, approximately 57% of the examinees were found to have received an external dose of less than 1 mSv and 94% less than 5 mSv, with a maximum dose of 25.1 mSv. Most of the external doses were from the ground shine pathway. The internal doses were not sufficiently evaluated primarily because the atmospheric radioactivity concentrations were not measured at spatio-temporal resolutions enabling the internal dose evaluation of inhabitants.

7.2 Radiation Monitoring Facilities

HIROMI YAMAZAWA

It was expected that the dose rate monitoring system would provide radiological information that was indispensable for formulation of emergency countermeasures. This section describes how the monitoring system was prepared before the accident and how it functioned during the accident. Improvements in the monitoring system since the accident will also be briefly reviewed.

7.2.1 Function of Monitoring Facilities

Stack monitors, which typically consist of a combination of such different radiation detectors as a scintillation detector and an ionisation chamber, are installed on the stacks of a nuclear power plant (NPP) to identify and quantify anthropogenic radioactivity in gaseous effluent released into the atmosphere. The stack monitors at the FDPNS site did not function from the very early stage of the accident due to the station blackout. Thus, no information on the release through the stacks was available from the stack monitors. In addition, because there were hydrogen explosions that seriously damaged the reactor buildings and caused substantial atmospheric releases by routes other than through the stacks, it would have been difficult to assess the atmospheric release from the plant even if the stack monitors had been functioning.

Monitoring stations are deployed around an NPP to identify any atmospheric release of anthropogenic radioactive materials at a detectable level (Figure 7.1). They are alternatively referred to as monitoring posts, with the choice of terms depending on the respective function. Their main function is to continuously measure the dose rate, which is transmitted through a telemeter system on a

Figure 7.1 Photo of a monitoring station.

real-time and online basis around the clock. In addition to radiation monitoring, many of the monitoring stations also observe wind direction and speed, precipitation, temperature and humidity.

Figure 7.1 shows a typical monitoring station with two different types of detectors mounted on the roof of the monitoring house. The thinner cylindrical detector with a round tip on the left side is an NaI (Tl) scintillation detector, and the thicker one on the right is an ionisation chamber. The former can provide spectral information on γ-rays, which can be used to estimate the composition of radionuclides. Although it can measure low dose rates at a high accuracy, it is inappropriate when the dose rate exceeds approximately 10 μSv/h. The latter detector is intended to be used under high dose rate conditions. The regular deployment of this type of detector means that, at the time of the FDNPS accident, preparation had been made for the measurement of the high dose rates caused by a large-scale nuclear accident. The right half of the monitoring house holds a diesel generator to supply electricity to the monitoring system when the supply through the grid fails. It is common for monitoring stations to have two types of detectors and a backup power supply. Therefore, with the exception of a relatively sparse spatial distribution, the preparedness of the monitoring system was regarded to be sufficient before the accident.

However, in reality, many of the monitoring stations suffered devastating damage from the earthquake and the tsunami. In addition to these direct effects, the loss of general infrastructure, such as electricity and communication lines, resulted in a fatal breakdown of the monitoring function as a centralised, online, real-time system, although some of the stations measured and locally stored dose rate data for a few days with the assistance of their backup generators (Investigation Committee on the Accident at the Fukushima Nuclear Power Stations, 2011).

7.2.2 Deployment and Operation of Monitoring Facilities

Routine environmental monitoring around an NPP is performed by the power company that owns the NPP and the prefectural government hosting the NPP in

Table 7.1 *Numbers of monitoring stations in prefectures in which NPPs are located, as of the date shown.*

Prefecture	July 2012	December 2015
Hokkaido	9	31
Aomori	15	26
Miyagi	7	40
Fukushima (Hama-dori)	23	1219
Ibaraki	25	101
Niigata	11	29
Ishikawa [Toyama]	10	30 [7]
Fukui [Kyoto, Shiga, Gifu]	24	29 [47]
Shizuoka	15	34
Shimane [Tottori]	11	29 [9]
Ehime [Yamaguchi]	8	25 [6]
Saga [Nagasaki, Fukuoka]	7	12 [22]
Kagoshima	7	13

Names and numbers of monitoring stations in square brackets are for neighbouring prefectures.

its territory. The monitoring by the power company is usually conducted at 5–10 points on the site boundary, approximately 0.5–1 km from the reactors and stacks. These monitoring stations are as important as the stack monitors for detecting an accidental release of radioactive materials and for estimating the composition and amount of radionuclides.

At the time of the FDNPS accident, monitoring by prefectural governments was performed at approximately 10 points, mainly within a 10 km arc of each site. The monitoring stations as of August 2011 are summarised in Table 7.1. The difference in the numbers among prefectures reflects the differences in the number of NPPs, topography and population distribution. The monitoring results were publicised on a real-time basis on webpages by each organisation and by governmental organisations, such as the Nuclear Safety Technology Center, which, under the auspice of the Ministry of Education, Culture, Sports, Science and Technology (MEXT), operated the Disaster Prevention and Nuclear Safety Network for Nuclear Environment to provide real-time dose rate information for all NPPs in Japan. Therefore, radiological information could be obtained by inhabitants provided they had access to the internet and that the monitoring system worked properly.

Within a few years following the accident, the monitoring capability was significantly enhanced by prefectural governments with the guidance of and support from the Nuclear Regulation Authority (NRA) of the national government. The monitoring stations for which dose rate data are available on a real-time basis on the NRA's website (Monitoring Information of Environmental Radioactivity

Level; http://radioactivity.nsr.go.jp/map/ja) are listed in Table 7.1. The drastic increase in the number of stations reflects both the increase in the density and the extension of areas of monitoring station deployment, the latter being from a 10 km circle to a 30 km circle from each NPP site. Although not listed in Table 7.1, several additional monitoring stations have also been deployed in each prefecture as part of a nationwide monitoring network. However, in contrast to this drastic change in the dose rate monitoring capability after the accident, the capability for air concentration measurements has not been substantially improved.

At the time of the accident, MEXT (now the NRA) had for many years been conducting a nationwide environmental radiation and radioactivity monitoring project, in which dose rate data were measured at one point per prefecture in addition to concentration measurements for environmental media such as soil, plants and environmental water. Although the original purpose of this project was to trace the consequences of radionuclide release from past atmospheric bomb tests, such observation data as the dose rates and deposition obtained by this project served to effectively capture the spatial extent of the impact from the accident. Moreover, this network provided the public with useful information, that the impact of the accident on the dose rate and the activity concentrations in environmental media at many locations were well below detectable levels or, even when detected, at insignificant levels.

7.2.3 Usefulness of Monitoring Data

The usefulness of the monitoring data in an emergency situation must be discussed from several perspectives. One aspect is the monitoring system's capability to capture the spatial extent of the radiological dose distribution. At the time of the accident, the number of monitoring stations was obviously insufficient in the sense that an accidental radioactive plume might have been detected at only one point or even missed due to the low density of the network. Exceptions would be the areas with multiple nuclear facilities (e.g. Ibaraki Prefecture), where the spatial extent and density of the monitoring sites were considered to be fair. However, in an analysis of the dose rate data observed in the criticality accident at Tokai-mura in 1999, it was found to be very difficult to depict contour maps of the dose rate distribution because the dose rate fluctuated due to variations in wind direction, even with a relatively large number of monitoring points (Hirao and Yamazawa, 2010). The aforementioned reinforcement of the dose rate monitoring network is reasonable and effective in improving the usefulness of the monitoring data.

Another point of discussion is the communication network. At the time of the FDNPS accident, MEXT was in charge of supervising off-site emergency monitoring and integrating the results, while METI was responsible for on-site information, including dose rate data and stack monitor data. Although the SPEEDI network system collected and displayed the dose rate monitoring data, it was mainly used for off-site data from prefectural governments, and the data from other organisations and the on-site data were displayed only if a special arrangement had been made in advance. For these reasons, the monitoring infrastructure, including the data communication scheme, was insufficient to integrate and analyse all of the available monitoring data to provide them to the relevant governmental bodies in charge of off-site countermeasures. From the perspective of the need for more comprehensive environmental monitoring data in a nuclear emergency, the necessity of and readiness to monitor data other than the dose rate should also be thoroughly reviewed in the restructuring of the nuclear regulatory system. This topic will be revisited in the following section.

7.3 Information Necessary for Off-Site Countermeasures

HIROMI YAMAZAWA

Up to this point in the chapter, we have mainly discussed dose rate monitoring. However, it is also necessary to monitor information that can enable the evaluation of internal doses in the public. Especially in the early stages of an accident, it is necessary to monitor the pathways of internal doses due to the inhalation of air in a radioactive plume and the intake of contaminated food. Thus, measurements of activity concentrations in environmental media such as air, water and agricultural products are essential. They cannot reasonably be inferred from dose rate data.

The most common equipment that has been used for dose rate monitoring is scintillation detectors, such as an NaI (Tl) detector. This type of detector has also been used in the monitoring network, which was fortified within a few years following the accident. The advantage of using this type is its ability to obtain pulse height distribution data, which correspond to γ-ray energy spectra. The dose rate is a secondary quantity calculated from the pulse height distribution data. It is becoming common to store the pulse height data and to analyse contributions from key natural and anthropogenic nuclides to the dose rate. Therefore, information on the ^{131}I concentration in the air, which is the largest contributor to internal doses, can be obtained. However, the relation between the concentrations of radionuclides in the air and the pulse height distribution is not simple because of factors affecting the γ-ray transfer, such as the geometrical distribution of obstacles with shielding effects. Technical considerations must be given to separating the contribution of the airborne radionuclides (cloud shine) to the pulse height distribution from that of the deposited radionuclides (ground shine). Recently, methods for estimating the concentrations of radionuclides in the air and on the ground surface have become available and have been applied to estimate the concentrations in the very early stages of the accident (Hirayama *et al.*, 2014; Terasaka *et al.*, 2016).

Another aspect is the direct measurement of concentrations in environmental media. Although analytical methods have been established and standardised in guidelines and manuals, it is a serious problem that it takes many hours to carry out the procedure, which includes sample collection, delivery to laboratories, sample preparation and analysis. The necessity for experts and specialised equipment, such as dust samplers and HPGe detectors, is the limiting factor for ongoing reinforcement of the monitoring capability. From this perspective, it should be noted that MEXT's nationwide environmental radiation and radioactivity monitoring project, which had been operational for several decades at the time of the accident, had fostered the prefectural governments' capability to analyse radioactivity in environmental samples and that the capability largely contributed to capturing the severity and spatial extent of the impacts of the accident. Moreover, this capability cannot be obtained in a short period of time because experts require proper education and training. This capability of the prefectural governments should be enhanced and organised in such a way that a quick and systematic response can be made to a nuclear emergency.

Based on the lessons learned in the accident that there was little information on atmospheric concentration of radionuclides, the NRA has revised their guideline for the emergency monitoring to include online and real-time monitoring of atmospheric concentration of radioiodine and other nuclides at several points within the 30 km arc. This revision will substantially enhance the effectiveness of the monitoring system as a key part of emergency preparedness.

7.4 Other Infrastructure

HIROMI YAMAZAWA

There are several points to be considered in the discussion on the infrastructure of off-site nuclear disaster control. These include the platform of radiation and radioactivity monitoring, the system for data collection and communication and the facility and organisation for data integration and analysis. Expertise of research institutions and universities can also be regarded as a kind of infrastructure of emergency preparedness.

7.4.1 Related Infrastructure in Japan

Based on the abovementioned infrastructure, existing and/or anticipated impacts of accidents should be comprehensively evaluated to formulate a list of effective and feasible options for countermeasures, which is to be presented to relevant decision-makers for selection and implementation. Other infrastructures that support the effective and efficient implementation of decided countermeasures are also essential, which include an emergency public communication system and a transportation system for evacuation. In the rest of this section, the monitoring platforms, as well as facilities and organisations for monitoring data analysis, will be discussed.

Depicting spatial distributions of accidental impacts in terms of dose rates or concentrations is of crucial importance. From this aspect, manual dose rate monitoring at fixed points is ineffective. For example, during the first few weeks following the FDNPS accident, car-borne monitoring was performed at fixed points every day to obtain data on the γ dose rate, which was caused by ground shine from radionuclides deposited on the ground surface. Thus, the dose rates measured only showed smooth temporal variations due mainly to the radioactive decay of nuclides at points too sparsely distributed in the area of interest to draw dose rate contour maps. In contrast, the greater effectiveness of airborne monitoring was clearly demonstrated by the results obtained by the airborne monitoring of the US DOE during the very early stages of the accident, the monitoring jointly conducted by MEXT and the US DOE for the 80 km range during the first several weeks and the extensive efforts put forth by MEXT over the following months. It was not and will not be difficult from either a technical or a financial perspective for Japan to develop and deploy airborne monitoring systems that can be readily used in an emergency situation.

A good example of car-borne monitoring systems is the GPS-aided radiation monitoring system, KURAMA (Kyoto University RAdiation MApping system), developed by the Kyoto University Research Reactor Institute. Multiple sets of this system were extensively used to measure special distributions of the dose rate because of its easy-to-use and compact configuration, which enables operation in an ordinary car. Monitoring systems with similar configurations were also developed and used in environmental monitoring in the area affected by the accident from the early stage. One such system was developed by the group from the High Energy Accelerator Research Organization (KEK). The monitoring cars deployed by national and prefectural governments were usually of the multipurpose type, having multiple detectors and sampling devices. However, they were expensive. Although they served to provide monitoring data to a certain extent, the monitoring data provided by simpler systems such as KURAMA were much more useful in understanding the spatial distribution of contamination. The purposes and the configuration of the present mobile monitoring system for emergency response should be reviewed in this regard.

The present accident resulted in huge amounts of radioactivity being released into the ocean directly from the site and indirectly via deposition from the atmosphere. Preparation for real-time ocean monitoring had not been made at the time of the accident except for the monitoring of effluent. Although the amount of radioactivity released during the periods with offshore winds has been estimated, the uncertainty in the estimation is large due to the lack of monitoring data obtained in the vicinity of the site from which deposition into the ocean can be directly estimated. It is necessary to increase the preparedness of the infrastructure for maritime monitoring.

The local headquarters of disaster control stationed at the off-site centre was supposed to be in charge of integrating and analysing the monitoring data to formulate off-site countermeasures. However, this scheme did not function for several reasons. Due to the direct impacts of the earthquake on the transportation system, only a limited number of the persons dispatched could get to the centre. The electricity failure and the breakdown of logistics, in addition to the elevated level of radiation around the centre, were also causes for the failure of the emergency scheme. The hardware and logistics of the centre must be robust enough to keep functioning under such conditions. Expertise in radiology and environmental radioactivity is also essential for the centre. Therefore, the centre must have the capability to collect and integrate various types of monitored and predicted data that support the activities of experts. The centres to be constructed must not only be robust under disaster situations but also capable of enhancing the effectiveness and efficiency of the activities of experts in the centres.

7.4.2 Expertise and Research Institutions as Infrastructure

In the early stage of the accident, the national and local governments, relevant specialised agencies and institutions, societies of related scientific fields and researchers could not properly cope with the accident, leaving inhabitants in the affected area and both national and international communities poorly informed. However, it was also true that many activities were performed to evaluate and mitigate the damage from the accident. Many of these valuable activities were performed by institutions and researchers with expertise and resources that were not a part of nuclear emergency preparedness.

Research activities in the nuclear environmental safety field, such as research on radionuclide transfer in the environment, had been declining due to budget and manpower issues for a few decades, although there was a temporary increase after the Chernobyl accident in 1986. The criticality accident in Tokai-mura in 1999 resulted in the enhancement of preparedness and research in the radiation emergency medicine field. The lesson from the criticality accident did not lead to the enhancement of infrastructure that could effectively cope with a full-scale nuclear emergency. It is obvious that related basic research activities in universities and research institutions must be maintained to foster their expertise as true infrastructure for environmental safety of nuclear energy and that such organisations with expertise should contribute their best efforts. It was regrettable that environmental safety research had long been declining.

7.5 Monitoring of Rivers

YUICHI ONDA

With the clarification of the inventories and behaviour of radionuclides on the ground surface, predicting the transfer and accumulation of radionuclides related to future exposure is possible. If we can determine the radiocaesium inventory in river waters, riverbeds and suspended sediments collected in upper catchments and the mean radiocaesium concentration at certain sampling points, we can estimate radiocaesium concentrations in the riverbed at different locations with various particle size characteristics after correcting for the particle size distribution.

The results of the survey and analyses shown in Chapter 6 require that the following tasks be performed for the long-term prediction of radionuclide fluxes. At the moment, radionuclide concentrations of riverbeds are the only data available to the public from the monitoring survey conducted by the Ministry of the Environment. To calculate the flux of radionuclides flowing down a river, continuous measurements of turbidity at each monitoring location and consecutive monitoring of radioactive concentrations in suspended sediments using samplers (e.g. suspended sediment collectors) are necessary.

Radiocaesium attached to organic matter in suspended sediments is likely to migrate to algae and fish. Therefore, a survey focusing on the organic matter content in riverbed sediments and suspended sediments is essential in the future to clarify the mechanism of radiocaesium transfer. In addition to the periodic monitoring of suspended sediments and riverbed sediments being conducted by individual ministries, an overall monitoring linking the two surveys is required for a comprehensive calculation of the radionuclide fluxes. Thus, continuous and thorough monitoring must be conducted that involves the Ministry of Land, Infrastructure, Transport and Tourism, which is in charge of the management of rivers. These tasks call for a new framework led by specialists replacing the current liaison meetings on monitoring among ministries to study the transfer of radionuclides in the environment more precisely and concretely.

References

Hirao, S. and H. Yamazawa (2010). Release rate estimation of radioactive noble gases in the criticality accident at Tokai-mura from off-site monitoring data. *J. Nucl. Sci. Technol.*, **47**(1), 20–30.

Hirayama, H., M. Kawasaki, H. Matsumura, *et al.* (2014). Estimation of I-131 concentration using time history of pulse height distribution at monitoring post and detector response for radionuclide in plume. *Trans. Atom. Energy Soc. Japan*, **13**(3), 119–26.

Investigation Committee on the Accident at the Fukushima Nuclear Power Stations, 2011. Interim report, chapter 5. www.cas.go.jp/jp/seisaku/icanps/eng/interim-report.html (accessed 19 September 2018).

Terasaka, Y., H. Yamazawa, J. Hirouchi, *et al.* (2016). Air concentration estimation of radionuclides discharged from Fukushima Daiichi Nuclear Power Station using NaI (Tl) detector pulse height distribution measured in Ibaraki Prefecture. *J. Nucl. Sci. Technol.*, **53**(12),1919–32, doi:10.1080/00223131.2016.1193453.

8

Dispersion Modelling of Radioactive Materials

8.1 Overview of SPEEDI

HARUYASU NAGAI AND HIROMI YAMAZAWA

SPEEDI, the System for Prediction of Environmental Emergency Dose Information, is an emergency response system to predict the atmospheric dispersion of radioactive materials and radiological doses in the case of an atmospheric release of substantial radioactive materials from nuclear facilities in Japan. It has been operated by the Nuclear Safety Technology Center on consignment from the Ministry of Education, Culture, Sports, Science and Technology (MEXT) and local governments (Suda, 2006). After the accident at the Fukushima Daiichi Nuclear Power Station (FDNPS) of the Tokyo Electric Power Company (TEPCO) due to the Great East Japan Earthquake on 11 March 2011, which caused a substantial discharge of radioactive materials into the atmospheric and oceanic environments, SPEEDI became recognised by not only the nuclear emergency community but also the public at home and abroad, and the issue of its utilisation was discussed by the government, the Diet and independent accident investigations (Independent Investigation Commission on the Fukushima Nuclear Accident, 2012; Investigation Committee on the Accident at Fukushima Nuclear Power Stations of Tokyo Electric Power Company, 2012; National Diet of Japan Fukushima Nuclear Accident Independent Investigation Commission, 2012).

In this chapter, an overview of SPEEDI is provided in the context of its development, functions and role in the framework of nuclear emergency management. Thereafter, we examine how it was used and how it should be used from a system development perspective. We believe that our review can provide lessons or tasks for improving the prediction system and for considering better utilisation of the system; it is also beneficial to consider reconstructing the framework of nuclear emergency management. Furthermore, we hope this review will prove useful in understanding and effectively using the atmospheric dispersion predictions from the system in the case of a similar accident in the future.

8.1.1 Background of Development

The development of SPEEDI was begun by the former Japan Atomic Energy Research Institute (presently the Japan Atomic Energy Agency (JAEA)) in 1979 after the Three Mile Island (TMI) nuclear power station accident, and the basic system was completed in 1984 (Imai *et al.*, 1985). Thereafter, SPEEDI was transferred to the Nuclear Safety Technology Center in 1986 and was consolidated as a network system. Since then, the Nuclear Safety Technology Center has been accumulating operational experience and improving the system. Moreover, computational models of SPEEDI have been continuously improved to increase the prediction accuracy and enhance functions by developers in the former Japan Atomic Energy Research Institute (Nagai *et al.*, 1999). Having incorporated these new developments, the advanced SPEEDI has been operating since 2005.

8.1.2 Functions of SPEEDI

SPEEDI has several key functions, including data acquisition and monitoring, meteorological prediction, atmospheric dispersion prediction and dose estimation, and visualisation and distribution of the predictions (Suda, 2006).

The first function, data acquisition and monitoring, executes processes to obtain weather forecast data and surface observation (AMeDAS) data from the Japan Meteorological Agency (JMA), to collect meteorological and radiation data from environmental monitoring posts located around the nuclear facilities in Japan, and to automatically monitor radiation data for the detection of abnormal conditions.

The second function, meteorological prediction, utilises a local-scale meteorological model to provide meteorological forecasts up to 44 hours into the future within a 100 × 100 km area around the nuclear facilities in Japan using weather forecast data from the JMA. Furthermore, detailed local-scale wind fields are calculated within a 25 × 25 km area, which includes the Emergency Planning Zone (EPZ) within a 10 km radius from a nuclear facility.

Third, the atmospheric dispersion prediction and dose estimation function calculates the atmospheric dispersion of radioactive materials to predict distributions of radioactive materials and radiological doses due to a nuclear accident using detailed local-scale wind fields around a nuclear facility and the source term, which includes the released radionuclide composition, release rate and release duration. Details of the meteorological prediction, atmospheric dispersion and dose calculations are described in the following section.

Finally, the visualisation and distribution of the predictions are conducted using maps with geographic information that are distributed to the emergency response centres of both the national and local governments.

8.1.3 Prediction Models

The prediction flow of SPEEDI is as follows. First, the meteorological fields within a small area around a target nuclear facility are calculated. Next, the atmospheric dispersion calculation is executed to predict the environmental distribution of radioactive materials released from the facility based on the local wind field. Finally, radiological doses are estimated from the environmental distribution of radioactive materials. Therefore, the prediction accuracy depends primarily on the meteorological prediction. The advanced SPEEDI currently operated by the Nuclear Safety Technology Center consists of a local-scale meteorological prediction model (PHYSIC), a mass conservation wind field model (WIND21) and a particle dispersion and dose calculation model (PRWDA21), as shown in Figure 8.1.

The local-scale meteorological prediction model calculates local meteorological conditions by numerically solving conservation equations of momentum, heat, turbulence, physical processes and surface boundary conditions for a 100 × 100 km area around a nuclear facility using numerical weather forecast data (GPV) provided by the JMA as the initial and boundary conditions. The calculation domain is divided into 30 vertical layers with 50 × 50 horizontal grids at a resolution of 2 km. In this calculation, meteorological observation data around the nuclear facility can be assimilated to improve the prediction accuracy.

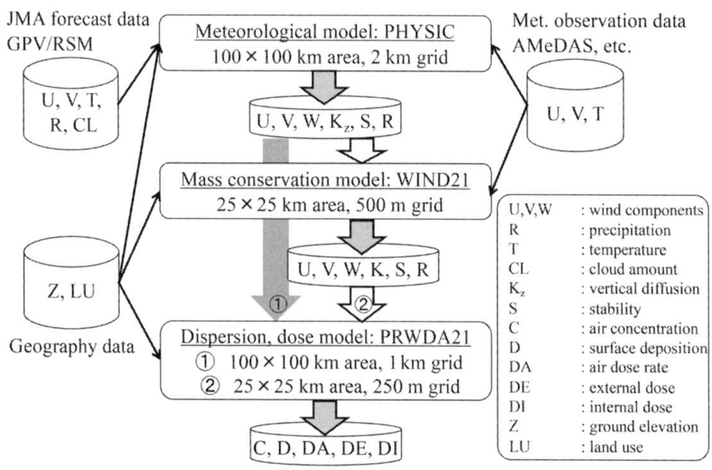

Figure 8.1 SPEEDI model structure and calculation flow.

The mass conservation wind field model calculates detailed local-scale wind field information. The calculation domain is set to cover the EPZ with a radius of up to 10 km, where disaster prevention countermeasures are intensified as specified in the guidelines for disaster prevention countermeasures titled 'About the Disaster Prevention Countermeasures for Nuclear Facilities' (hereafter 'Countermeasures Guidelines') by the former Nuclear Safety Commission (Nuclear Safety Commission, 1980). Therefore, meteorological prediction data for a 100 × 100 km area with 2 km resolution are interpolated to higher resolution for the wind field calculations over a 25 × 25 km area in the vicinity of a nuclear facility to consider the effects of local topography by variational analysis with a 500 m resolution grid containing 50 × 50 points.

The particle dispersion and dose calculation model calculates the atmospheric dispersion of radioactive materials discharged from the nuclear facility using the wind field data calculated by the local-scale meteorological prediction model or mass conservation wind field model. For the atmospheric dispersion of radioactive materials discharged from a nuclear facility, the emission source is generally regarded as a point, and a Lagrangian particle dispersion model is utilised to precisely evaluate the radiological impact of locally high concentrations of radioactive materials near the nuclear facility. Radioactive materials discharged into the atmosphere are represented by numerous marker particles, and the air concentration, surface deposition and radiological doses are calculated by tracing the trajectories of these particles. These values are calculated for 100 × 100 grids to resolve detailed local-scale distributions of radioactive material in the vicinity of the source point; a 25 × 25 km area at a resolution of 250 m is used for the output from the mass conservation wind field model, and a 100 × 100 km area at a resolution of 1 km is used for the output from the local-scale meteorological prediction model. In the radiation calculation, three-dimensional distributions of the radioactive plume are considered by integrating the radiation emitted from each marker particle. Radiation from deposited radioactive materials and internal doses due to inhalation are calculated from the surface deposition amount and air concentration, respectively, of radioactive materials based on the corresponding dose conversion factor.

8.2 Role of SPEEDI in the Emergency Response Framework

HARUYASU NAGAI AND HIROMI YAMAZAWA

At the time of the FDNPS accident, the role and utilisation of SPEEDI were described in the Countermeasures Guidelines (Nuclear Safety Commission, 1980) and 'The Guideline for Environmental Radiation Monitoring' (hereafter, 'Monitoring Guidelines') by the former Nuclear Safety Commission (Nuclear Safety Commission, 2008).

In the Countermeasures Guidelines (Nuclear Safety Commission, 1980), the consolidation of SPEEDI and a system to estimate conditions of nuclear facilities (ERSS, the Emergency Response Support System), networking of various systems and the construction of a cooperative framework are described. Additionally, it is stated that the personal dose expected without any protective action (projection dose) is estimated using SPEEDI and other information. Furthermore, 'Chapter 4 Emergency environmental radiation monitoring: (1) The first step monitoring' states that environmental monitoring data and information from SPEEDI are combined to estimate the projection dose, and any decision about protective action is made based on this information. These descriptions only indicate the basic directions for the utilisation of SPEEDI in the emergency response framework; a more specific method is described in the Monitoring Guidelines.

In the Monitoring Guidelines (Nuclear Safety Commission, 2008), the specific method for applying SPEEDI is described in 'Chapter 4 Emergency environmental monitoring'. The sections titled '4-3 Plan and operation' and '4-3-1 Construction of emergency response framework' specify that the radiation team at the Nuclear Emergency Response Headquarters, which is in charge of the collection and assessment of environmental monitoring data, is responsible for the operation of SPEEDI and conducts estimates of the radiological doses for the public based on the SPEEDI network system and other information. The specific method to use SPEEDI is described in the section titled '4-3-2 Method of implementation: (1) The first step of environmental monitoring', as follows. The first step of environmental monitoring should be launched immediately after a nuclear emergency. The environmental monitoring results, source term, meteorological data and information from the SPEEDI network system should be used to estimate the projection dose, and any decision about protective action can be made with this information. Furthermore, the utilisation of SPEEDI to select places for measurements of air dose rates and air concentrations of radioactive materials and environmental sampling is also described.

The method to estimate the projection dose is described in the section titled '4-4 Estimation of doses' in the Monitoring Guidelines as follows. To determine if protective action is necessary, the concentration of radioactive materials and the projection dose for the public around a nuclear facility are first predicted by simulations. Next, the environmental monitoring results are used to evaluate the actual concentration of radioactive materials and doses for the dominant radionuclide species and radiation released from the nuclear facility. Additionally, the estimated projection dose should be comprehensively implemented using the SPEEDI prediction map and measurements of actual air dose rates. Furthermore, the exposition section entitled 'K About SPEEDI network system' precisely describes the utility of SPEEDI in the context of accident progression and the correction of prediction results by considering measurements.

The above descriptions indicate that predictions by SPEEDI (or in combination with ERSS) are not independently used to determine whether protective action should be taken; instead, the system is utilised in cooperation with environmental monitoring data in a mutually complementary form. The selection of environmental monitoring points based on the atmospheric dispersion prediction, the interpolation of discretely distributed environmental monitoring data and the correction of prediction results using environmental monitoring data are examples of complementary cooperation for the comprehensive implementation of an emergency response. Therefore, SPEEDI alone (or in combination with ERSS) is not sufficient to obtain the necessary information, and the prediction results with some uncertainties must be utilised as much as possible by considering errors, uncertainties and limitations due to external factors, such as a lack of necessary input data.

8.3 Response to the Fukushima Daiichi Nuclear Power Station Accident

HARUYASU NAGAI AND HIROMI YAMAZAWA

The utilisation of SPEEDI for the FDNPS accident is precisely described in the report by MEXT titled 'Review of Response by MEXT Concerning the Recovery from the Great East Japan Earthquake' (MEXT, 2012a). It is summarised briefly below.

The Great East Japan Earthquake occurred at 14:46 on 11 March 2011, and one hour later TEPCO sent an Article 10 Notification (the facility's entire AC power supply had failed) of the Act on Special Measures Concerning Nuclear Emergency Preparedness. Next, MEXT ordered the Nuclear Safety Technology Center to operate SPEEDI in emergency response mode and carry out a calculation with a unit release condition around the clock according to the manual for nuclear emergency response (MEXT, 2008). The Nuclear Safety Technology Center started SPEEDI calculations on the hour every hour, and sent the first result to MEXT at 17:00, and they then started sending results to the Nuclear and Industrial Safety Agency, Nuclear Safety Commission, JAEA and the off-site Nuclear Emergency Response Headquarters after 17:40. The prediction results were faxed to the Nuclear Emergency Response Headquarters in Fukushima Prefecture after 10:05 on 12 March. Thereafter, around-the-clock SPEEDI operations and calculations were performed every hour for several years.

MEXT utilised the SPEEDI predictions to select monitoring points during the early phase of the accident, which enabled the timely dispatch of a monitoring team to the high air dose rate zone to monitor the air dose rate. The Nuclear Emergency Response Headquarters, Nuclear and Industrial Safety Agency and Nuclear Safety Commission ordered SPEEDI predictions with various source term assumptions. The Nuclear Safety Commission cooperated with the JAEA to estimate the source term via a reverse estimation method using environmental monitoring data and the SPEEDI predictions (described in '2.1 Atmospheric release of radioactive materials' of the MEXT report); they evaluated the integrated dose using the estimated source term in the SPEEDI calculations. These applications of SPEEDI were novel uses of the system that were not included in the emergency response manual. However, SPEEDI calculations using the source term from ERSS as described in the manual were not performed because of the ERSS outage.

SPEEDI was operated steadily according to the emergency response manual as described above. However, the SPEEDI predictions were judged to be unreliable because they did not simulate actual conditions, there was a lack of essential input and source term data for quantitative estimation, and it was not utilised in the protective action (MEXT, 2012a).

In the previous section describing the role of SPEEDI, we noted that SPEEDI predictions need to be utilised for the comprehensive implementation of an emergency response by considering errors, uncertainties and limitations of the system and in cooperation with environmental monitoring data and knowledge of nuclear experts. However, SPEEDI predictions were not utilised in this way but were excluded from the information for the evacuation planning because they were not based on the source term from the ERSS. It is indeed regrettable that SPEEDI had been always used in combination with the ERSS in comprehensive nuclear disaster preparedness drills and that this usage has long been considered the only way. We look forward to the realisation of an integrated utilisation of SPEEDI in combination with environmental monitoring.

8.4 How Should We Have Utilised SPEEDI?

HARUYASU NAGAI AND HIROMI YAMAZAWA

The previous section showed that SPEEDI was not utilised for protective actions, such as evacuation and sheltering, after the FDNPS accident. However, the prediction results assuming a unit release condition were utilised for the environment-monitoring plan (MEXT, 2012a). Additionally, the accident investigation by the government (Investigation Committee on the Accident at Fukushima Nuclear Power Stations of Tokyo Electric Power Company, 2012) expressed their viewpoint as follows. Prediction results were obtained through SPEEDI that assumed unit emissions; if that information had been distributed, the respective local governments and residents could have selected a more appropriate timing or direction of evacuation. Even if the emission source information could not have been obtained from the ERSS, it is believed that SPEEDI could have been better utilised. Possible protective actions based on the SPEEDI predictions, especially for the case on 15 March, were further examined.

After a loud explosion was heard around 6:00 on 15 March, a high radiation dose rate of 11 930 μSv/h was measured near the main gate of the FDNPS at 9:00. This monitoring information clearly indicated a substantial release of radioactive materials, and this condition should have been considered, especially for taking preventive actions. While the SPEEDI predictions showed that the flow direction of the radioactive plume during this period was towards the southwest, it rotated clockwise to the northwest at night. Thereafter, the radioactive plume was predicted to flow towards the Pacific Ocean after noon on 16 March, and this condition continued for several days. By combining monitoring information and the SPEEDI predictions, it was expected that the radioactive plume would pass through a wide area on 15 March (this information was used to successfully measure the plume using monitoring cars in Namie Town, Fukushima, from 20:40 to 20:50 on 15 March; MEXT, 2012a). This information was available for protective action consideration and to reduce radiation exposure (internal dose by inhalation and external dose by direct radiation from the radioactive plume). Because the radioactive plume continued to flow towards the Pacific Ocean after noon on 16 March, high air dose rates observed at many places in Fukushima Prefecture were identified not as direct radiation from the radioactive plume but as radiation from deposited radioactive materials (ground shine). Due to this information, protective action shifted from sheltering indoors to evacuation to reduce the long-term integration of external doses from ground shine since there was no risk of radiation exposure from the radioactive plume.

The utilisation of SPEEDI predictions described above may be difficult to implement without researchers regularly examining the results of atmospheric dispersion simulations. Moreover, it is important that these prediction results are obtained in a timely manner for protective action. Concerning the SPEEDI prediction results that were available to the public on the MEXT website (MEXT, 2012b), the SPEEDI calculations assuming a unit release condition were visualised only for times up to two hours into the future until 7:00 on 16 March. Thereafter, this interval was lengthened to three hours. These results were designed to indicate the condition of the plume at that time, although they were not sufficient to grasp changing trends. Additionally, the meteorological prediction and reproducibility are always dependent on the variability of the meteorological conditions. Although the changing meteorological conditions are reproduced well, a slight time lag in the prediction may cause a large difference in wind direction predictions. Therefore, a prediction result at a specific time, which may have large discrepancies in terms of the wind direction, gives an image that is not reliable for protective action. A method of displaying images to show long-term trends in changing meteorological conditions should be used to obtain useful information for protective action.

As described in Section 8.1.2, SPEEDI can predict atmospheric dispersion up to 44 hours into the future and can output prediction results at one-hour intervals. A long-term SPEEDI prediction was also ordered by the Nuclear Emergency Response Headquarters (Nuclear Regulation Authority, 2011). According to the prediction results sent early on 15 March, a distribution map of the accumulated air concentration was constructed from the 24-hour calculations after 1:00 on 15 March based on a continuous unit release condition. This showed that higher values were predicted around areas towards the south and northwest of the Fukushima Daiichi Nuclear Power Station. This result indicated that the flow direction of the radioactive plume was towards the south, after which it rotated clockwise to the northwest, causing a significant impact around this area. This result was consistent with the results assuming a unit release condition published hourly on the MEXT website (MEXT, 2012b). Therefore, by constructing distribution maps of the air concentration for not only the 24-hour accumulated value but also all prediction results at one-hour intervals, the trend in the flow direction of the radioactive plume on 15 March could have been determined. Although this type of prediction results in large datasets, it should be provided as useful information to make protective action decisions, regardless of the time needed to send this information.

8.5 Lessons and Tasks for SPEEDI from the Accident

HARUYASU NAGAI AND HIROMI YAMAZAWA

By reflecting on the problems in the emergency responses to the accident, MEXT arrived at SPEEDI improvements (MEXT, 2012a). The major issues were effective utilisation of the prediction results without a definitive source term, improvement of infrastructure for delivery of prediction results and sharing of information, and prompt and adequate release of prediction results. In particular, a review of the emergency response manual, improvement of the method to deliver and release prediction results, prompt collection of environmental monitoring data and development of functions to estimate the source term were considered. However, when the NRA developed new guidelines for nuclear emergency countermeasures titled 'The Nuclear Emergency Response Guidelines' (Nuclear Regulation Authority, 2012), they opted not to use SPEEDI for the judgement of protective measures in emergencies (Nuclear Regulation Authority, 2014). This occurred although the Meteorological Society of Japan had pointed out the necessity of atmospheric dispersion predictions and proposed their own effective utilisation method

(Meteorological Society of Japan, 2014). Furthermore, in response to a strong desire within local governments to use SPEEDI for emergency countermeasures, the Council of Ministers on Nuclear Energy decided that local governments may use atmospheric dispersion predictions for emergency countermeasures and receive support needed for implementation from the government (Council of Ministers on Nuclear Energy, 2016). Issues on the revitalisation of SPEEDI are still being considered. In addition to these considerations, we propose a revision of our understanding of the prediction system.

To make the best use of a prediction system like SPEEDI for protective actions, the director of such actions should have sufficient knowledge about the prediction system and adequate orders for operating the system and displaying the results. Although SPEEDI has been operated as directed and has provided prediction results continuously for years after the FDNPS accident, the critical issue is the failure to effectively utilise the results for protective actions. Even if the improvement of SPEEDI or the development of a more advanced prediction system is achieved in the future, these systems might not be utilised for protective actions, as was true in the Fukushima case, thus not making the best use of the system as mentioned above. Although this issue may be considered in the review of the emergency response manual, the director of protective actions must also change his or her perspective on the prediction system. That person must not be a user of the prediction system but a commander of the prediction system who gives orders to operate the system effectively and thus provide usable results. With this way of thinking, the prediction system can be operated with a sense of responsibility to best utilise it for protective actions under any and all circumstances.

In addition to the improvement of the prediction system, we must consider and promote the training of personnel who use the system and the construction of a framework to support the director of protective actions.

8.6 Recent Status of Atmospheric Dispersion Modelling

HARUYASU NAGAI AND HIROMI YAMAZAWA

SPEEDI is an example of a system to model the atmospheric dispersion of trace substances. In general, atmospheric transport models or atmospheric dispersion models are developed for various purposes and in different ways. Most are intended for large-scale atmospheric transport from Asia to conduct impact assessment of air quality or climate based on chemical substances such as photochemical oxidants or particulate substances like yellow sand and other aerosols. These models, including SPEEDI, have similarities and differences.

One common point is that these models calculate the air concentration and surface deposition of target substances by combining three-dimensional meteorological field calculations and dispersion calculations. In meteorological field calculations, general circulation models or regional meteorological models are used to predict the three-dimensional distributions of wind, turbulence, convection and precipitation. Most models have a function to correct the prediction results by minimising the difference between the calculated and observed values with four-dimensional data assimilation; moreover, they can reproduce past or present conditions with a high degree of accuracy.

The differences between the models are the degree of detail in the physical processes and spatiotemporal scales, which is attributed to the target substances and phenomenon. For example, MASINGAR (Meteorological Research Institute) and SPRINTARS (Kyushu University) are intended for use on a global scale. They use general circulation models for meteorological calculations and have detailed models for the generation and elimination processes of aerosols on a global scale, although their performance is limited for regional-scale atmospheric dispersion in Japan. On the other hand, regional-scale chemical transport models have been developed in combination with regional meteorological models; examples include WRF and CMAQ (National Institute for Environmental Studies), Chem (Japan Agency for Marine-Earth Science and Technology) and CAMx (Central Research Institute of Electric Power Industry). These models use detailed data for turbulence, cloud and rain from regional meteorological models and geographical data to precisely reproduce the wet and dry deposition processes.

WSPEEDI-II, which has been developed by advancing SPEEDI, uses the same framework as the above regional-scale models in combination with the regional meteorological model MM5 and the dispersion model GEARN. The difference from the above models is that WSPEEDI-II uses a rather simple scheme for deposition processes, although it can treat detailed distributions of air concentrations in the vicinity of a source point. The deposition process of WSPEEDI-II calculates wet and dry deposition using a simple scavenging coefficient and deposition velocity, respectively, which have specific values without considering the chemical forms of the radioactive materials, based on the spatial distribution of precipitation. On the other hand, to assess the impact of the accidental release of hazardous substances, WSPEEDI-II uses a Lagrangian particle dispersion model, which is suitable for examining the atmospheric dispersion phenomenon from a point source. There is a clear difference from the chemical transport models that use the Eulerian method to calculate the transport of already widespread substances. The Lagrangian particle dispersion model, which has no numerical diffusion, performs well in regard to the reproducibility of the spatial distributions of the air concentration in the vicinity of a source point. For example, surface air concentration can be realistically reproduced in the case of an elevated release point.

Although recent developments in regard to atmospheric dispersion models have made significant progress, there are many problems in reproducing the atmospheric dispersion processes for the FDNPS accident. One of the problems is the large uncertainty in the source term of the radioactive materials. However, even if we have a precise source term, models cannot reproduce the conditions. It is necessary to evaluate the model performance on the advection and diffusion process by atmospheric motion, dry deposition by the interaction with ground surface and wet deposition by cloud and precipitation. Model validation has not been conducted for the case with a varying release from a point source. It is necessary to conduct tracer experiments for this purpose and to perform validation and further improvement of models.

References

Council of Ministers on Nuclear Energy (2016). Way of thinking for the enhancement of nuclear emergency countermeasures: response to requests from the Association of Prefectural Governors based on the lessons learned from Fukushima, http://bit.ly/2BVt5Gw (accessed 19 September 2018) (in Japanese).

Imai, K., M. Chino, H. Ishikawa, *et al.* (1985). *SPEEDI: A Computer Code System for the Real-Time Prediction of Radiation Dose to the Public due to an Accidental Release.* Tokyo: Japan Atomic Energy Research Institute.

Independent Investigation Commission on the Fukushima Nuclear Accident (2012). Investigation: investigation report, 28 February (in Japanese).

Investigation Committee on the Accident at Fukushima Nuclear Power Stations of Tokyo Electric Power Company (2012). Final report, 23 July.

Meteorological Society of Japan, (2014). Recommendations on strengthening of atmospheric dispersion monitoring and prediction technologies for radioactive materials discharged due to an accident at a nuclear facility. www.metsoc.jp/2014/12/17/2467 (accessed 19 September 2018) (in Japanese).

MEXT (2008). *The Manual for Nuclear Emergency Response* (in Japanese).

(2012a). Review of response by MEXT concerning the recovery from the Great East Japan Earthquake, 27 July (in Japanese).

(2012b). Prediction results by the SPEEDI network system. http://radioactivity.mext.go.jp/ja/list/201/list-1.html (accessed 31 August 2012) (in Japanese).

Nagai, H., M. Chino and H. Yamazawa (1999). Development of scheme for predicting atmospheric dispersion of radionuclides during nuclear emergency by using atmospheric dynamic model. *J. At. Energy Soc. Japan*, **41**(7), 53–61 (in Japanese).

National Diet of Japan Fukushima Nuclear Accident Independent Investigation Commission (2012). Investigation report, 28 June (in Japanese).

Nuclear Regulation Authority (2011). List of the SPEEDI prediction results by the Nuclear Emergency Response Headquarters (from March 14 to May 5, 2011). http://bit.ly/2VqIGFo (accessed 19 September 2018) (in Japanese).

(2012). The Nuclear Emergency Response Guidelines (revised in 2017) (in Japanese).

(2014). About operation of the System for Prediction of Environmental Emergency Dose Information (SPEEDI). www.nsr.go.jp/data/000027740.pdf (accessed 19 September 2018) (in Japanese).
Nuclear Safety Commission (1980). About the disaster prevention countermeasures for nuclear facilities (revised in 2010) (in Japanese).
(2008). The guidelines for environmental radiation monitoring (revised in 2010) (in Japanese).
Suda, N. (2006). The present status and future prospects of the SPEEDI network system. *Japan. J. Health Phys.*, **41**(2), 88–98 (in Japanese).

9

Off-Site Decontamination

The Fukushima Daiichi nuclear disaster released massive amounts of radioactive substances into the environment. For isotopes with short half-lives, the primary safety concerns relate to the impacts on human health by initial exposure relatively soon after the disaster. However, isotopes with long half-lives remain in the environment for a long time; they move through the environment by many pathways and raise concerns regarding the threats to human health and everyday life. To reduce these impacts from the radioactive substances that have been dispersed widely throughout the environment, it is necessary to remove and isolate these substances from the places where people live and work, which is the purpose of decontamination. This chapter discusses the concept of decontamination, the methods used in decontamination, the institutional frameworks for decontamination and the processing and disposal of the soil and waste that is generated by decontamination.

9.1 Concept of Decontamination and Its Application

YUICHI MORIGUCHI

Internationally, the removal of radioactive substances from the environment is expressed primarily by using the term 'cleanup'. In Japan, the word *josen* is used, which means decontamination or literally the 'removal of contamination'. However, radioactive substances cannot be destroyed, and the contamination that they cause does not disappear. Critics have indicated that the 'removal of contamination' is in fact no more than the 'relocation of contamination' from one placc to another. However, the essence of decontamination consists of distancing radioactive substances from people's living environments and other objects that the substances may impact, and of managing these substances. The term decontamination also refers to the removal of radioactive substances from the clothing and personal belongings of evacuees who flee a disaster and the workers who respond to a disaster, but in this chapter decontamination is used to refer to the cleanup of a contaminated off-site environment.

Decontamination is sometimes necessary in situations other than nuclear disasters, such as when an unsealed radioactive source is mishandled, but in these cases the contamination is typically localised. Decontamination also occurs when the soil at former factory sites or illegal dumping sites is contaminated by harmful materials other than radioactive substances, but here, too, the geographical extent of the contamination is far smaller than the geographical extent of contamination that is caused by nuclear disasters.

There are two typical decontamination locations and goals. The decontamination of city and town streets is implemented to reduce external exposure to radiation among people who must spend a long time in contaminated areas because they live, work or attend school there. The decontamination of agricultural land is implemented to reduce or eliminate contamination in agricultural products.

By contrast, how to approach the decontamination of forests has become a major issue. Japan's forests cover two-thirds of the country's land area and an even higher proportion of some districts that remain heavily contaminated with radioactive substances from the Fukushima Daiichi nuclear disaster. Furthermore, rain washes the radioactive substances that are deposited on land into the ocean through rivers and lakes. Amid concerns that the contamination of marine and freshwater products may persist for many years, the bottoms of rivers, lakes, reservoirs and the ocean may also become targets of decontamination.

Decontamination represents attempts to remove radioactive substances from the places where they were deposited or where they were subsequently transferred by natural phenomena. The further extraction and separation of radioactive substances from the soil, plant matter and the types of waste that are generated by these removal actions, as well as from the sewage sludge and incinerator ash that is generated by artificial processes that concentrate or accumulate these radioactive substances, are also a type of decontamination, because they involve separating and distancing radioactive substances from people. In addition, without the final disposal of these materials, decontamination cannot truly be considered complete. As shown in Figure 9.1, this chapter interprets the narrow definition of decontamination to include the removal of materials from contaminated locations, and the broader definition to include the entire process, from post-removal concentration, separation and other processing to storage and disposal. These processes are described in detail in the following sections. Although decontamination is not limited to specific isotopes, the composition of the emissions from the Fukushima Daiichi disaster has meant that on land, the primary decontamination focus has been on the removal and isolation of radioactive caesium.

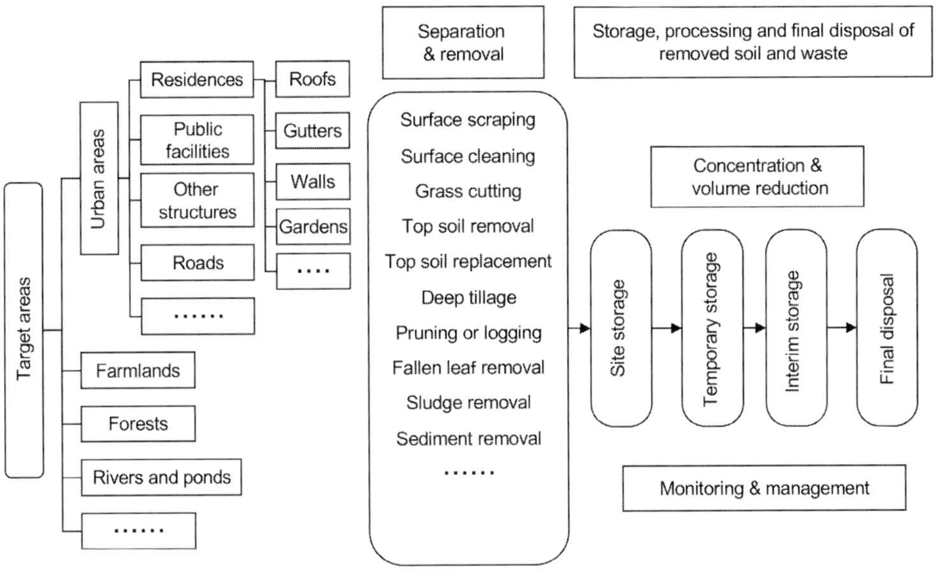

Figure 9.1 Overview of off-site decontamination.

9.2 Decontamination Techniques Used at Contaminated Sites

YUICHI MORIGUCHI

The techniques for removing the contaminants from contaminated sites can be broadly divided into the following two categories: (a) the removal, in full, of the materials that contain radioactive substances, such as the excavation and transport of topsoil to which caesium has adhered; and (b) the selective extraction and removal, to the maximum extent possible, of the radioactive substances at the contaminated site. The isotopes in the fallout that occurred over land adhered to everything on the Earth's surface, including the soil, paved roads, the roofs and walls of buildings, trees and the bottoms of bodies of water. The decisions concerning which of these two types of removal action to use are made according to the place or item that is to be decontaminated. The ease of separating isotopes from the surfaces to which they have adhered varies depending on the nature of the surface and the type of isotope, and in cases such as roofs and walls, fully removing the contaminated material can be difficult.

Information regarding these removal techniques was collected prior to the Fukushima Daiichi nuclear disaster in technical documents that were published by the International Atomic Energy Agency (IAEA) (IAEA, 1999) and elsewhere. This existing knowledge was also reflected in the first guidelines that the Ministry of the Environment compiled at the end of 2011 to ensure that decontamination was implemented in accordance with the legal framework (Ministry of the

Environment, 2011), which is described in the following. However, because of the characteristics of the soil and the locally distinctive materials that are present in the landscape, accurately predicting which techniques would be most effective concerning the Fukushima Daiichi disaster proved to be difficult. Faced with the clear need to make improvements based on the experience that was gained through implementing decontamination, the Ministry of the Environment released a second edition of its decontamination guidelines in May 2013 (Ministry of the Environment, 2013a) and has added further supplements since then.

9.3 Timeline of the Decontamination-Related Events Following the Disaster

YUICHI MORIGUCHI

9.3.1 The Period of Chaos Prior to the Enactment of the Act on Special Measures

This section examines the timeline of the events that occurred after the Fukushima Daiichi nuclear disaster. After the disaster, experts soon recognised the need for decontamination. However, the scale of the release of radioactive substances into the off-site environment and the resulting contamination were not foreseen, and legal frameworks and systems for responding to this disaster had not been established. I (Yuichi Moriguchi) personally visited the Ministry of the Environment soon after the disaster and argued that action had to be taken to address the radioactive contamination of the soil. However, the response was unsatisfactory because radioactive substances and the items that were contaminated by them had been entirely excluded from the existing legal framework for environmental protection, which was based on the Basic Environment Law.

For example, Article 2 of the Soil Contamination Countermeasures Act, which establishes the definition of soil contamination, defines 'designated hazardous substances' as 'substances (excluding radioactive substances) that are designated by a Cabinet Order as likely to have harmful effects on human health when present in soil'. Because of this explicit exception, environmental administrators were unable to address radioactive contamination. Similarly, although the sewer sludge and ash from incinerating waste were contaminated by radioactive substances, the Waste Management Act excluded the contamination that is caused by radioactive substances.

Against this backdrop, individuals in the fields of environmental contamination and waste disposal used personal connections to exchange information with people in the nuclear power and radiation fields, and individuals from all of these fields participated in efforts to establish unofficial study meetings. Nuclear and radiation experts did not wait for the official response, which was delayed by legal restrictions, but instead travelled to the contaminated regions of Fukushima Prefecture on their own under the auspices of non-profit organisations. In some cases, these experts made recommendations to local government leaders or conducted experimental decontamination actions. Reports emerged that some local governments were receiving many recommendations regarding radioactive contamination, including some recommendations with a questionable scientific or technical basis.

9.3.2 Establishment of the Act on Special Measures and the Designation of Zones

For several months following the disaster, decontamination measures were limited to these unofficial responses, and the government's inability to address contaminated waste under existing legal frameworks also remained unresolved. To address these two problems, the government enacted the Act on Special Measures Concerning the Handling of Environment Pollution by Radioactive Materials Discharged by the Nuclear Power Station Accident Associated with the Tohoku District – Off the Pacific Ocean Earthquake that Occurred on March 11, 2011 (which is abbreviated as the Act on Special Measures Concerning the Handling of Pollution by Radioactive Materials and is further shortened in this chapter to the 'Act on Special Measures'). The Act on Special Measures was offered as lawmaker-initiated legislation, was approved and enacted on 26 August 2011 and promulgated and partially put into effect on 30 August 2011.

The Act on Special Measures established measures to be taken in the following two broad areas: decontaminating the soil and other materials that are contaminated by radioactive substances, which comprises the main topic of this chapter, and processing the waste that was contaminated by radioactive substances. The Act took effect in full on 1 January 2012. During the intervening span of just less than four months, work was performed at a frantic pace to establish basic policies, cabinet and ministerial orders and decontamination guidelines. The Committee on Environmental Remediation, in which I participated, was established at the Ministry of the Environment as a forum for discussing decontamination.

Under the Act on Special Measures, the national government directly implements decontamination in special decontamination areas, while in intensive contamination survey areas local governments conduct surveys of contamination, establish decontamination plans and implement decontamination. In the aftermath of the Fukushima Daiichi disaster, the special decontamination areas comprised restricted areas and deliberate evacuation areas and covered 11 municipalities. The restricted areas and deliberate evacuation areas were later reorganised into three categories based on air dose, namely, difficult-to-return-to zones, restricted residence zones and zones in preparation for the lifting of the evacuation order (hereafter shortened as derestriction preparation zones). In difficult-to-return-to zones, where radiation levels were highest, decontamination was limited to model projects only, while in restricted residence zone and derestriction preparation zones, wide-area decontamination was implemented. Evacuation orders were derestricted in stages, and all orders for the restricted residence zones and derestriction preparation zones were lifted by 1 April 2017 as planned decontamination for these two zones was completed. Remediation for difficult-to-return-to zones is underway, focusing on specific bases designated by another Act on Special Measures for the Reconstruction and Revitalization of Fukushima as of April 2018.

Meanwhile, a total of 41 municipalities in Fukushima Prefecture and 63 municipalities in Iwate, Miyagi, Ibaraki, Tochigi, Gunma, Saitama and Chiba Prefectures were designated intensive contamination survey areas, where air dose rates exceeded 0.23 μSv/h, the annual equivalent of more than 1 mSv of radiation exposure above background levels. Subsequent surveys resulted in this designation being lifted from 12 municipalities, leaving a total of 92 municipalities with that designation. By the end of March 2018, decontamination in intensive contamination survey areas was completed for all 36 municipalities in Fukushima Prefecture and 56 municipalities outside of Fukushima Prefecture.

9.3.3 Basic Decontamination Principles and Targeted Land Use

The Basic Principles under the Act for Special Measures to address environmental contamination from radioactive substances were established by a cabinet decision on 11 November 2011. The introductory portion of this policy is as follows:

- Decontamination measures should target soil, built structures, roads, rivers, lakes, coastal areas, harbours, agricultural land, forests, etc. However, because these targets cover an extremely wide area, it is necessary to prioritise the formulation of special-area decontamination implementation plans and other decontamination implementation plans in areas where these measures are necessary for the protection of human health. It is also necessary to implement measures that are carefully adjusted according to radiation levels.
- Within these prioritised areas, it is necessary to further prioritise decontamination implementation in areas that are routinely used by children, who are more sensitive than adults to the effects of radiation.
- The areas that are contaminated by radioactive substances from the Fukushima Daiichi nuclear disaster include large tracts of agricultural land and forest.

- The decontamination of soil and other measures that are implemented on agricultural land shall consider the restoration of the conditions that are necessary for the resumption of agricultural production.
- Regarding forests, decontamination shall be prioritised in areas near residences, etc.

This basic policy acknowledges that the targets of decontamination measures are varied and geographically widespread, including various types of land use as well as marine and freshwater areas, and prioritises the decontamination of residential areas and agricultural land.

Under the Basic Principles, the government stipulated that forests near residential areas should be prioritised. However, the highly contaminated areas from the Fukushima Daiichi disaster had a high proportion of forests in them, and in some areas these forests were closely interwoven with inhabited areas, which caused local stakeholders to strongly desire that the forests be decontaminated. Subsequently, forest decontamination became an important subject of further discussion. The goals of forest decontamination can be summarised as follows: (1) in residential areas that are close to forests, reducing residents' exposure to radiation from the radioactive substances that are deposited in forests; (2) reducing the redistribution of radioactive substances that are deposited in forests, either through airborne redispersal to surrounding agricultural or residential land or through rainwater runoff to the lower reaches of rivers and the ocean; and (3) enabling the resumption of forestry production, recreation and other uses of the forest.

9.4 Demonstration Tests and Demonstration Model Projects for Decontamination Technologies

YUICHI MORIGUCHI

Concurrent with institutional responses, technical responses to the contamination were also increasingly offered starting in mid-2011. The Cleanup Subcommittee of the Atomic Energy Society of Japan was one of the central fora of the scientific and technical discussions. Meanwhile, the Japan Atomic Energy Agency (JAEA) was primarily responsible for implementing trials with government funds. Among these trials was the Decontamination Technology Demonstration Test Project, which was commissioned by the Cabinet Office. Separately, the Ministry of the Environment also implemented decontamination technology test projects, and JAEA's Fukushima branch handled the selection and evaluation of the technologies for this programme. In addition to these basic technology development projects for each type of place or object that was targeted for decontamination, decontamination trials also occurred in the areas that were designated for cleanup through the

Restricted Areas and Deliberate Evacuation Areas Model Decontamination Test Project, which began with a call for participants in October 2011. Under this project, 12 participating municipalities were divided into three groups; each of these groups included forest, agricultural land, residential areas, buildings, roads and other elements that were targeted for decontamination, and a range of radiation levels.

According to Ministry of the Environment documents (Environment Management Bureau, Ministry of the Environment, 2012), these three groups included 16 districts that were targeted for decontamination, and these districts ranged in area from 3 to 37 ha, comprising a total of 209 ha. The insights that were gained from this on-site work regarding each type of place or object that was targeted for decontamination are summarised in this document. For example, this document reported that the effectiveness of decontamination in reducing radiation levels on building roofs varied according to the material that the roof is made of. The results were achieved more easily with galvanised iron and clay-tile roofs than with cement tiles, while baked steel and slate roofs had low levels of residual radioactive substances prior to decontamination. The quantitative results and other insights that were gained from the trials are also summarised regarding gutters, home gardens, the roofs and walls of large buildings, agricultural land, roads, parks, forests, trees and other targets of decontamination. The trials also yielded information on the effectiveness of processing the wastewater that was contaminated during the decontamination process by methods such as filtration, adsorption, coagulation and sedimentation; the dust generation that is caused by the crushing of leaves and branches to reduce their volume; and the measurements of radioactive caesium concentrations in the emissions and ash that are generated by incineration.

The quantity of soil that was removed during decontamination work was also measured, and air doses at the temporary storage sites for this soil were measured as well. This project also measured the radiation exposure among workers who were engaged in decontamination. When exposure was projected by assuming decontamination workers would work continuously for five years in the types of work that caused the highest exposure and in the model project areas with the highest radiation levels, the total projected exposure exceeded the limit of 100 mSv in some cases. Because the existing Ordinance on Prevention of Ionizing Radiation Hazards (Ionising Ordinance) did not anticipate a need to manage the exposure among decontamination workers, a new Decontamination Ionising Ordinance (a shortened name) was enacted prior to the full implementation of decontamination under the Act on Special Measures.

Following this series of trial projects that lasted from the latter part of fiscal year 2011 into 2012, full-scale decontamination began. The results were reported in a summary on the effectiveness of various decontamination methods that were used in the programme (Ministry of the Environment, 2013b) and a report on model decontamination trials in the difficult-to-return-to zone (Ministry of the Environment, 2014a). Experiences on decontamination for the living environment conducted by the Ministry of the Environment until fiscal year 2014 were compiled as a 'Decontamination Report' (Ministry of the Environment, 2015). The target date for completing the wide-area decontamination outside of the difficult-to-return-to zone was set for March 2017 to accelerate the reconstruction of Fukushima.

9.5 Contamination Levels Required to Trigger Intensive Survey for the Necessity of Decontamination Work and of the Goals of Decontamination

YUICHI MORIGUCHI

9.5.1 Designation of Target Areas for Decontamination and Reduction Goals

The aim of decontamination in evacuated regions is to lower radiation to a level that allows residents to return home. Similarly, in the regions that were not evacuated but had radiation levels above a certain level, the aim is to lower radiation levels to create a living environment that is safer and causes residents less concern. However, opinions vary widely regarding how low to set targets for decontamination and the degree of contamination that should trigger survey for the necessity of decontamination work.

These questions became one of the main points of contention during the approximately four months between the passage of the Act on Special Measures and its full-scale implementation. Contamination levels were discussed in terms of yearly radiation exposure that was above background levels, which was estimated based on air dose rates. This calculation assumes that individuals spend eight hours a day outside and 16 hours a day inside and that indoor doses are equal to outdoor doses multiplied by 0.4 because some radiation is blocked. Given that $(8 + 16 \times 0.4)/24 = 0.6$, the yearly radiation dose equals 0.6 times the radiation dose that is calculated simply by multiplying the hourly air dose rate by the number of hours. By using this calculation method, 1 mSv/year is 0.19 μSv/h, and adding a background dose of 0.04 μSv/h to this number yields 0.23 μSv/h. This value was used as the criterion to designate intensive contamination survey areas.

However, the target outcome levels for decontamination were not always clearly expressed in terms of this annual dose above background levels. The November 2011 Basic Principle for the Act on Special Measures stated that actual target levels should be established based on the results of model projects and other factors, with the goal of shrinking the areas where doses exceeded 20 mSv/year rapidly and in phases.

Meanwhile, in areas where the yearly dose was below 20 mSv, goals were established as follows.

1. In the long term, lower radiation doses to below 1 mSv/year above background levels.
2. By the end of August 2013, reduce the yearly above-background doses for ordinary residents by approximately 50% compared with the doses at the end of August 2011, including reductions from the natural decay of radioactive substances.
3. Because it is important to restore the environments where children can live safely, reduce the yearly above-background doses for children by approximately 60% compared with the doses at the end of August 2011, including reductions from the natural decay of radioactive substances, by prioritising the decontamination of schools, parks and other places where children spend time.

Regarding the areas with yearly doses between 1 and 20 mSv, achieving goals 2 and 3 in the areas on the lower end of the spectrum would also lower yearly doses below 1 mSv, but in the areas with relatively high dose levels this reduction was not the case. Although a yearly dose of 1 mSv or less was expressed as a long-term goal, it was sometimes difficult, given the technical and financial constraints on achieving this goal, for the national and local governments to reach agreements with local residents who demanded the rapid achievement of this level.

9.5.2 Wide-Area Decontamination and Localised Decontamination

As described above, under the Act on Special Measures, areas with annual doses over 1 mSv were designated intensive contamination survey areas, but wide-area decontamination had been neither assumed nor anticipated in these areas. In September 2011, the government indicated that it was considering an approach in which the annual dose level that was required to apply wide-area decontamination would be set at 5 mSv/year (equivalent to approximately 1 μSv/h), while districts with doses of between 1 and 5 mSv/year would undergo localised decontamination of the spots with high radiation levels. Ultimately, the government did not officially announce this figure of 5 mSv/year, but the principle of prioritising decontamination at spots and districts with high radiation levels seems reasonable.

This value of 5 mSv/year for above-background exposure (an air dose rate of 1 μSv/h) was used in some earlier cases as a reference value that indicated the government's recognition that decontamination was necessary. In October 2011, when the above-noted discussions were occurring, the Cabinet Office, Ministry of Education, Culture, Sports, Science and Technology (MEXT) and Ministry of the Environment issued 'Guidelines for Responding to Sites with Radiation Levels Higher than Those of Surrounding Areas in Prefectures Other than Fukushima'. These guidelines stated that when local public bodies or private-sector organisations discovered spots with radiation levels that were higher than the levels in surrounding areas, they should report this information to MEXT, and established procedures would be initiated for the national government or local public bodies to respond by monitoring and decontaminating the spot. The standard that was required to trigger this process was defined as the 'spots where air dose rates 1 m above ground exceed those in surrounding areas by 1 μSv/h or more'.

Several days after these response guidelines were released, doses that far exceeded the above-noted requirement were recorded on public property in the city of Kashiwa, Chiba Prefecture, and this was the first report submitted to MEXT. Based on the procedures that were outlined in the response guidelines for handling situations in which straightforward decontamination was not possible, the case was reported to the Ministry of the Environment, and at the ministry's request I participated in supporting the further analysis and decontamination of the site. The immediate cause of the contamination in this case was a broken street gutter that caused rain to seep into the soil, but the broader cause was the fact that rainwater drained from a large roof onto this location. For this reason, in March 2012, the Ministry of the Environment released 'Guidelines for Responding to Localized Spots Contaminated by Radioactive Substances', which listed large roofs and parking lots as examples of hotspots. Additionally, in places where wide-area decontamination occurred and post-cleanup monitoring identified localised radiation hotspots, the system called for follow-up decontamination.

9.6 Temporary and Interim Storage, Processing and the Final Disposal of Soil and Waste Generated by Decontamination

YUICHI MORIGUCHI

As noted above, it is not possible to extinguish radioactive substances through decontamination; therefore, the need arises for new locations at which to place the materials that are removed from the sites where they were impacting daily life and production activities. Even before the decontamination that was based on the Act on Special Measures began, measures such as the removal of the topsoil in schoolyards and at public facilities were being taken in areas with high radiation levels. From the start, the question of where to take the contaminated soil after removal posed a problem. In addition to localised responses, such as burying the soil or piling it in inaccessible places, specific sites were established where soil was brought for temporary storage. Without these temporary storage sites, it is difficult to conduct large-scale decontamination, but it is not easy to gain the understanding of residents regarding the collection and storage of large amounts of radioactive soil and waste near their homes. In the areas where the national government was directly in charge of decontamination activities, these activities were performed only after temporary storage locations had been secured. However, in areas where local governments were responsible for decontamination, there were many cases in which a temporary storage site could not be secured, and contaminated waste was therefore temporarily stored or buried at other sites that were designated for decontamination. The difficulty of securing temporary storage sites was one factor that slowed the progress of decontamination overall.

On the same day that the Act on Special Measures was enacted, the Nuclear Emergency Response Headquarters released its Basic Policy for Emergency Response on Decontamination Work. This policy stated that for the time being, 'it would be more realistic that municipalities or local communities have designated temporary repository sites'. Furthermore, this policy affirmed that the government would 'develop and disclose' a roadmap as soon as possible for constructing disposal sites that require long-term management and for ensuring the safety of these sites. In response, on 29 October 2011, the Ministry of the Environment released a roadmap that detailed the waste-processing flows for inside and outside Fukushima Prefecture.

In this roadmap, the item that attracted the most attention was the so-called 'interim storage facilities.' These facilities were intended to collect, safely store and manage, for a set period of time, both the large quantities of soil and waste that were generated by decontamination and the highly contaminated materials such as sewage sludge and incinerator ash that, independent of decontamination work, had already begun to accumulate. Initially, this roadmap indicated that its basic approach was to establish one such facility per prefecture or similar administrative district. At the end of September, senior officials at the Ministry of the Environment announced that these facilities would be established in a total of eight prefectures, including some prefectures in the Kanto region. However, the roadmap indicated that these interim storage facilities would be established in Fukushima Prefecture only. Furthermore, the roadmap stated that waste would be stored at interim storage facilities in Fukushima Prefecture for not more than 30 years, after which it would be transferred to a final disposal site outside the prefecture. The location for the interim storage facility was to be determined in consultation with local communities and was to be situated within a 16 km^2 area surrounding the Fukushima Daiichi nuclear power plant. It was anticipated that the volume of waste that required strict long-term management could be reduced by separating and concentrating the radioactive substances in soil and other waste (Ministry of the Environment, 2014b).

In contrast, the roadmap indicated that outside Fukushima Prefecture – where the quantity of decontamination-related soil and waste and highly contaminated waste is comparatively small – the national government would establish facilities for disposing of the waste that was highly contaminated by radioactive substances from the nuclear accident (designated radioactive waste), while most of the other materials would be transferred to existing waste disposal sites for final disposal. Subsequently, the national government announced that in Miyagi, Ibaraki, Tochigi, Gunma and Chiba Prefectures, where it proved to be difficult for existing disposal sites to handle all of the waste, it would establish designated waste disposal sites. In the autumn of 2012, several locations in Ibaraki and Tochigi were proposed as potential sites for these waste disposal facilities, but local communities strongly opposed the move. This opposition led to a reconsideration of the process that was used for selecting sites, and as of early 2018 no potential sites had been determined.

The initial roadmap proposed a stepwise process of storing waste first at decontamination sites, then at temporary storage sites in each community or municipality, then at prefectural interim storage facilities, and then at final disposal sites. The key to the smooth implementation of this plan lies in attaining local approval for each of these facilities. Government policy calls for transferring waste from interim storage facilities in Fukushima Prefecture to other prefectures for final disposal, and this policy has been stipulated in legislation, but the practical discussions that are required to implement this policy have yet to begin.

9.7 Conclusion

YUICHI MORIGUCHI

The IAEA's report on its October 2011 fact-finding mission to Fukushima advised Japan to consider the most rational priorities for decontamination from the perspective of lowering exposure while also focusing on forests (IAEA, 2011). However, the residents of the areas that were impacted by the nuclear accident demanded the restoration of their living environments and the return of their previous lives, and they forcefully demanded the government and corporations assume responsibility for the disaster and cleanup. Decontamination has proceeded in this context, but there is a need to dispassionately ascertain whether it will be possible to restore the environment in all contaminated areas to a state in which residents can resume their previous lives. Some parties have begun to express the opinion that in addition to responding to evacuees' desire to return home even if it takes a long time and requires the best decontamination work possible, preparations should also be made to support the option of evacuees to relocate permanently in cases where decontamination is not anticipated to reduce the radiation below a certain level within a certain number of years.

The path to the final disposal of the soil and waste that is generated by decontamination is long. However, if we fail to start on this path for fear of its length, it will be impossible to reduce the impact of the current contamination. In some cases, local citizens have been motivated to launch their own decontamination efforts by their desire to do something regarding the contamination that surrounds them; with the support of experts, these efforts have led government bodies to take a more proactive stance. Decontamination is one method to reduce the impact of radioactive substances, but work should not stop there. We must also endeavour to restore our surrounding environment in a broader sense, including restoring the trust in government and experts that has been lost because of the disaster. Without question, this path is also long.

References

Environment Management Bureau, Ministry of the Environment (2012). Summary report of the decontamination model projects for Restricted Areas and Deliberate Evacuation Areas (in Japanese, English edition not available).

IAEA (1999). Technologies for remediation of radioactively contaminated sites. IAEA-TECDOC-1086.

(2011). *Final Report of the International Mission on Remediation of Large Contaminated Areas Off-site the Fukushima Dai-ichi NPP, 7–15 October 2011, Japan*. www.iaea.org/sites/default/files/final_report151111.pdf (accessed 1 July 2018).

Ministry of the Environment (2011). *Decontamination Guidelines*, 1st edition (in Japanese, English edition not available).

(2013a). *Decontamination Guidelines*, 2nd edition. http://bit.ly/2VpJHxu (accessed 1 July 2018) (tentative translation).

(2013b). Effects of decontamination methods in decontamination projects undertaken to date by national and local governments (in Japanese, English edition not available).

(2014a). Report of the decontamination model projects for difficult-to-return areas (in Japanese, English edition not available).

(2014b). Booklet of Interim Storage Facility (ISF) for soil and waste. http://josen.env.go.jp/en/storage (accessed 1 July 2018) (in Japanese, English edition not available).

(2015). Decontamination report: a compilation of experiences to date on decontamination for the living environment conducted by the Ministry of the Environment (MOEJ). http://bit.ly/2VpyzRl (accessed 1 July 2018).

Part III

Lessons and Future Issues from the Fukushima Accident

It is necessary to analyse how people acted at the time of the accident and what we should do in the future to understand the environmental transport of the radioactive materials to improve future readiness and countermeasures for urgent and serious disasters such as the Fukushima accident. Chapter 10 discusses the relationship between science and society, while Chapter 11 introduces snapshots of activities of researchers and academic communities in order to provide guidance to the coming generations.

10

Urgent Actions by Scientists

10.1 The Gathering and Distribution of Information Required for Applying Countermeasures at the Disaster Site

TOKUSHI SHIBATA

The things we learned from the responses to the Fukushima Daiichi Nuclear Power Station (FDNPS) accident are as follows. (1) If experts had been properly placed in the Nuclear Emergency Response Headquarters and the Fukushima Prefectural Disaster Response Headquarters, the data calculated by SPEEDI or the results obtained from the airborne survey conducted by the US Department of Energy (DOE) might have been more effectively utilised, and people would likely not have been evacuated to areas with high dose rates. (2) If experts had been properly placed in the Fukushima Prefectural Disaster Response Headquarters, the distribution and administration of stable iodine preparations could have been performed in a similar manner by different local governments. (3) In the case of an emergency disaster, if special budgetary actions had been taken for the investigation of radioactive contamination, the collection of soil samples could have started earlier, and detailed maps could have also been created. These situations occurred because the government asked us to follow procedures to budget according to a normal situation. They requested a list including all items with unit prices and exact numbers of necessary items for soil sampling for about 11 000 samples from about 2200 locations. The list also had to include travel expenses for persons who took part in the soil sampling project, taxi fares for the transfer of persons from the headquarters to the sampling points, etc. This job took about one month because we had to get information from participants from 98 organisations. If we could have used the budget to pay for necessary items with receipts by drawing money from a special account, we could have started the soil sampling project one month earlier.

The measures against disasters that significantly affect the environment were distributed through communications from the national government headquarters to the headquarters of local governments. It became obvious in this accident that the proper system to communicate information had not been established for disasters in which accurate judgement required highly professional knowledge. Therefore, in such a case, it is desirable that many researchers make a collective effort to gather necessary information and then provide appropriate information to headquarters. To take such an action, a leading group must be organised, and the members of the leading group should be chosen from a variety of fields. The Science Council of Japan (SCJ) is a candidate to become that leading group as it includes specialists from a variety of fields.

However, the SCJ is not an organisation that can provide a rapid response for emergency disasters. The required items for such a leading group are:

1. The leading group should conduct regular training on collecting information and communication.
2. Sufficient communication and trust among the leading group and government, ministries and local governments should be established.
3. The leading group should carefully consider necessary information for various disasters and contact concerned organisations to collect information in normal situations.
4. The leading group should secure sufficient budgetary resources for information collection by the concerned organisations.
5. The leading group should incorporate the latest knowledge into the measures for disaster relief.

The following sections discuss several aspects of this problem, including the necessity of interdisciplinary research, the scientific explanation of the situation and uncertainty, the importance of verification, lessons from the IPCC, a scientist's message for presenting measures against disasters with a group voice, and the ideal nature of the autonomous distribution of information from scientists. In these reports, there are many clues for establishing an organisation in Japan that utilises solid information for emergency disasters.

10.2 The Need for Interdisciplinary Research

TOSHIMASA OHARA

Although eight years have now passed since the FDNPS accident, we are still looking for ways to resolve the issue of environmental contamination from the emitted radioactive substances. Radioactive contamination is a severe type of pollution that constitutes an extremely serious environmental problem from various perspectives, including its spatiotemporal spread, the scale of its impact on human health as well as on other organisms and ecosystems, and the fact that various environmental media have been contaminated. This issue can only be resolved by marshalling the many different scientific capabilities available in Japan while also seeking the cooperation of other nations. In this section we discuss the need for interdisciplinary research that has been brought to light by the Fukushima Daiichi radioactive contamination and consider the scientific approach required to tackle environmental contamination issues and to rehabilitate future disaster areas.

10.2.1 The Research Challenges Exposed by the Accident

Japanese scientists and scientific organisations need to directly tackle the serious environmental problems posed by Fukushima Daiichi radioactive contamination. As explained in Chapters 2–6, and also in Chapter 11, immediately following the accident many scientists dedicated themselves to ascertaining the status of the radioactive contamination in air, soil, forests, the ocean, organisms and ecosystems. The researchers involved showed tremendous mettle in the face of such a serious environmental crisis, and their actions demonstrated the true value of environmental research. As described in Chapter 11, these initial activities began spontaneously as small-scale grassroots initiatives launched by individual researchers and laboratories that gradually became more organised. Conducted at a time when Japanese society was enveloped in gloom, these forward-looking endeavours have played a significant role in the reconstruction of the affected areas. However, given the extent and seriousness of the environmental contamination, research efforts still generally lack intensity; much more work needs to be done. One of the key shortcomings is the paucity of interdisciplinary research efforts.

10.2.2 The Importance of Interdisciplinary Research to Elucidate Dynamics

Current environmental issues are very complex and diverse; they require a research approach informed by this multifaceted structure. Radioactive contamination caused by the FDNPS accident is a typical example of such an issue. The following three structural axes are particularly important:

1. the multilayered nature of the spatiotemporal scale of the phenomenon (multiple scales);

2. interactions between and among various environmental media (multiple media); and
3. the multifaceted nature of impacts (multiple effects).

With respect to (1), in spatial terms, radioactive substances were deposited on the ground after spreading in the atmosphere from the nuclear power station to a broader scale encompassing much of eastern Japan; locations throughout the Northern Hemisphere were also affected (spatial multiscale properties). In terms of time, there were impacts from short-term radiation exposure to radioactive substances released into the atmosphere immediately after the accident and long-term radiation exposure to radioactive substances such as ^{90}Sr and ^{137}Cs with long half-lives (temporal multiscale properties). With respect to (2), ascertaining the status of radioactive substances released into the environment, elucidating their dynamics and predicting their movements requires an understanding of the behaviour of radioactive substances in diverse environments, including air, soil, forests, rivers, lakes and oceans. This requires integrating multimedia environmental modelling with multimedia environmental dynamics measurement. For (3), radiation can affect human health, production and everyday life in various ways, with different radionuclides also having different impacts and exposure occurring via various routes. This type of multifaceted structure is characteristic of general problems related to material cycles in the environment and various types of environmental pollution.

Tackling problems related to complex environmental systems requires an integrated research approach and an interdisciplinary research structure. An 'integrated research approach' means a style of research that combines observation, laboratory experiments, data analysis, modelling, impact assessment, development and evaluation of technology, economic analysis, policy evaluation and other research methods to achieve objectives through discussion and information sharing among different research groups. This integrated research approach also requires an interdisciplinary research structure composed of researchers in the earth sciences, nuclear physics, social sciences and various other fields.

This type of bottom-up approach to research has been extremely uncommon in Japan due to its scientific traditions, compartmentalised bureaucracy and existing academic divisions. Although the FDNPS accident has resulted in some interdisciplinary research, such as the integration of atmospheric and ocean models performed jointly by the Central Research Institute of Electric Power Industry, the National Institute for Environmental Studies (NIES), JAMSTEC, JAEA, universities and other organisations, and the integrated research on multimedia modelling and environmental dynamics measurement conducted by NIES and JAEA, such efforts in the initial stage after the accident have not yet achieved nationwide momentum. However, 2012 saw the launch of the Grant-in-Aid for Scientific Research on Innovative Areas project 'Interdisciplinary Study on Environmental Transfer of Radionuclides from the Fukushima Daiichi Nuclear Power Plant Accident (ISET-R)' (head investigator: Professor Yuichi Onda, University of Tsukuba). This research project is being conducted by researchers specialising in observations, laboratory experiments, data analysis, modelling and various other methodologies in a variety of disciplines, including radiation chemistry, geochemistry, atmospheric sciences, marine sciences, hydrological geomorphology, ecology and forest sciences. The project has laid the foundation for (2) above – the elucidation of multimedia environmental dynamics through an interdisciplinary and integrated research approach. This project has produced many scientific products (see www.ied.tsukuba.ac.jp/hydrogeo/isetr/ISETRen/resultEN.html) and concluded successfully at the end of FY2016. Additionally, in FY2015, the Environment Research and Technology Development Fund of the Environmental Restoration and Conservation Agency, 'Interdisciplinary Study for Exposure Assessment and Risk Assessment of Airborne Radionuclides Released by the Nuclear Power Plant Accident' (head investigator: Professor Yuichi Moriguchi, University of Tokyo) started. This project aims to accumulate the knowledge required to reduce the uncertainty in early exposure doses by integrating two approaches: experimental science to measure radionuclides in airborne microparticles collected during the early phase after the accident; and atmospheric transport modelling, exposure and dose assessment to describe processes from environmental emission to health impact. In the spring of 2016, the Centre for Environmental Creation was constructed in Fukushima Prefecture and its inter-institutional research system for environmental restoration and creation was established through collaboration among the Fukushima prefectural government, JAEA and NIES. It is expected that the centre will be a base of interdisciplinary research and a research network hub of many domestic and international organisations related to the FDNPS accident.

In the event of a crisis situation such as the FDNPS accident, the emergency response of the central government and related organisations is of vital importance, although the activities of academic bodies are also important. As is described in Chapter 11, the Japan Geoscience Union and other academic societies called for scientists to act urgently to respond to the crisis, setting in motion a wide range of activities. However, almost all of these activities were volunteer efforts, and the lack of any mechanisms for linking these efforts sufficiently even with the government's emergency investigations was a major problem. It is necessary for the SCJ to reinforce the function for supervising and coordinating such efforts to avoid this situation. Emergency research funding proposals were also submitted by scientists, although there was no category under which such research could be implemented, and it was only one year later that the Scientific Research on Innovative Areas category described above was adopted and a research organisation established. The only emergency research funding open to public subscription is the Japan Science and Technology Agency's J-RAPID programme for supporting urgent collaborative research with other countries. In the USA, the National Science Foundation (NSF) established research funding that included private donations, based on which emergency ship-based observations were carried out. To prepare for future crisis situations, a research funding category under which scientists can submit urgent research proposals should be re-established.

10.2.3 The Importance of Interdisciplinary Research for Reconstruction

The radioactive contamination caused by the FDNPS accident makes a convincing case for the need for the academic and research communities to cooperate on interdisciplinary research initiatives designed to contribute to the reconstruction of affected areas. Figure 10.1 shows the types of specialised fields that would need to be involved in research on the radioactive contamination issue, with the research process represented in the form of a Plan–Do–Check–Action (PDCA) cycle.

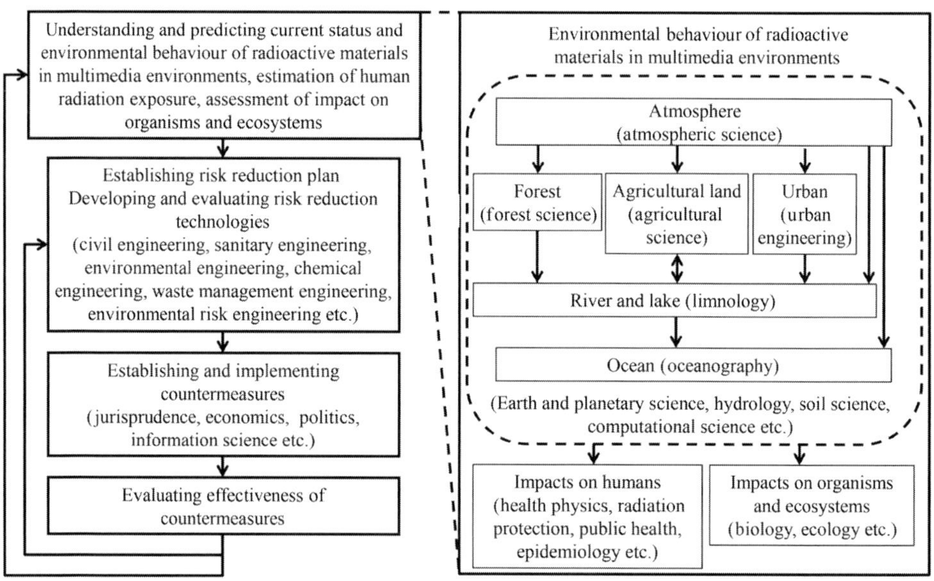

Figure 10.1 Research flowchart for tackling radioactive contamination, and some of the research fields that would be involved in the research.

Developing an understanding of the status and dynamics of such environmental contamination would require the involvement of a wide range of research fields in science and engineering in addition to radiation science. Assessing impacts and formulating measures for reducing radiation would furthermore require cooperation with specialists in medical fields such as radiology and public health; engineering fields such as nuclear reactor, civil, river, environmental and waste engineering; and social science fields such as sociology, economics and law. Sociology, information science and other disciplines would also need to be leveraged to share information with the general public. Thus, tackling this issue clearly requires cooperation among almost all fields of science. Integrating these diverse research fields is the task of environmental science and research. However, even today, there is a paucity of cooperation in interdisciplinary research. This situation perhaps reflects the conservative nature of Japan's academic and research communities or the nascent status of its environmental science community. It was against this backdrop that the Society for Remediation of Radioactive Contamination in Environment was launched in November 2011 to conduct interdisciplinary activities as an internationally oriented, integrated academic organisation devoted to the remediation of radioactively contaminated environments with the participation of specialists in many different fields of basic and applied science. I fervently hope that such research activities spanning different disciplines will be facilitated throughout Japan.

The 1986 Chernobyl accident resulted in a great amount of knowledge and taught us many lessons regarding the impacts of radioactive substances released into the environment as a result of a nuclear power plant accident. Japan also has a duty to scientifically elucidate the various impacts of the FDNPS accident, share the resulting knowledge with the world and pass on the lessons learned to future generations. Therefore, pursuing the type of research described here and elucidating the totality of environmental impacts is critical.

In the medium to long term, we must prepare for future nuclear power-related accidents and other natural or man-made disasters and minimise sudden environmental impacts. This can be accomplished by enlisting the cooperation of both natural and social scientists to systematise various findings for contributing to measures to prevent or mitigate environmental impacts triggered by disasters. Other important challenges related to protecting people in the event of a disaster include the building of monitoring networks, forecasting systems and emergency response systems for dispatching experts and developing a knowledge infrastructure related to risk communication. Finally, environmental emergency research needs to be established to explore the relationship between disasters and the environment, provide a solution for rapid environmental recovery, and aim to build an environmentally resilient society.

10.3 Explanation of Scientific Phenomena and Uncertainties: The Importance of Validation –Lessons from the IPCC

TERUYUKI NAKAJIMA

A serious problem at the time of the Fukushima accident was identifying reliable information. The Office of the Prime Minister did not regard the SPEEDI calculation results as reliable; thus, they did not order a quick release of the results. The first publication was delayed until 23 March, and daily distribution of the results started in April, almost one month after the accident. In this period, residents who evacuated in the same direction that the radioactive materials were transported received radiation exposure.

This decision of the Office of the Prime Minister was made based on political consideration of preventing panic and harmful rumours caused by the flow of unreliable information. Some meteorology professionals knowledgeable about numerical model uncertainty were also cautious about releasing the model simulation results.[1] The SPEEDI simulation results without a definitive source term (see Section 8.3) are more erroneous under complex meteorological conditions. Publishing unreliable information might cause confusion in society and induce secondary disasters. Based on long-term experience with problems in weather forecasting operations, there is a system in meteorological operation law that prohibits publication of weather forecasting results without permission from the Japan Meteorological Agency.

On the other hand, although ex post facto, it might have reduced the amount of exposure if the evacuation routes had been designed based on the SPEEDI results as discussed in Sections 8.4 and 8.5. SPEEDI was developed based on scientific research knowledge; simulations reflected the meteorological condition at that time, and the probability of misjudgement of the situation could have been reduced if the simulation results were used with interpretation by professionals, instead of judgement having been made without knowledge. Moreover, it is true that the public needs timely scientific information when faced with large volumes of unreliable knowledge on the internet.

Then, what is the best way to disseminate information? One idea is to establish an emergency response committee under a scientific organisation such as the Science Council of Japan (SCJ) at the time of a serious disaster, consisting of professionals and academic society representatives in related fields, who would issue scientific information that is as accurate as possible with enough explanation

[1] See www.metsoc.jp/others/News/message_110318.pdf.

of uncertainties to the public. The SCJ established action guidelines in emergencies, as described in Section 11.2.

Dissemination of the uncertainty information is especially important. Weather forecasting provides a useful example. People may become angry if there is rain when the weather agency has provided information that 'It will be fine today'. Hence, the meteorological agency introduced probability weather forecasts in 1980. In this method, the weather forecast includes the rainfall probability with an explanation that the weather forecast has uncertainty. Namely, scientific knowledge is used for decision-making after people understand that scientific knowledge includes uncertainties. Currently, people will make decisions regarding whether to bring an umbrella based on the forecasting probability. They will bring an umbrella if they go outside with a precious dress that they do not want to get wet even when the probability is low. They may not bring an umbrella for high-probability conditions if they do not care about getting wet. Nobody can become angry about the weather forecasting when there is rain.

It has become common to add an uncertainty index to disseminated information. It is said in physics that a hypothesis needs a 99.99% accuracy rate to be recognised as a first-law physical principle. There are larger uncertainties in the earth sciences, for which validation is very difficult.

In the global warming problem, how to express uncertainty has been discussed in the drafting process of the assessment report of the Intergovernmental Panel on Climate Change (IPCC). Several concepts to express the uncertainty have been introduced. Namely, in the case that the uncertainty can be qualitatively evaluated, it is defined by a relative sense regarding the amount and quality of the evidence (the amount of information to indicate the truth or validity of the idea obtained from theories, observations and models) and degree of consensus among estimates (degree of agreement between a certain finding and past references)[2]. The IPCC uses two factors: the grade of evidence as sufficient (A), medium level (B) and insufficient (C); and degree of consensus among estimates as high (1), medium (2), and insufficient (3). From these two factors, a level of scientific understanding (LOSU) is assigned one of five levels: high, medium, medium–low, low, very low (not evaluated).

[2] When a quantitative evaluation of the validity of the data, model and analysis results is possible based on the judgement of professionals, the level of confidence is defined as very high confidence (at least 90%), high confidence (approximately 80%), medium confidence (approximately 50%), low confidence (approximately 20%) and very low confidence (less than 10%). When the uncertainty of a result is statistically evaluated based on evidence and statistical analysis by professionals, we use the likelihood as virtually certain ($>$ 99%), extremely likely ($>$95%), very likely ($>$ 90%), likely ($>$66%), more likely than not ($>$ 50%), unlikely ($<$ 33%), very unlikely ($<$ 10%) and extremely unlikely ($<$ 5%). The confidence range is given by 5–95%.

The IPCC assessment reports must be approved through two reviews by professionals and governments before reaching the final draft stage. The process opens the draft to professionals and governments, collects criticisms and comments and reflects them in the subsequent draft. Based on this process, errors in the manuscripts are reduced, and missing information is added. Moreover, drafting rules require the presentation of only objective scientific knowledge and not subjective opinions to suggest proposals to affect public decision-making for global warming mitigation measures.

Even with such efforts, unsuitable statements were included in the Fourth Assessment Report published in 2007, and criticisms were raised. Responding to this situation, the IPCC conducted an international review of the assessment system in 2010 to introduce a more rigorous system. They added careful selection of evidence and clarification of references for assessment, introduction of the concept for uncertainty, enhancement of suitable communication with the media and more dialogue with the public, and exclusion of authors who have conflicts of interest.

The abovementioned story of the IPCC suggests the importance of increased validation and explanation of uncertainty when scientific knowledge of high public concern is disseminated. It is difficult for scientific knowledge with insufficient validation to be accepted by society. The IPCC took 20 years to provide validation and establishment of transparency for the assessment system. A social movement towards taking actions against global warming phenomena began.

Such a system, which takes time to implement, may not be relevant to the information dissemination problem of the SPEEDI results under the situation of an emergency and the possibility of an unpredictable, intense nuclear explosion. However, the above discussion includes important propositions for the preparation of future accidents. First, the Office of the Prime Minister should use the SPEEDI results if the uncertainties are well understood. In the weather forecasting system, decisions should be made through an analysis of observed and model-simulated meteorological fields. At the time of the Fukushima accident, it was possible to find optimum evacuation routes via a synthetic analysis of wind conditions and SPEEDI calculation results utilising satellite data and weather forecasting model results, assuming a professional team from the meteorological agency was formed, even though there was an excuse that the validation of the SPEEDI result was not possible in the situation in which the ground-based meteorological and emission monitoring networks shut down. Today's weather forecasting is based on ensemble weather forecasting in which several tens of model runs are used to account for the effect of errors involved in the initial conditions of the wind and temperature fields so that the uncertainty of the forecast can be evaluated. Therefore, a future system should be developed for providing numerical computation results and uncertainties from these data analyses.

Moreover, the multi-model ensemble method has been introduced in recent climate predictions; model biases are also evaluated objectively. Model results other than those from SPEEDI could serve as a second option. The JAEA has developed models succeeding SPEEDI (WSPEEDI and SPEEDI-MP) that have been successfully applied to various events, such as the fire explosion of the Power Reactor and Nuclear Fuel Development Corporation (March 1997), the criticality accident of the Tokai-mura JCO Uranium Processing Plant Agency (September 1999) and the transport phenomenon of volcanic gases from Miyake Island (August 2000). Modelling results were obtained within one to two weeks for the Fukushima accident from these models and other emergent modified air pollution transport models. It was therefore possible to make use of these results as second-opinion information for the government.

10.4 Proposal for *Group Voice:* Going beyond the Limits of *One Voice* and Making Information Provided by Scientists Available to the Public in Emergency Situations

HIROMI YOKOYAMA

After the Great East Japan Earthquake (GEJE), much discussion arose over how scientific information should be provided to the public in emergency situations. Among other issues, much criticism and misunderstanding was directed at the Meteorological Society of Japan's request for member scientists to refrain from making the results of simulations conducted by respective members available online and via other sources in order to avoid interfering with or even contradicting official government announcements, which have been termed its 'one voice'. At the same time, data from the System for Prediction of Environmental Emergency Dose Information (SPEEDI), which was supposed to predict the dispersion of radioactive material, were not disclosed because the system had not been prepared for an evacuation of citizens. Although discussion of how researchers should act in such situations has continued to the present day, few concrete proposals have been made.

The need for one voice, representing a consensus of scientists in an emergency situation, has also been discussed. However, if researchers' views are different to each other's, more than a few people noted that useful opinions may be excluded by a one voice strategy or that, in the first place, it is impossible to bring diverse opinions from a research community together into one voice.

This section discusses the conditions in which researchers can provide information that they believe is useful to the public and better contribute to society in emergency situations, especially the diffusion of radioactive material from the FDNPS accident. The author also proposes an approach to achieve the goal of making information provided by researchers available to the public in such situations.

10.4.1 Two Hurdles: Legislation and Clarification of Responsibility

After the GEJE, as the nuclear plant accident continued, providing information on the direction of the plume's flow was particularly problematic as two hurdles remained. First, legislation must be enacted that enables researchers to provide information in emergency situations; second, the locus of responsibility for researchers' provision of information must be clarified.

The first hurdle is legislation. Concern arose that the provision of information by the Meteorological Society of Japan might conflict with the Meteorological Service Act. Article 17 of this law, which prescribes basic systems concerning meteorological services in Japan, stipulates that 'Any person other than the Japan Meteorological Agency who intends to perform the services for forecasting meteorological phenomena, terrestrial phenomena, tsunamis, storm surges, high waves, or floods (hereinafter referred to as 'forecasting services') shall obtain a license from the Director-General of the Japan Meteorological Agency.' When the nation is in crisis, it would be absurd for no action to be taken only because such action might be illegal. This certainly is an issue that must be addressed.

Second, clarification of responsibilities is required. Responsibility here has two meanings, the first of which is scientific responsibility and the second of which is political responsibility. Normally, scientists are responsible for the content of science announced. Here, we refer to this as 'scientific responsibility'. In 1988, climatologist Jim Hansen made a statement to the US Congress in which he fulfilled his scientific responsibility by explaining about the increase in carbon dioxide. In 2004, seismologists were criticised because they did not refute or correct the government member who explained that no major earthquake had occurred in L'Aquila, Italy. The safety declaration was not scientific and after the explanation a big earthquake occurred. The seismologists attended the same meeting but they did not point out the mistake and did not object to the safety declaration. They did not fulfil their scientific responsibilities. Basically, scientists should respond to the situation of society and fulfil their scientific responsibility.

Meanwhile, it is the role of politicians to make judgements and evaluations based on scientific knowledge. For example, risk *assessment* of food is the role of scientists, but risk *management* is the role of the agency or politician in charge. This is because politicians are responsible for risk judgement and evacuation instructions are made by politicians. That is not a role that scientists are expected to fulfil. Regarding science, it is important to enhance the debate that is the basis of democracy, so political decisions reflecting the argument are desirable. Providing easy-to-understand explanations from scientists is important in terms of ensuring access to discourse. The maturity of democratic debate, together with the concept of the public sphere, were established as the basis of discussion by Hannah Arendt and Jürgen Habermas in the field of philosophy. However, what is different from the usual time is the technical problem in the event of an emergency of how to transmit information when there is no time to convert that information into an easy-to-understand article.

After the GEJE, some Japanese scientists thought they should release the data in order to fulfil their scientific responsibility. But in an emergency, the boundary between scientific and political responsibility becomes ambiguous. Other scientists thought they should not impede the government's one voice. Because the political responsibility is attached to the disclosure of data that affects the lives of people, those data should not be disclosed easily, but rather were discussed very honestly rather than with an eye towards self-preservation. Those who dealt with data from a typhoon in which many people died thought that data that affect human lives should be dealt with extremely carefully. In the Meteorological Agency, data are handled carefully. In this era when many people use a social networking service (SNS), disaster information and how to use the information of scientists who were in the public sector in the past are problems.

Individual researchers or a group of researchers might simulate radioactive material dispersion beyond the usual scope of their research publications and then make the simulation results available to the public. Of course, scientists do not issue evacuation orders, but this boundary is ambiguous. Researchers will be asked to interpret the published data and explain evacuation based on it. If it is not an emergency situation, like food safety, the results can be submitted to the government for judgement. Scientists were discussing whether to disclose the data to the public or not. Of course, free speech is always enshrined and individual scientists freely made remarks on SNS which were not vetted.

After the GEJE, a group led by nuclear physicists at Osaka University and the University of Tokyo was established to measure radiation levels in soil. This volunteer group immediately took action to take measurements in Fukushima while they continued to coordinate with local governments and MEXT; their actions resulted in a radiation dose map that was published by MEXT.

The continuous coordination with MEXT facilitated the group's ability to both take radiation dose measurements and to publish their measurement data. Although it took time to coordinate with MEXT, their efforts resulted in the successful release of official data by MEXT. Thus, neither individual researchers nor research groups were responsible for publishing the information; the relevant ministry or agency, MEXT in this case, was responsible.

In such emergency situations, it is essential to publish data with a clear indication of responsibility. It is also important for the government and the relevant ministries and agencies to discuss platforms for publishing data from researchers for which they (the government and relevant entities) take responsibility. On the other hand, researchers should keep in mind that individual researchers, groups, universities, research institutes or academic societies should not take responsibility but that the researchers should delegate responsibility for publishing data to the government and the relevant ministries and agencies.

10.4.2 Group Voice Rather than One Voice

The need and possibility for a so-called one voice or unique voice representing a consensus of opinions from experts has been discussed since the earthquake. Even the SCJ, an organisation representing various scientific communities in Japan, reportedly found it difficult to obtain a consensus of opinions ('one voice') from its committee members when they tried to express their views on low dose exposure. Because it was difficult even for the SCJ to build a consensus, they decided to release their view as a comment by the chairman.

Based on the abovementioned information, the author hereby makes the following proposal for information provision processes. Specifically, the author proposes the following two-step process.

First, information should be provided as a group voice rather than as one voice. In contrast to one voice representing the consensus of an entire scientific community, a group voice represents information and proposals provided by a group of researchers who have a common intention.

This approach should demonstrate well that prompt voluntary action is possible when taken by a research community that is in regular communication and is accustomed to working together. Such organic cooperation and the emergence of different voices from different groups in a community will yield desirable results. Data may vary from group to group; however, the same data provided by different groups make those data more credible. In the case of the nuclear physics community, which has few members, we did not see many groups established to release data; however, depending on the scientific field, several separate groups might be started.

Of course, a group voice does not prevent individual researchers who do not participate in the group from expressing and giving their opinions. Rather, free opinions from a number of individual researchers will accelerate discussion, which may lead to good proposals. However, what can be accomplished by an individual researcher is limited, and data analysis and interpretation by one person may be biased. To maintain the objectivity of data to be released, it is more desirable for researchers to be united and work as a community rather than as individuals.

If such researcher groups proliferate, the relevant ministry or agency would not be able to handle the number of requests for data release from the various groups, and the characteristics and credibility of each group would not be clear to the public. Therefore, a prudent second step for information provision would be setting up an organisation that evaluates the credibility of each group and helps the relevant ministry or agency to make judgements. Representatives of the communities the groups belong to, such as academic societies and associations, would be appropriate for this role. If data released by various groups, as well as their interpretations, were compiled in a comparison chart and posted on the website of the relevant academic society or association, they could be widely accessible and useful to society.

It is difficult to precisely predict the type of emergencies that will occur in the future. Of course, the preparation of an optimal system is necessary, although the ability to address each emergency case with flexibility is also necessary. The model proposed in this section is only an ideal concept, yet it anticipates that a bottom-up release of information will be promptly carried out by researchers in future emergency cases.

10.4.3 Proposal for the Release of Uncertain Information with Explanation on an 'as Needed' Basis

Under normal circumstances, researchers will not be able to release information unless two conditions are satisfied: first, the information is submitted as a manuscript to a journal and reviewed, after which the results are then announced to society. However, in an emergency there is no time to write and publish a paper. Scientists do not normally send data that have not undergone peer review directly to wider society. When it is necessary in the event of an emergency to disclose data without receiving a peer review, I believe that being a group, not a single person, contributes in a manner that is good for maintaining the quality of the data and devising a way to communicate.

The study of risk communication is advancing in the fields of social psychology, science and technology and society and science communication. What has emerged is a gap in critical thinking skills between specialists (i.e. researchers) and citizens. Researchers naturally care about the accuracy of information, while not only citizens but also journalists care more about researchers' transparency. For example, the Japan Aerospace eXploration Agency (JAXA) will hold a press conference every few hours when a signal from a spacecraft arrives, and information on the mission is dispensed from time to time. It is important that researchers distribute the information they have at the moment when the public's interest is high and information can be provided about, for example, an accident. Even if the information provided a while ago is wrong, they can maintain the trust from society by making corrections from time to time. Roughly correct and timely information is more socially meaningful than extremely correct and late information. If researchers are unwilling to disclose information, they may be thought to be hiding something, and confidence in them will be lost. That is why the Japan Meteorological Society was denigrated by journalists. In emergency situations, for researchers to play their roles as specialists, they should immediately disclose information needed by citizens at risk with proper explanations, even when such information may be incomplete or difficult to disclose. Furthermore, information released on an 'as needed' basis should be updated according to the expectations and needs of citizens at risk. If the government takes responsibility for disclosing data as proposed above, this proposal would immediately become more feasible. Stallen and Coppock (1987) gave four reasons for communicating risk to the public. One of them, the practical reason, mandates the provision of information to people at risk so they can avoid harm. In emergency situations, researchers should not hesitate to disclose information for fear of the bad effects of such disclosures. It is not researchers in laboratories, but the citizens themselves facing risk, who will make judgements based on the disclosed information.

Even if researchers disclose no information in emergency situations, information will still be available everywhere in today's online society. However, it is likely that most of that information will be of lower quality than that possessed by researchers. If so, disclosing data, even with the explanation that they may contain a high degree of uncertainty, will still be useful to citizens facing risk.

Furthermore, there is another important point in emergency situations. The prime minister, the leader in risk management, should have a competent scientist with whom to work directly, which means one who is trusted by the prime minister. After the GEJE, then Prime Minister Kan called one specialist after another from the Tokyo Institute of Technology because, as a graduate of that institute, he had absolute faith in them.

10.4.4 Summary for Group Voice

After the GEJE many researchers felt a sense of urgency because no system existed through which they could put to best use the relevant information they had for society. If there had been a system through which researchers could do as much as possible as researchers under the aegis of legislation that delegated responsibility for data disclosure to the ministries and agencies, these situations would have likely turned out differently. It is still difficult to judge to what extent researchers can directly contribute to society; however, if citizens had been able to observe the many researchers trying to contribute to society, at least their level of trust in scientists and researchers might have been greater.

Things that were not possible at the time of the GEJE may still not be possible when another major disaster happens. It is more productive to prepare to achieve the possible than to push to achieve the impossible. The idea of group voice is useful as a system for providing information by small groups of scientists. I hope that this idea will be discussed as a practical model by research in the small public sphere in SNS in the coming years.

10.5 The Autonomous Dissemination of Information from Scientists

MASATOSHI IMADA

When the 2011 earthquake and the nuclear power station accident occurred, the public experienced great distrust and disappointment due to the silence of scientists in areas such as weather and earthquake science, who were expected to provide useful information and ideas regarding the developing situation. One could say that keeping silent is strict and faithful to the norm of science in a situation in which no reliable knowledge had been obtained, although the problem was never that simple.

Another norm to be followed by scientists is a philosophy of speaking one's thoughts solely based on one's convictions after unrestricted studies through close investigation of all possibilities that can be listed. Unfortunately, not only operational organisations such as the Japan Meteorological Agency and the Nuclear Safety Commission, but also researchers in the areas of weather, earthquakes, nuclear power and others were not able to impress people, at least those outside the community, with their enthusiasm for the full use of research community collaborations and the networks to grasp the situation and thus work towards resolving the crisis. However, many people felt that a strange silence and inexplicable self-control dominated the expert community.

The situation that occurred in Japan after the 2011 earthquake and the nuclear power plant accident and the silent response of the scientists have revealed some of the large defects and contradiction between the science communities in Japan and the social system and its financial support that they rely on. Many people felt that organisations such as SPEEDI and the Japan Meteorological Agency did not provide useful information in the unprecedented disaster and emergency (which may occur at any time, as history has taught us; the history of science contains repeated cautions on the limitations of the human imagination). Moreover, people also assumed that the organisations did not release information because agile actions and information disclosure were not regarded as their duty. There may be an opinion that the routine work of SPEEDI and the Japan Meteorological Agency was performed within a range of regulations and assumed duty, but people had naively expected agile action at the highest level of science and technology in response to the emergency, even beyond their routine duties, from these organisations, which had been receiving considerable financial support. The majority of the people now know that an agile correspondence cannot be accommodated by the operational organisations in charge of routine work and that they are easily affected by politics. It thus appeared to be difficult for them to work purely within the scientific spirit to expand beyond their statutory mission.

It is disappointing that we still do not see any cases of autonomous testifying and/or reflection by the operational organisations involved in nuclear energy and weather. We are also concerned that there has been no change, even after the earthquake, in the constitution of the communities that have been complacent regarding the unfounded assumption of the 'safety myth'.[3] Although eight years have passed since the earthquake and the nuclear power plant disaster, we still have to be concerned with the problem of the apparent inability of the operational organisations to adequately reflect on their actions. However, the main subject of this section is not this point.

A more serious problem is that self-control and an odd silence were also found for the scientists and researchers who did not belong to the operational organisations. This attitude is far from the basic spirit of science. Although easily misunderstood, the reality is that personnel in the operational organisations, including governmental committees, are not necessarily scientists. A crucial problem, which people outside the fields have felt, is that even scientists not belonging to the operational organisations as a whole were unduly influenced by the same atmosphere as the operational organisations of Japan, in which those scientists worried about having an excessively large impact and also hesitated or avoided the dissemination of important information because it fell outside the purview of their primary duty. Such silence was contrary to the public good, and such behaviour led to distrust of and disappointment in scientists.

How can we create a mechanism and robust social structure that can flexibly address such an unexpected and serious disaster that has a large impact on society? Can scientists play a role in such a mechanism? Do scientists have no other way than to be silent because the professional operational organisations cannot flexibly cope with an emergency?

On the contrary, when the operational organisations cannot or choose not to handle serious problems, the only possible way to handle them, if any, must involve state-of-the-art science and any research group related to that disaster. A body that can handle the crisis must be an organisation that brings scientists in

[3] It had been advocated mainly by the Japanese government and electric power companies that Japanese nuclear power plants are all completely different from those that caused serious accidents such as at Chernobyl and Three Mile Island, and Japanese nuclear power plants are absolutely safe against earthquakes and tsunamis. After the Fukushima accident, this turned out to be a myth.

the related fields together, uses cutting-edge ideas and makes close contact with, yet maintains independence from, the political world.

Will scientists escape from such involvement by claiming that it is not their business? Of course, scientists should act on the basis of self-belief as the central spirit of science, and avoid imposing military discipline with controls, especially in an emergency. However, at the same time, persons outside the community and, even more, the administrator of the upper organisation should not interfere or suppress the activities of the scientists to voluntarily challenge problems not addressed by operational organisations.

On the other hand, confusion will occur in the emergency response if many activities are randomly organised by scientists without any coordination or control. Not only to avoid such confusion but also to respond efficiently and flexibly to the situation, it is essential for experts in each field to share knowledge about the reliability of scientific approaches as well as on uncertainties in forecasting, observations and experiments. Unprepared actions without such coordination would only amplify the confusion.

Given the need for coordination, the question arises whether academic societies, where members of the operational organisations are involved as well, can lead such activities and preparation. The public will expect those academic societies to take the lead. However, through their experiences after the earthquake, where sincere reflection and remorse about the silence has not yet occurred, people will reach the conclusion that most of the academic societies cannot take the lead in the scientific activities necessary for an emergency response. In fact, there exist the following opinions: academic societies must appreciate the majority opinions such that the society should represent the absolute consensus of researchers in the field; or it should not commit beyond their 'own business'; or it should not work on uncertain problems which can be shared with members of the operational organisations.

While keeping these opinions and the real situation in mind, what types of robust and flexible organisational form and structure can allow scientists to play a role in creating a society that can cope with urgent problems and crises that extend beyond the capabilities of operational organisations? Moreover, scientists attend to their ongoing business of promoting and advancing their own research. Hence, we must build an organisation to achieve the two seemingly irreconcilable tasks of maintaining the independent freedom of science while providing an emergency response that can handle a crisis. One way is to build a volunteer group that is independent of academic society but consists of researchers as reservists or volunteers in the basic science field related to the operational organisations and to develop a protocol so that they can work together in daily activities in preparation for crises. In addition, this group should be established with a mechanism to gain the trust of the public and constant support from society. To support this strong mechanism, we need the integration of infrastructure, organisation, system construction and an authority structure that secures the prompt and efficient use of rich data and networks owned by the operators and operational organisations. Furthermore, a mechanism to review the response performance of volunteers and volunteer groups is needed on a daily basis. Verifying the reliability of the information that originates from the group is also crucial.

Some heroic activities were undertaken by individuals after the 2011 earthquake and nuclear power station emergency, even in the absence of effective mechanisms. These activities should be highly appreciated. However, because most of these activities were not developed into organised activities, emergency responses were ill-timed and often ineffective. Moreover, with the excuse of avoiding panic and confusion, counterproductive measures to control and even suppress a wide range of activities and actions to overcome the emergency were implemented. As a consequence, hesitation or self-restraint in regard to disseminating information was widespread among individual researchers. Behind much of this short-sighted behaviour was a lack of strong and rich human networks supported by scientists with an effective information dissemination protocol.

It is necessary to establish a neutral and independent organisation to aggregate and disseminate sound scientific and technical information endorsed by people of great wisdom to make the contributions from science and scientists at the highest level of expertise to society without atrophy, self-regulation and control, both in times of emergencies as well as under normal conditions. It is necessary to build a mechanism that allows this organisation to strengthen research cooperation, to support and advise researchers and scientists in the field, to protect researchers and information dissemination, and to provide advice to operational organisations and industries from a different perspective from the government. The problem is complex and requires a holistic vision based on the perspectives of scientists and researchers with interdisciplinary viewpoints, which is difficult to implement via individual groups.

Science is viable only when two conditions are satisfied. The first is preserving diversity even after reaching consensus through the confrontation and sublation cycle, which enjoys development enabled by coexisting freedom and competition after surviving experimental tests. The second is the claims and activities based solely on the scientific conscience of each researcher. Following this philosophy, this neutral and independent organisation for information dissemination should not control or suppress scientists who transmit information, maintaining freedom and diversity. It is important to convey a variety of ideas rather than a single voice when there are uncertainties and conflicts, yet reliability should also be maintained at a high level. If uncertainties exist, as is often true in emergencies and crisis problems, the new organisation should clarify the distribution of scientific opinions to the public with estimations of the reliability of the information. It is crucially important to continue such efforts at normal times as well.

However, in a situation in which an urgent response is needed or for a problem that includes the national interest, it is important to quickly provide reliable information and show how experts and information are used in the decision-making process, allowing for the free dissemination of information by each scientist. In the earthquake and the nuclear power plant accident, it is unclear whether there were efforts to address the situation using the best experts, knowledge, cooperation and mobility; this was considered to be the main reason for the public's distrust. It is essential to establish an organisation that brings together scientists on the front line to analyse reliable information and then react flexibly by taking advantage of the scientists' expertise. Without daily and constant activities, those actions will not work in times of crisis.

Of course, the establishment of links between information publicity and administrative protocols is also needed. However, for issues requiring a political decision, the administrative and political level must make a responsible decision based on the multiple views that have been presented by scientists, while the scientists must make their best efforts without being held responsible for the results and consequences of the political decisions. It is also desirable to form a mechanism to review the validity of political decisions later by independent scientist organisations for the sake of making better decisions in the future.

In addition, it is desirable to provide appropriate advice from the independent perspective of scientists to operational organisations that might become somewhat rigid in their thinking due to their own interests.

The neutral and independent organisations engaged in the abovementioned activities need to play a role to protect scientists from attacks and control when they provide opinions and conduct activities based on their scientific conscience. The opinions of scientists at the highest level are believed to contribute to the resolution and mitigation of problems with profound societal impacts, even when faced with uncertainty. However, presenting these opinions from a scientific perspective becomes impossible if the scientists are burdened with unlimited responsibility for inadvertent outcomes due to uncertainty in the problem. Moreover, the opinions from experts should be accepted by the public only after sufficient legal consideration, forethought of the social impact and effort to obtain the public's trust.

The SCJ should take the leading role in such a neutral and independent organisation of volunteer scientists and researchers. However, the current system of the SCJ has a strong constraint due to its activity wholly relying on volunteer work and lacks the leadership to lead the science and technology communities in Japan. One of the major causes is the financial situation of the SCJ. Given this situation, it is necessary to form an autonomous organisation that regularly examines the reliability of, evaluates, promotes and disseminates scientific information supported by scientific and technical wisdom to obtain greater insight in the field. Unlike the current situation of the SCJ, this organisation must have solid financial support; otherwise, it will be difficult to effectively work by incorporating the rapidly growing network to capture cutting-edge research. In Japan, there are almost no independent organisations supported by the government for the scientific community to consider scientific evidence and express their opinions. This situation is different from those of the USA, the United Kingdom and others, reflecting the shallowness of the state of science in Japan to date and its short tradition.

To ensure the transmission of highly reliable information, it is necessary to create a flexible and robust information dissemination environment that can function even during and after an emergency situation. In such an environment, scientists would be able to freely access a large amount of conflicting information and transparently collect and share high-quality scientific data. In particular, such an environment is indispensable for evaluation of the reliability of various data for issues of the national interest and to enable professional researchers and the general public to access and refer to them. This problem is closely related to the long-term issue of developing science and technology activities, which already occupy a substantial portion of the national budget, and the resulting explosive increase in scientific information, which should be conveyed in a meaningful way backed by the understanding of the public who financially support them. There have been some attempts, such as the DIAS (Data Integration and Analysis System), although they have not worked well and have not been helpful in a disaster emergency. Spanning the boundaries of ministries and agencies, there also needs to be a mechanism for the transparent and reliable rapid dissemination of information regarding disaster and emergency events that focuses on public interests.

As discussed above, the academic societies that are only linked with the present SCJ and operational organisations have not solved the underlying problems because of the limitations arising from their structures and principles. Development of possible activities of the SCJ and strong cooperation with new organisations is desired. At the same time, it is also necessary to pay attention to the limitations of the SCJ as a federation of societies. Beyond that, the development of a new organisation and structure, as well as efforts for autonomous dissemination of information by collecting wisdom from scientists, are also sorely needed. The new organisation responsible for this scientific information and transmission should make independent and critical inspections of the closed science communities, which have been created by the operational organisations and their related academic societies, and check their activities as a coordinator, separate from all stakeholders. The scientists in Japan, who experienced an unprecedented earthquake disaster and the nuclear power station accident, have a mission to create a new trend in the scientific community that provides us with a new framework for information dissemination, which also delivers messages from Japan to the world. This framework is the core to support society in the response to a disaster.

Finally, the first-response team in the event of an emergency should of course be a professional force comprising disaster and emergency response experts. At this stage, there is little that scientists can do, and it would not be suitable to organise an emergency response team comprising solely of scientists. It is the scientists who should continue to monitor and check whether the crisis response system can function, whether it can fully incorporate outcomes of state-of-the-art science and technology and whether the information is reliable. Thus, the scientists should verify, examine and advise on whether the crisis correspondence system for emergencies is highly reliable and reflects the latest science.

In addition to presenting the personal opinion of the author, this section summarises the discussions of the Information Dissemination Subcommittee of Computational Science Simulation of the SCJ and the interim report of the Working Group for Information Transmission from the Computational Science to Society of the Liaison Committee in the Advanced Institute for Computational Science, to which the author contributed. The author would like to express his gratitude for the lively discussions in both of these committees.

11

Emergency Actions and Messages Related to the Fukushima Accident

In this chapter, we introduce emergency activities undertaken by researchers and academic communities. The cross-field collaboration arose from a series of small emergency activities. We describe the stories of these activities and what the researchers felt in the development of the accident correspondence as well as thoughts for the future that might be useful for the next generation.

11.1 Reports from Fukushima University

AKIRA WATANABE

Fukushima University is the only national university in Fukushima Prefecture that is located near the earthquake and nuclear accident. The university provided much of the correspondence about the earthquake and the nuclear accident. Moreover, the highest dose rate from the radioactive material released into the environment by the nuclear accident was measured at the public facilities of the university. Putting aside the reality of crisis management, here we describe several thoughts for a safer society.

On the afternoon of 11 March 2011, while drinking a cup of coffee after the conclusion of a conference, a large shake like I had never experienced occurred. At 14:46 my remaining coffee was spilled into the carpet and I nestled under a table and waited for the shaking to subside. I had experienced many earthquakes since 1978, but this one was extraordinarily long.

My position as vice president requires me to coordinate crisis management of the university as soon as possible to protect the Facilities Division. However, when asked for safety confirmation, I advised not staying inside due to the magnitude of the aftershocks, instead suggesting that people should go outdoors. Determining whether people were trapped in elevators amid the subsequent aftershocks took approximately one hour. There were about 100 people on the campus. On 12 March, parents and children who had been on campus for evening exams gathered in the central square. We were instructed to ensure safety around 15:30. In addition, I was instructed to open the shelter for stranded commuters, and we began to distribute food and water.

Although the apartment complex near the university typically housed nearly 2000 people, only approximately 150 people who had not gone home were there due to the spring break period; it was still necessary to ration food and water. Although the university water supply and sewerage system were broken, electricity was available, which allowed people to remain abreast of the news on television. In addition to the surreal appearance of the tsunami disaster, much like in a movie, information from the Tokyo Electric Power Company was released regarding the first nuclear power plant accident. However, due to radiation leaks, I could not focus. During the nuclear power plant accident, I believed that before the cooling function was activated the crisis headquarters had performed earthquake response procedures. After having finished the immediate safety checks, I was instructed to stay at home until 15 March unless needed at the crisis headquarters.

However, safety cannot be confirmed by appearance. Although the ceiling of several buildings in the university had collapsed, there were several places with no major damage. The recovery lasted until the end of April. In addition, the supply of gasoline had also been disrupted, which meant customers had to wait up to eight hours in line through late March.

11.1.1 The Seriousness of the Fukushima Daiichi Nuclear Power Station Accident

The Fukushima Daiichi Nuclear Power Station (FDNPS) lost its cooling function, which was followed by the news that the situation had become far more serious. At 21:23 on 11 March, evacuation orders for the area within 3 km of the nuclear power plant were issued. Gradually, the nuclear accident had become more serious than the earthquake. While I corresponded with evacuees through the night, the disaster headquarters meeting occurred the following morning to confirm the safety of the university members; similar meetings were held twice in the afternoon. However, in the later correspondence, the nuclear power plant accident became the major issue.

Unit 1 exploded at 15:36 on 12 March. We could not understand the hydrogen explosion of the nuclear reactor. People were concerned about whether radioactive material was leaking. First, the evacuation order was issued covering the area within 10 km of the FDNPS, then the area expanded to 20 km that evening. When the hydrogen explosion first happened at 15:36 it was not then thought that radioactive material was released. The release of radioactive materials was not clear even in the statement of the Japanese government. We prepared to evacuate from the university. The cooling function of the nuclear reactor remained stopped, and it was not a cooling that could be safely secured. We had concerns about re-criticality. Unit 3 also suffered a hydrogen explosion at 11:14 on 14 March. Cooling of unit 2 was also inadequate, and we were in a crisis situation. Water vapour containing high-concentration radioactive substances was released from the exhaust port of unit 2 at 02:00 on 15 March. At 6:10, a loud noise came from unit 2. The pressure of the containment vessel decreased from 7.3 atm to 1.5 atm in 25 minutes. The leaked water vapour was reported to contain radioactive material, but the Japanese government continued to announce that there was no health impact. People who work at the FDNPS reported by phone or email from early in the morning of 15 March that they should evacuate to their families and friends. As a result, traffic congestion occurred early in the morning on National Route 114. Meanwhile, plumes containing high concentrations of radioactive material passed along Route 114. I was contacted with rumours of a high dose in the vicinity of Iitate along the Route 114 evacuation path, which originated from the Fukushima mitigation measures. I began to conduct a brief transport study using the NOAA HYSPLIT model. The surface emissions moved to a height of 50 m above sea level and towards the sky over Fukushima City, where the radioactivity was expected to increase by a factor of 19. The display of the GM dosimeter in my office at the university indicated a similar result. However, this did not explain the observed high dose at Iitate, which advanced to a height of 300 m and passed through the centre of the village.

I held a crisis task force meeting at the university on 16 March, and reported the prediction of the transportation forecast of the radioactive plume on 17 March. Radioactive substances were expected if the situation continued. Instructions were provided to help to avoid indoor exposure. However, residents were already concerned due to the high dose rates. It was unclear whether the best approach was to reduce the number of students at the campus by sending them home. Meanwhile, the trains from Sendai (about 70 km north of Fukushima), Niigata (about 170 km west of Fukushima) and Nasushiobara (about 100 km south of Fukushima) were restored. To help students return home, we carried out home support to these three points.

On 17 March the cooling state entered a serious situation, and water injection by helicopter began. However, there was no relief from these efforts. In the afternoon, the US government issued a recommendation to evacuate Americans within 80 km of the accident site. Fukushima University is within 60 km of the accident site, and we had to consider the evacuation of the university itself. I was asked whether I had to wait for the evacuation recommendation from the Japanese government. The water supply to the nuclear fuel pool finally began in the evening with the support of the Self-Defense Forces; the use of special vehicles similar to those of the Tokyo Fire Department indicated the seriousness of the situation. However, other than the government's statement that there were no immediate health effects due to exposure, there was no information on the environmental radioactivity.

On 19 March, a Fukushima University Graduate School of Symbiotic Systems Science meeting decided to measure the dose rate with a 2 km grid of Fukushima City. In coordination with Fukushima Prefecture, five dosimeters were obtained. On 20 March the successful injection of cooling water by special vehicles to the unit 4 fuel pool resulted in some relief. On 23 March, five LP gas taxis were chartered, and the measurements began. The high dose rates from the village of Iitate had already been reported by the university, and an evacuation order was issued based on different data from the International Atomic Energy Agency (IAEA). Another research institution announced its own radiation dose measurement results, although the values were different. The cause was related to differences in instrumentation and altitude. In lieu of this situation, it was decided to limit the monitoring height to 1 m over the 2 km grid.

On 23 March, the US DOE published the US military unmanned aircraft results of the radiation dose around the FDNPS for 17–19 March. The observation results agreed with the prediction results of SPEEDI, assuming quantitative discharge. It became obvious that the contamination distribution predictions by SPEEDI were not being effectively utilised. However, the observation results of the DOE also showed that the high dose range extended northwestward from the accident site, but consistency with the air dose rate on the ground had not been confirmed.

On 27 March, the Fukushima radiation measurement team presented their results, which are shown in Figure 11.1. Those not only coincided with the high dose rate range that the DOE found through its aircraft observations, but also caught the emergence of a high dose rate range in the indoor evacuation recommendation area of 20–30 km from the accident site. We decided to first use this observation result for evacuation behaviour because the high dose range existed in the area that Namie Town residents had evacuated. Residents of Namie-machi felt distrust of the government's uncertain evacuation order. However, residents who evacuated to the Namie-machi Tsushima branch more than 30 km from the accident site accepted the measurement results of Fukushima University. The measurement result indicated that the dose rate of the evacuated Tsushima branch was high, so they understood the necessity to evacuate again and evacuated to Nihonmatsu. It also helped to review the evacuation plan of the Japanese government. That is because Fukushima University was near the accident site and there was an established trust in the relationship.

Although the FDNPS cooling pump began to work again from electricity supplied by a commercial power company, it was not until April that this information brought peace of mind to people in the area. Moreover, dosimeter data

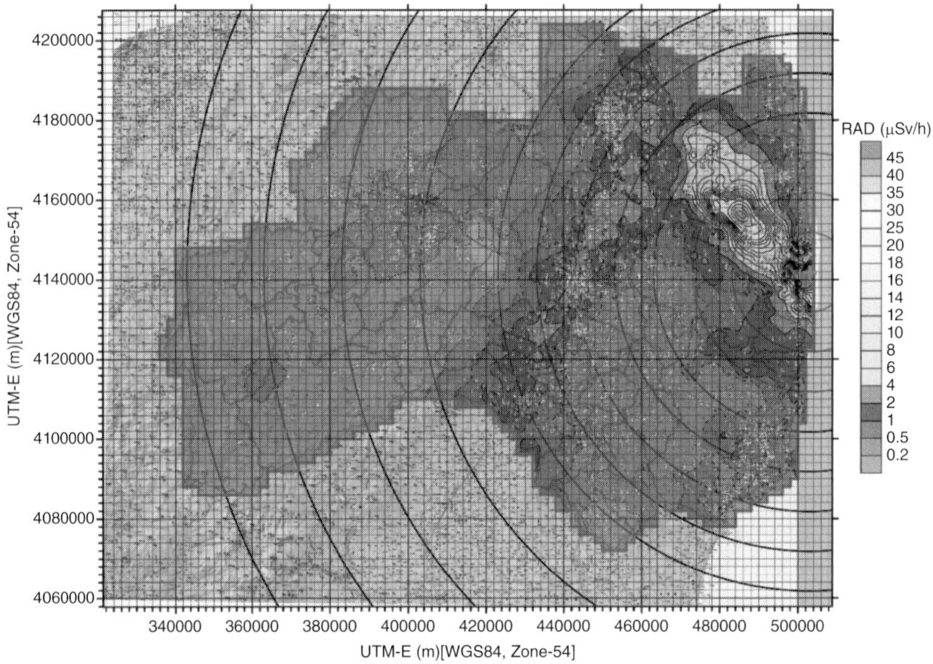

Figure 11.1 The air dose rate distribution at a height of 1 m at the end of March 2011 observed by the Fukushima University radiation measurement team is shown. This result was announced on 17 June 2011. The figure also includes the radiation monitoring results conducted at elementary and junior high schools over 12–16 April; the measurement of the dose rates within 20 km from the FDNPS was conducted by the Ministry of Education, Culture, Sports, Science and Technology. A black and white version of this figure will appear in some formats. For the colour version, refer to the plate section.

gradually became more widely available, indicating a value about 3 μSv/h around the university. In order to start classes at the university, it was necessary to reduce the dose rate and restore the transportation network. The decay curve was calculated from the change in the dose rate up to that point, and it was confirmed by extrapolation that the dose rate would become 1 μSv/h or less in May. However, at that time the composition of radioactive material was not known. We did not understand what was being done. At the end of April, the rate of decay of the dose rate gradually declined. The dose rate did not decrease as initially calculated. In addition, because the 1 μSv/h level was not achieved, many colleagues' faculty opposed the start of classes.

Both Vaisala, Inc. and Tokyo University offered assistance; it was possible to perform various observations such as those described in Section 3.6. The installation of a high-volume sampler was performed by Dr H. Tsuruta of the University of Tokyo. This effort was to give the appropriate information to those who thought they would not be able to open windows in the classroom to prevent radiation exposure, or who thought they had to put on a raincoat or mask in the hot weather. In addition, the time-consuming measurement support and high-volume sampler for radioactive material in rainwater at Osaka University were underwritten by Professor Shinohara and the University of Tokyo isotope laboratory under Dr Hiyama. These efforts further expanded the installation of high-volume samplers using impact filters, which, according to the Tokyo Institute of Technology's Professor Yoshida, provided powerful data showing the relationship between the radioactive material and pollen, which had become a problem of interest.

Classes in elementary, junior high and high schools in Fukushima City began as scheduled in mid-April. However, parents were worried about radiation exposure while students were going to and from school and exposure due to the opening and closing of windows of the school building. Therefore, outdoor physical education lessons were cancelled, and students began to move out of Fukushima Prefecture. The measurement of radioactive materials in the general environment progressed, and it became clear that most of the radioactive caesium was on the surface of the soil; however, it was shown that surface peeling of the soil was effective for reducing exposure. Meanwhile, at Fukushima University, an experiment on the effectiveness of 'deep ploughing' and 'soil exfoliation' was carried out, and results showed that removing the top 10 cm of topsoil reduced exposure to one-tenth of previous levels. Disposing of exfoliated soil was an issue, but this was used as a standard radiation reduction measure at schools in Fukushima Prefecture. In some cases, miscellaneous soil exfoliation work was done and the dose rate only decreased to half, but the dose rate fell to about 0.2 μSv/h. However, once decontaminated there was a high dose rate at a height higher than the ground surface, and the need to decontaminate buildings and trees became clear. Decontamination of a wide area became an issue, and the cost of decontamination was not clear. We realised that we could not simply or easily return to the environment as it was before the accident.

11.1.2 Lessons from the Fukushima Accident

Seven years have passed since the FDNPS accident. At the accident site, the number of tanks for the accumulation of contaminated water has reached 1000 (1000 tonnes each), meaning the green area has turned into tank forest. Approximately 7000 people work daily on treatment of contaminated water and rubble. Seventy per cent of the people working at the accident site are from Fukushima Prefecture; they have not only experienced radiation contamination disasters but are also responsible for the treatment of their contaminated hometown. How many of our activities since the accident were useful as safety measures for the residents and measures to prevent future accidents? The more time goes by, the more we think about our unpreparedness as a society regarding nuclear disaster prevention. In the eight local governments close to the accident site, only 13.5% of the residents have returned, and many of the returning residents are elderly people over 65 years old. The numbers of returning people under the age of 40, who will support the future of the municipality, are quite low, even in the low dose rate area. The reason is that the FDNPS, where the aftermath of the accident continues, is nearby and the damage to health remains unknown, even at low dose rates. In some cities, towns and villages, it is said that the transition to an ageing society has been accelerated and become a reality as much as 20 years early due to this accident because returnees are limited to the elderly. The accident separated neighbouring families, destroyed workplaces and devastated local communities. Even in municipalities far away from the accident site, troubles occurred in decontamination storage and decontamination methods, which had a great influence on various regional clusters. Based on lessons learned from this accident, how much was the safety management of nuclear power plants strengthened? Certainly the revetment work corresponding to the prediction of the height of the tsunami was reinforced, and the activity situation of the active fault was also reconsidered. Safety testing of and personnel training at nuclear power plants were implemented, and evacuation areas were also changed. The affiliation of the Nuclear and Industrial Safety Agency (NISA) changed from the Ministry of Economy, Trade and Industry to the Ministry of the Environment, and its name was changed to the Nuclear Regulatory Authority. However, there are 1573 fuel rods in nuclear reactor buildings of units 1, 2 and 3, and they have not been collected yet. In addition, the melted condition of 1496 fuel rods in the operating nuclear reactors was recently confirmed. Not only is the cooling space for those fuels insufficient, the final disposal method and location are uncertain.

Fukushima's lesson is that we cannot accept that accidents occur only due to greater disasters than are accounted for. In the 40 years since the start of the facility's operation, exhaust operation training to lower the pressure of the containment vessel and inspection of the ancillary steel structure supporting the transmission line of the commercial power supply were not conducted even once. The operating system supported by the safety myth has been exposed. Safety myths cannot be implemented by only one company or country. The safety myth grows as people become indifferent to risks while believing they are safe. Fukushima's lesson is to build a society in which all activities have various risks, people understand the risks properly and then people share those risks. What I learned from the accident at Fukushima is the importance of building a safety management system that goes beyond generations. Safety management is impossible if there are no manufacturing-experienced persons after 40 years, and events that cannot be dealt with by reading manual alone occur. One of the lessons learned from the accident at Fukushima is that it is important for experienced manufacturing personnel at the nuclear power plant to inherit all the technologies and connect them to the next generation. The lifetime of a nuclear power generator is not the life expectancy of a machine; it is important to decide it based on the life span of a person who has safety management and manufacturing experience.

For the past 1300 years, Ise Jingu Shrine has held an event in which the shrine is rebuilt every 20 years. By rebuilding the shrine once every 20 years, we inherit the skills and technology to make the shrine and 714 kinds of clothes and furniture. This is certainly one way to convey culture and technology to the following generations. As a result, people have been able to visit the unchanged shrine for 1300 years. I think that such a periodic transmission of skills and technology is one way in which the management of nuclear power plants can avoid accidents. Furthermore, it is unlikely that lessons learned from the Fukushima accident will be effectively used, such as extending the life of the nuclear reactor from 40 years to 60 years. Even if the Fukushima accident had not happened, I would be keenly aware that there are structural problems in safety measures specific to Japan that will not be solved. The social structure that creates a safety myth does not change easily. It is a task to build a society that can understand risks by gradually accumulating facts based on science and technology.

11.2 Efforts of the Science Council of Japan and Scientific Societies and Unions

TERUYUKI NAKAJIMA, TOKUSHI SHIBATA AND TOMOYUKI TAKAHASHI

This section and the following three sections introduce the activities performed by each academic community.

At the time of the Fukushima accident, without information on how to contribute to the countermeasures of the government, studies and activities were conducted quite haphazardly. Soil investigations in Fukushima Prefecture began on 14 March at Fukushima University, indicating the importance of bottom-up activities to complement top-down activities undertaken by the government. However, the decision-making system in Japan was slow to include such bottom-up efforts in governmental countermeasures.

The role of the SCJ as the headquarters of the academic community and scientific societies and unions was important. The SCJ held an emergency workshop titled ‘Today, What Can We Do?’ to begin exchanging ideas for emergency actions, which culminated in the SCJ statement titled ‘Disaster in Northeast Japan and Nuclear Emergency’ on 18 March. They then established the Great East Japan Earthquake Committee and issued an SCJ recommendation titled ‘First Emergency Recommendation Regarding the Response to the Great East Japan Earthquake’ on 25 March, which was followed by many other recommendations and reports.

Beginning in October, the twenty-second term of the SCJ established its Committee on Supporting Reconstruction after the Great East Japan Earthquake to plan activities for deliberation, investigation and recommendations. The Nuclear Accident Response Subcommittee of the General Engineering Committee established an investigation sub-subcommittee for environmental contamination by nuclear plant accidents and working groups for atmospheric modelling and early data excavation and collection. The modelling working group made an international call to solicit model calculation results for the transport of radioactive materials. Research groups contributed results from 9 regional atmospheric models, 6 global atmospheric models and 11 ocean models (Science Council of Japan, 2014). A data collection working group was established to collect data to better evaluate the public exposure dose. They reorganised the working group into a data archive working group for radiation and radioactivity to archive data and records of measurement activities from the perspective of societal implications for the next generation. In preparation for data excavation, they conducted a questionnaire of existing observational data types in cooperation with 22 academic societies and unions. The main issues were found to be the methods of announcement and collection to prevent the loss and burying of data, sustainable methods for managing the collected data and system management.

The various academic societies and unions also conducted emergency gatherings and investigations and issued messages. On 15 March the Japan Radiation Chemical Society called for volunteer measurements. On 17 March the president of the Atmosphere, Ocean and Environment Science Section of the Japan Geoscience Union issued a message to solicit emergency cooperation for emergency support from both national and international communities. On 20 April, a joint meeting, later dubbed the 'Joint Board of Environmental Radiation Nuclear Physics and Earth Science', was held with nuclear physics and earth science researchers. These extended research groups appealed to MEXT for a large-scale investigation and research study. Regarding marine observations, the Oceanographic Society of Japan organised a disaster response working group in mid-April and started a marine survey using research vessels. From April to July, workshops and lectures were held by the Japan Geoscience Union, the Meteorological Society of Japan, the Oceanographic Society of Japan, the Japan Society for Analytical Chemistry, the Geochemical Society of Japan, the Japan Society of Nuclear and Radiochemical Sciences and other organisations. In August, a Fukushima session was programmed in the Goldschmidt Conference, and a joint communiqué was announced by Mitsuru Ebihara, the president of the Geochemical Society of Japan, Bernard Bourdon, the president of the European Geochemical Society, and Samuel Mukasa, the president of the International Geochemical Society.

Through these activities, a national network of the earth sciences gradually emerged. As a result, a project proposal to the Grant-in-Aid for Scientific Research on Innovative Areas (principal investigator: Yuichi Onda, Tsukuba University) was submitted and approved in June 2012. A year had passed from the first movements of the researchers. Another bottom-up programme was an emergency funding programme from the US National Science Foundation (NSF-RAPID) and the corresponding programme of the Japan Science and Technology Agency, the International Emergency Joint Research and Research Support Program (J-RAPID), which was announced on 18 April 2011.

Reflecting on these events, we observed that each activity was connected with the others and that gradually the combined force led to larger effects. The academic community took a large role in taking care of distributing messages to the international communities, many of which were beyond the control of governmental activity. In this process, discussions began between academic communities and the MEXT Emergency Operation Center (EOC). Today, many measurement data and model simulation results have been collected through these efforts. This lesson indicates that although researchers' movements can be slow in an emergency situation, this does not mean capitulating because they are capable of providing a large amount of information through persistent research activities (Nakajima *et al.*, 2011).

Activities continued. A workshop titled 'One Year from the Fukushima Daiichi Power Station Accident: Validation of Monitoring Data for Environmental Radiation and Radioactivity' was held in March 2012 by the Chemical Society of Japan. An open workshop titled 'Reconstruction of Environmental Emission and Diffusion Process from the Fukushima Daiichi Power Station Accident' was held in March 2012 by the JAEA; an international workshop titled 'Fukushima Ocean' was held in November 2012; and the 'Fukushima Special Session' was held in January 2013 by the American Meteorological Society.

In February 2014, the SCJ established action guidelines for emergent large-scale disasters to increase the SCJ function of collecting and disseminating scientific knowledge by the following mechanisms, as also schematically illustrated in Figure 11.2:

1. declaration and termination of emergency;
2. call for an emergency response committee consisting of presidents, vice-presidents, division officers, representatives of committees relating to the emergency situation, professionals in fields related to the emergency, etc.;
3. actions taken by SCJ, including:
   - the president issues discourses, statements and recommendations;
   - communication of opinions with and information provision requests from governmental agencies;
   - information sharing in the SCJ and release of information to public;

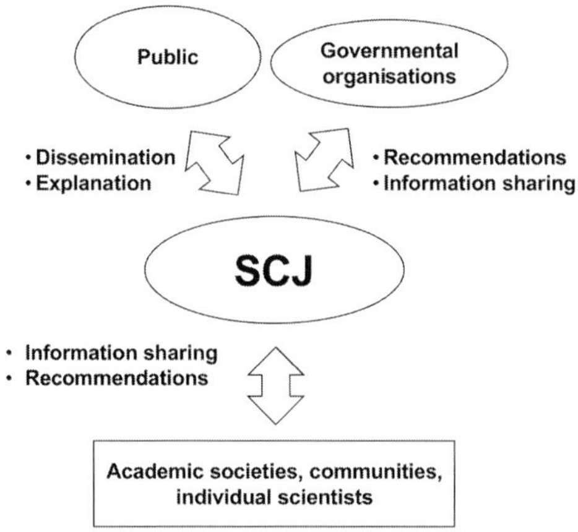

Figure 11.2 Guidelines on activities of the SCJ in emergency situations (as of February 2016).

4. cooperation with disaster research academic organisations; and
5. contact and collaboration with academic organisations abroad.

By this mechanism, part of the chaotic situation of information dissemination in the Fukushima accident will be resolved. Professionals representing an academic authority, the SCJ in this case, collect the scientific information and can disseminate useful high-quality information to governmental agencies and the public with enough explanation and with information on the uncertainty. By this mechanism, young scientists can be protected by the authority for publicising their research results about the accident, which might be useful.

In parallel, the academic societies established the Japan Academic Network for Disaster Reduction in January 2014 and started close cooperation with the SCJ. To date, 56 societies have joined the network. The first emergency actions by the SCJ and the Academic Network were undertaken at the time of the Kumamoto earthquakes in April 2016. They made calls for emergency investigations, data collection and coordinated media presentations. Related academic societies provided their knowledge to the network to provide accurate and detailed explanations to the public. We feel the realisation of this improved situation of information dissemination reflects our having learned from the Fukushima accident.

11.3 Urgent Atmospheric Measurements Under Collaboration between Geoscientists and Radiological Chemists

HARUO TSURUTA AND TERUYUKI NAKAJIMA

Just after the Fukushima accident, air pollutant transport models were urgently improved by experts in the community of material transport modelling, and a simulation of the Fukushima Daiichi disaster was conducted. Radioactive substances that are released into the atmosphere are transported through the atmosphere as gases or particles, or while they transform from gases to particles, and the latest air pollution models reproduce well the process of their widespread dispersion (see Chapter 3). If experts had provided the results of multiple models and an expert team that comprised the staff of the Japan Meteorological Agency and other agencies had analysed these models with meteorological fields, the results would likely have been extremely useful for emergency countermeasures. However, although many modelling results that used bottom-up calculations were published internationally (e.g. Morino *et al.*, 2011; Takemura *et al.*, 2011), they were not adequately used to make an urgent plan.

At the time of the disaster, many earth scientists realised that they had to act quickly to collect materials and data that would otherwise be permanently lost, and as a result two groups launched parallel emergency action plans. In an emergency response to the Fukushima Daiichi accident, the Japan Society of Nuclear and Radiochemical Sciences (JNRS) called on its members belonging to different groups to monitor radioactive substances as soon as possible. Subsequently, the statements that were posted on the society's website by its president spurred a varied series of related initiatives. Starting on 15 March, calls went out to the relevant parties, and atmospheric radionuclide monitoring began at the High Energy Accelerator Research Organization (KEK), RIKEN, Kanazawa University, Niigata University, Nagoya University, Osaka University, Tokushima University, Kyushu University and elsewhere. Beginning on 22 March, a system was established for checking and releasing these data to the public; this was led by Noriyuki Momoshima of Kyushu University (the director of the JNRS). Simultaneously, a collaborative team among the JNRS, the Geochemical Society of Japan and the Japan Geoscience Union launched atmospheric radionuclide monitoring. An application was made to MEXT for emergency funding to provide the equipment and filters for the emergency collection of atmospheric (aerosol) samples, Ge detectors to measure radionuclides and participation in the soil project (described in the following). However, this funding was not granted, and accordingly most of this research was conducted by volunteers without compensation. The JNRS also received a request to measure radionuclides in the agricultural soil survey that was led by Fukushima's prefectural government and the Ministry of Agriculture, Forestry and Fisheries. Then, the JNRS responded through Gakushuin University and Tohoku University's Research Center for Electron Photon Science (which was Tsutomu Ohtsuki's group at that time). These grassroots activities evolved into the Japan Geoscience Union's Emergency Radioactive Substance Research Team (led by Kazuyuki Kita, Ibaraki University), which implemented multipoint observations of radioactive substances in atmospheric aerosols throughout a wide region

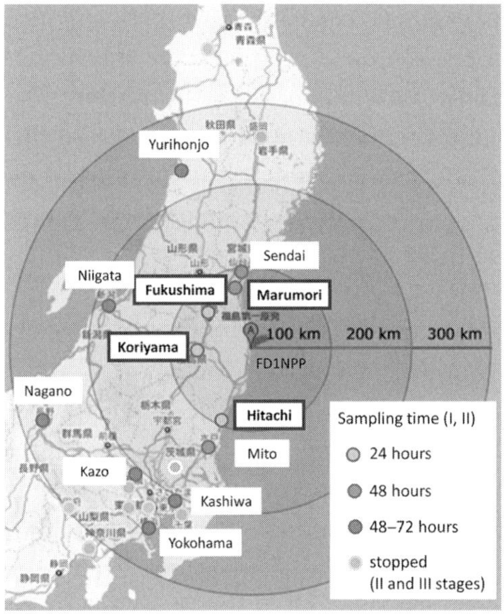

Figure 11.3 Emergency monitoring sites for atmospheric aerosols by the Emergency Radioactive Substance Research Team organised by the Japan Geoscience Union under Kazuyuki Kita. The first stage (I) was from late March to May 2011, with monitoring at about 20 sites. In the second stage (II, until August 2011), the monitoring was conducted at 11 sites (Yurihonjo, Sendai, Fukushima, Koriyama, Niigata, Hitachi, Mito, Kashiwa, Kazo, Yokohama and Nagano). In the third stage (III, from August 2011) the monitoring has been carried out at four sites (Marumori, Fukushima, Koriyama and Hitachi). The monitoring at Marumori was started in December 2011 when the monitoring stopped at Sendai. The original figure made by Kazuyuki Kita was modified.

surrounding the FDNPS (Figures 3.24 and 11.3). The early two-year results at Fukushima and Koriyama were described by Tsuruta and Nakajima (2012). Currently, the long-term monitoring of atmospheric radionuclides is continuing at four locations (Figure 11.3). The long-term variations of atmospheric radiocaesium are shown in Figure 3.14, which indicates some peaks still being observed even five years after the accident. A comprehensive analysis of these long-term observations will be summarised elsewhere. Efforts are underway to compile the data that have been separately gathered and monitored through these various initiatives so that they are not scattered and lost (see Section 11.2).

To evaluate internal radiation exposure, data on ^{131}I are essential. However, this isotope has a half-life of eight days, and data can be collected only through real-time monitoring. For this reason, ^{131}I data measured just after the accident are extremely limited (see Section 3.7). In this context, Nippon Hoso Kyokai (NHK) and Masaharu Okano, a former researcher of RIKEN, succeeded in collecting the monitoring post data on atmospheric ^{131}I in Fukushima Prefecture. Preliminary analysis of the data was conducted by him in cooperation with Teruyuki Nakajima and Haruo Tsuruta (University of Tokyo at that time). Masayuki Takigawa (the Japan Agency for Marine-Earth Science and Technology (JAMSTEC)) simulated the pathways of ^{131}I in the atmosphere by using an atmospheric transport model, and Hitoshi Sato (Ibaraki Prefectural University of Health Sciences) performed preliminary calculations of equivalent doses. Including a new member, Fumiya Tanabe, a former researcher of the JAEA and an expert on the events within the reactors, a team supported by the NHK was formed to discuss how the initiatives that are described above could use this type of data and other data that were released later (NHK, 2012; 2013).

A few years later, an estimate of atmospheric radionuclides released from the FDNPS just after the accident was produced by analysing the pulse height distribution dataset from an NaI(Tl) detector at monitoring posts in Fukushima Prefecture near the FDNPS (Hirayama *et al.*, 2015). In addition, an estimate of atmospheric radionuclides was also produced by analysing the pulse height distribution dataset from an NaI(Tl) detector at monitoring posts by Ibaraki Prefecture located in Tokai-mura (Terasaka *et al.*, 2016). Both estimates are useful for understanding atmospheric ^{131}I just after the accident, although there are uncertainties about analytical methodology. Furthermore, the atmospheric ^{131}I concentrations which were measured just after the accident by several institutions and open to the public were analysed (Tsuruta *et al.*, 2013; Lebel *et al.*, 2016). Although many researchers have endeavoured to clarify the behaviour of atmospheric ^{131}I just after the accident, the whole picture of atmospheric ^{131}I has not been understood yet. More comprehensive studies are expected under a collaboration of experts between atmospheric sciences and nuclear sciences.

11.4 Urgent Survey for the Disaster at Sea

MITSUO UEMATSU, TAKESHI KAWANO AND ATSUSHI TSUDA

Radioactive substances spread over the Pacific Ocean after the accident at FDNPS. It is more difficult to comprehend the situation of radioactive contamination in the sea than that on the land because well-equipped research vessels and experienced members must be prepared for marine surveys. The facilities where radioactive material can be measured and the experts to measure them are very limited in Japan. The behaviour of radioactive material induced anxiety in the people close to the Pacific Ocean and caused an international panic because the radioactive material spread from the nuclear plant to the sea through the air and by direct effluent disposal. A marine survey was needed immediately over a wider area. Therefore, international collaboration was necessary for this survey activity. The international science community, led by US and Japanese scientists, collaborated to investigate the impacts of the Fukushima radioactive material in the air and the sea. In this section the approaches of the survey by MEXT and international activities of scientists from the Oceanographic Society of Japan (JOS) are discussed.

11.4.1 Implementation of a Marine Survey Plan by MEXT

The Comprehensive Radiation Monitoring Plan was announced by MEXT on 22 March 2011, 11 days after the earthquake. Thereafter, the marine survey of radioactive material began. The Japan Agency for Marine-Earth Science and Technology (JAMSTEC) immediately dispatched the research vessel (R/V) *Hakuho Maru* to 30 km off the Fukushima site to monitor radioactive material. The research vessel was ordered to survey in the afternoon of 21 March, and departed the next day. On 23 March the survey began. The observation stations were determined on the evening of 21 March by the Marine Ecology Research Institute (MERI) based on past survey experience. The main tasks (seawater sampling, aerosol sampling and spatial radiation dose rate measurements) were fixed. There was not enough time to prepare for a new cruise task because a previous cruise of the research vessel ended just before the order; the main equipment for the observations was unloaded from the vessel. Therefore, the sampling of seawater was conducted using an onboard pump for surface water. Measurement instruments including a dosimeter and necessary consumables such as sampling bottles and chemicals were gathered on 21 March. They were loaded on 22 March at Yokosuka Port near the JAMSTEC main office. Experts including radiation protection supervisors were brought together and dispatched to cope with the developing situation. An immediate departure was executed because the main goal was to determine the status of radioactive contamination in the sea near the FDNPS for the safety of the Japanese people. Safety precautions, including a rule of a dose limit during the sampling operations, were laid out by experts and prescribed before conducting the tasks. During the cruise, the vessel called at Hitachinaka Port for a delivery of seawater and aerosol samples and restocked with provisions for the daily needs of the crew and researchers. In addition, a specialist in protection from radiation also boarded, and a briefing session regarding safety was held onboard. The air sampler and its filter holders were replaced with those used routinely by the JAEA. Thus, the first monitoring cruise began with considerable uncertainty about the safety of the observation team. Fortunately, the radiation was not dangerous and all tasks were carried out smoothly. In addition, the samples were brought to the JAEA daily, and seawater samples were immediately analysed, with the results being announced by MEXT.

The monitoring of the area within 30 km of the FDNPS was performed via four cruises until the beginning of May. During this period, four research vessels,

Hakuho Maru, *Mirai*, *Kairei* and *Yokosuka*, were rotated to avoid extra stress on the crew and researchers and to limit the long-term use of each vessel. During the R/V *Mirai* cruise, in addition to surface water sampling, water near the seafloor was also collected because the ship was equipped with water samplers. The number of sampling stations was increased to 16, and bottom sediment samples were collected at some stations. The early stages of the monitoring were an extraordinary experience for the researchers. First, massive amounts of marine debris were floating on the surface because the cruise occurred shortly after the disaster. The activities of the ship were limited to only daylight hours. Second, responding to a nuclear disaster was a new experience for many of the researchers. There were some anxieties. Was the seawater for flushing toilets dangerous? Could we handle the waste from the strainer of the engine cooling pump with an unknown radioactivity level? Could we escape from the area when highly contaminated water was released again? Would it be possible that there could be another explosion at the FDNPS?

Researchers typically plan based on their own interests to make onboard observations. However, these activities were not the case. It took much more time because new rules were needed for the surveys. How much radioactive material was contained in the seawater samples? What bottle would be suitable for sampling? Which acid should be added to the seawater sample, nitric or hydrochloric? What amount of acid should be added? How long would it take for aerosol sampling? There were also issues related to the procurement of daily needs. However, these problems were gradually resolved with time. Participating scientists were not satisfied with these activities because these cruises were not for research but for monitoring. However, the researchers believed they were making a contribution to society in the aftermath of the disaster.

After June, the activity shifted from offshore to the open ocean due to dilution and diffusion/advection of the radioactive materials in the ocean. In the 2011 fiscal year, 12 survey cruises were conducted and the data for low-level radioactive material collected in the open sea were also announced by MEXT. This project was commissioned by MERI.

11.4.2 Emergency Disaster Response Activities by the Oceanographic Society of Japan

On 14 April, Motoyoshi Ikeda (Hokkaido University) and his colleagues held a meeting at the Hongo Campus of the University of Tokyo on the environmental damage to the ocean by the earthquake. Unexpectedly, more than 100 participants joined the meeting. Preliminary reports regarding the earthquake, tsunami, and contamination of radioactive material were given by several speakers, which was followed by an exchange of opinions and discussion. At the end of the meeting, Ikeda stated that the JOS should disseminate information and recommendations in response to earthquake disasters. On the next day, 15 April, the recommendations from the meeting were discussed by the JOS executive committee, and an earthquake working group for correspondence was established. On 18 April, Kimio Hanawa, the president of the JOS, declared that the JOS would devote its efforts to clarification of the current situation and prediction of the marine environment. Through the recommendations and preparation of the research survey plans, the JOS could contribute to the earthquake disaster response to protect the public.

However, there are very few experts in the field of radiochemistry in marine science. From the 1950s to the 1960s, frequent atmospheric nuclear bomb tests resulted in a radioactive contamination problem in the environment. At that time, a survey and research system in Japan was established. Thereafter, the nuclear reactor accidents at Chernobyl and Three Mile Island occurred. In spite of such accidents, over time the number of researchers and departments related to environmental radiochemistry decreased gradually. There are few research resources to accommodate such an accident. In the case of oceanographic surveys, it is necessary to use research vessels equipped with water sampling systems, sediment samplers and biological sample collectors. Eagerness alone was not enough to accomplish the survey. TEPCO began measuring coastal waters, including the FDNPS drain, on 22 March. MEXT ordered R/V *Hakuho Maru* to survey the region off Fukushima in late March, and JAMSTEC and MERI were in charge of observations within 30 km. The radioactivity monitoring by the government was led by the Ministry of Land, Infrastructure and Transport (MLIT) and the Ministry of Agriculture, Forestry and Fisheries (MAFF), and MEXT was responsible for correspondence. Deep respect was granted to the crew and researchers investigating the area with large amounts of debris drifting on the surface of the ocean. However, it was evident that these efforts were not sufficient to determine the actual condition.

In response to these circumstances, the JOS earthquake correspondence working group announced recommendations for the monitoring system on 16 May and 25 July. The recommendations in May proposed that the reproduction and prediction of radioactive material spread using numerical models based on the data from the wide area observation were necessary. It was also noted that the data obtained from precise observations in the coastal areas where highly contaminated fish and shellfish were caught were insufficient for running numerical models that would represent the marine ecosystem.

In July, a broad marine observation survey was achieved by the government; however, analytical methods were reserved for emergency actions in response to high-level radioactive samples. Thus, many early data obtained from the seawater samples were below the detection limit, meaning they could not be used to verify the model. The concentrations of radioactive material are expected to increase in higher trophic-level organisms, yet the data below the detection limit in seawater were not suitable for determining the safety level of the biota. Under such circumstances, proposals to apply a very sensitive analytical method were strongly recommended. A proposal titled 'Concerning Marine Pollution Research on the Fukushima Daiichi Nuclear Power Plant Accident' was announced by the JOS working group. This proposal was not limited to only an announcement on their home page; it was also sent to other relevant organisations. Immediately after this, very sensitive analytical methods were applied in the government monitoring programme. However, it is unknown whether this proposal was an effective government programme.

Furthermore, bottom-up observations were collected by researchers over time. Academic research vessels, including R/Vs *Tansei Maru* and *Hakuho Maru* (JAMSTEC), were used to support the bottom-up research. Operation plans for the year were already finalised before the beginning of the fiscal year. Without changing the framework for the cooperative use of the schedule, the cooperative research vessel steering committee organised by the Atmosphere and Ocean Research Institute, the University of Tokyo, initiated the following processes with the marine research community. On 13 April, the research vessel operating group under the cooperative research vessel steering committee provided a questionnaire to the scientists and chief of the cruises for R/Vs *Tansei Maru* and *Hakuho Maru* to determine whether they could provide additional support or change the study area to the Sanriku coastal region as well as provide additional ship time. There were many positive responses; therefore, 20 days of ship time were available for the earthquake correspondence cruise. Subsequently, the three themes of the earthquake mechanism, diffusion of the radioactive substances and disturbance of the marine ecosystem by the tsunami were initiated for the earthquake correspondence cruises on 19 April. Within two weeks, 11 applications were submitted, and all proposals were accepted after the evaluation of the research vessel operating group under the cooperative research vessel steering committee. In mid-May, a new ship schedule plan, including the earthquake correspondence cruises, was set. In addition, many offers of cooperation to collect samples and surveys from scientists were provided. In the end, eight cruises (four with the R/V *Tansei Maru* and four with the R/V *Hakuho Maru*) were conducted over 202 days as cooperative cruises for the earthquake correspondence committee, and six earthquake correspondence cruises were conducted over 45 days with the R/V *Tansei Maru*. The bottom-up process to finalise the ship schedule took a long time due to the very complicated procedures. In this case, the cruise plan was quickly fixed without destroying the bottom-up process for cooperative activity. This was possible because the research community, including the members of the cooperative research vessel steering committee, possessed enthusiasm and determination due to the emergency situation.

The disaster correspondence cruises were performed by numerous research institutions. Cooperative samplings for the monitoring programme in addition to the cruises of the R/Vs *Tansei Maru* and *Hakuho Maru* were also conducted. If we had had a centralised decision-making mechanism, it would have been possible to perform more efficient monitoring, survey and research. Unfortunately, this was difficult to control under the complicated structures of the agencies and their missions. The role of the JOS earthquake correspondence working group was to collect cruise information related to the earthquake correspondence and to arrange efficient sampling and chemical analysis, as well as the cruises.

A list of the earthquake correspondence cruises in fiscal year 2011 is presented in Figure 11.4. Most of the cruises were conducted by universities, JAMSTEC and the Fisheries Research Agency. More than half of the cruises were to survey radioactive material. The subgroup of radiochemical analysis of the JOS earthquake correspondence working group published a manual for sampling, pretreatment and measurement of radioactivity for the various samples. It was a serious issue for safety management against irradiation exposure on the ship. Shigeyoshi Otosaka (JAEA) and Tatsuo Aono (National Institute of Radiological Sciences) developed radioactivity safety guidelines for the monitoring cruise. A counterplan for the prevention of radioactive contamination on the ship was established by Takashi Ishimaru, the principal investigator of the R/V *Umitaka-maru* cruise operated by the Tokyo University of Marine Science and Technology (TUMST), which was used in several subsequent cruises. The JOS earthquake correspondence working group had been effectively functioning as an information hub with only a few experts.

The climax of the earthquake correspondence surveys was the cooperative work with the Japan Broadcasting Corporation (NHK) and the JOS earthquake correspondence working group. A survey within 20 km of the FDNPS was conducted. TEPCO only monitored a few aspects, and these data were not good enough to clarify the contamination levels and the concentration processes of high-level radioactivity in fish bodies caught in this area. Tadashi Ikemoto, NHK chief director, negotiated with related organisations and finally obtained permission to survey from Okuma-machi (see *JOS newsletter* 2(1) for details). The joint survey plan was summarised by Jota Kanda (TUMST) and Takashi Ishimaru, and the survey was performed by volunteers of Sanyo Techno Marine Ltd, TUMST, the University of Tokyo and Tokai University. Invaluable samples of seawater, sediments and marine biota were collected from 32 stations in this area over nine months following the accident. This survey could not have been accomplished without the support of the fishermen forced to refrain from fishing due to the

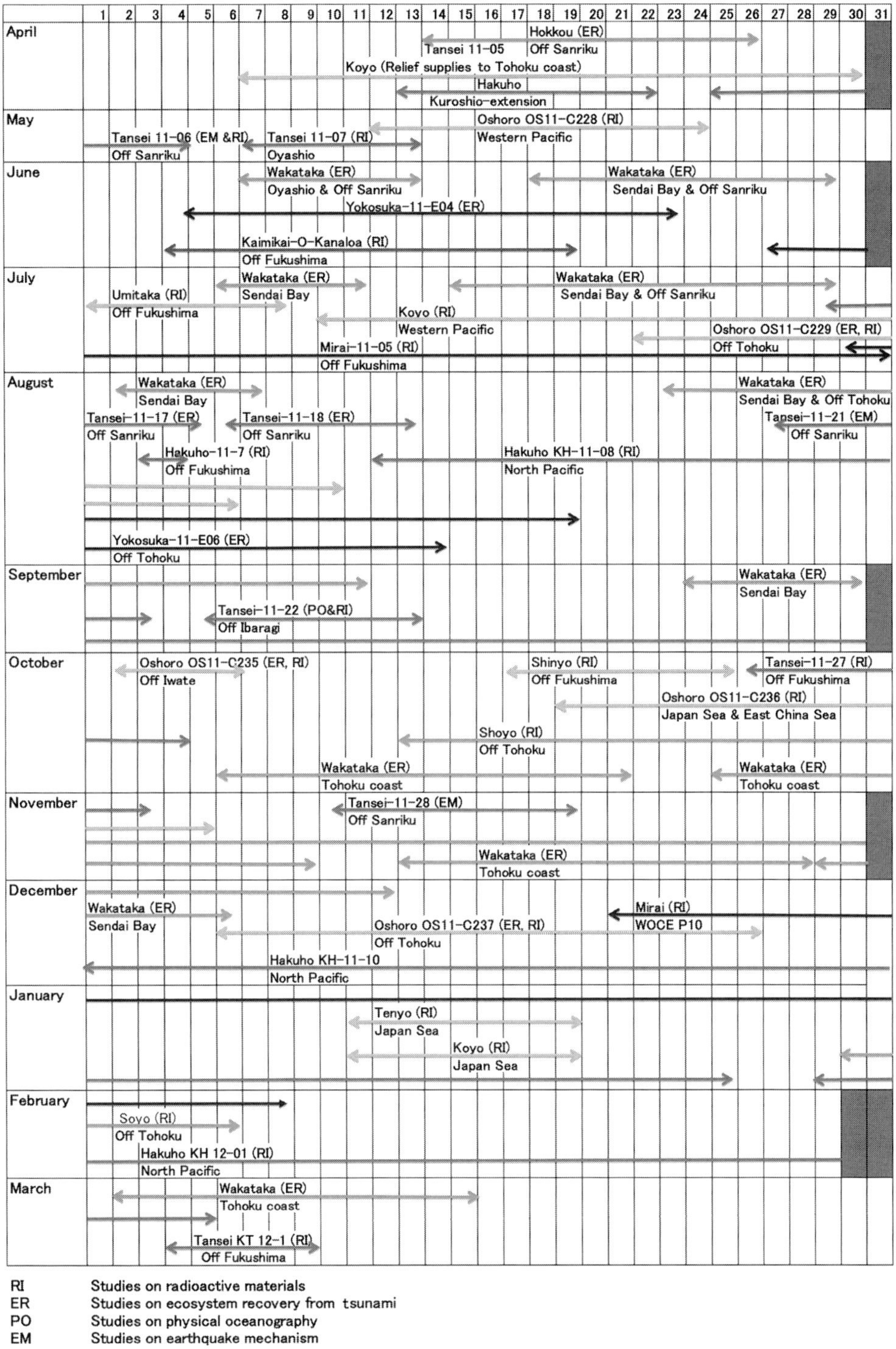

Figure 11.4 List of the disaster correspondence cruises (summary of the JOS earthquake correspondence working group).

unprecedented accident and the staff of the Fukushima Prefectural Fisheries Experiment Station. It was the goal to ensure that fishing activities returned to normal in the Fukushima waters and people would enjoy delicious, safe fish from Fukushima in the near future.

The disaster correspondence under the bottom-up volunteer framework by the society and researchers was conducted for more than a year and half. Scientists attempted to contribute for the betterment of the public beyond the traditional top-down framework. In the future, it needs to be verified whether these attempts were suitable or whether there were other important endeavours. Moreover, much effort was put forth by other institutions and individuals, yet some of these could not succeed because of various difficulties. It is disheartening to not have the resources to support these efforts, which thereby fall short of achieving their objectives.

11.4.3 Implementation of International Collaborative Radioactivity Survey Cruises

After the meeting about the earthquake on 14 April 2011, Mitsuo Uematsu of the University of Tokyo was requested to contact Ken Buesseler at Woods Hole Oceanographic Institution (WHOI) for collaboration on international radioactive survey cruises using a US research vessel through Michio Aoyama (Meteorological Research Institute) on 20 April. Incidentally, Ken and Mitsuo participated in the Liege Colloquium held in Denmark on 3–6 May. During the colloquium, Ken received notice that his proposal for the survey cruise had been accepted by the Gordon and Betty Moore Foundation. The plan was immediately set, and the R/V *Kaimikai-O-Kanaloa* (R/V *KOK*) of the University of Hawaii was reserved for the cruise. The port of call, transportation of equipment from abroad to Japan and scientists were determined on a day-to-day basis. A survey permit in the Exclusive Economic Zone (EEZ) of Japan was obtained. However, there is a very old law prescribed in the Meiji era (1899) that limits the transport of cargo by foreign ships between Japanese ports or a round trip involving a Japanese port, such as to or from Yokohama Port. The coastal transport licence was finally obtained after considerable time and money were spent on the preparation of additional documents. Jun Nishikawa and Hiroomi Miyamoto (both from the University of Tokyo) joined the R/V *KOK* cruise as members of the biology group to monitor the radioactivity in the zooplankton and micronekton over 4–19 June 2011. This cruise was planned to occur over the open ocean, where low-level radioactivity was expected to be found. Samples were collected at the request of Oxford University, the IAEA, the Tokyo Institute of Technology, Nihon University and the Meteorological Research Institute; these institutions could not join the cruise. Before the cruise, brief meetings were held in Tokyo with John Victor

Roos, the US ambassador to Japan. The USA voiced very strong concerns regarding marine radioactive contamination in the Pacific Ocean.

Various marine samples were analysed for the same radionuclides at the related institutions to verify the analytic results. Therefore, this collaboration also contributed to increasing the reliability of the data. However, the radioactive contamination survey was limited to the atmosphere and the land by MEXT. The marine radioactivity survey had a low priority. The JOS earthquake correspondence working group submitted a proposal to the international emergency joint research and research support programme (J-RAPID) by the Japan Science and Technology (JST) committee to obtain funding for the reagents needed to pretreat the various samples, the sample containers and the cost of sample transportation. After the cruise, the proposal was accepted. In addition, the shortfall was assisted by the University of Tokyo's general discretionary budget. The Grant-in-Aid for Scientific Research on Innovative Areas 'Interdisciplinary Study on Environmental Transfer of Radionuclides from the Fukushima Daiichi NPP Accident' (principal investigator: Yuichi Onda of Tsukuba University) from 2012 to 2016 was adopted, and a long-term survey research system has been established.

11.5 Participation of Nuclear Physicists in the Screening Survey

ISAO TANIHATA AND MAMORU FUJIWARA

Emails concerning the situation of the nuclear reactors began to circulate to the mailing list of the nuclear physics community, when on 12 March news was broadcast that the Fukushima Daiichi Nuclear Power Plant no. 1 had suffered a hydrogen explosion. Apart from that, Ryugo Hayano of the Faculty of Science at the University of Tokyo sent various messages through Twitter, and the circle of discussion widened rapidly. On 15 March, discussion started based on available information such as the atomic decay heat calculated by Kazuhiro Matsumoto of Aichi Shukutoku University. On the same day, researchers exchanged information that radioactive materials had begun arriving in the Kanto area; this was indicated by radiation level increases at the High Energy Accelerator Research Organization (KEK) in Tsukuba City, the JAEA in Tokai-mura, the Institute of Physical and Chemical Research (RIKEN) in Wako City and elsewhere.

'Terrible things are happening; what can we do in this situation as researchers of nuclear physics?' Consumed with this thought, Fujiwara and Tanihata of the Osaka University Research Center for Nuclear Physics (RCNP) conferred and issued an email at 02:00 on 16 March calling for an urgent meeting. Prior to this action, we (Tanihata and Fujiwara) had received and been motivated by an email from Hideto Enyo of RIKEN requesting the establishment of a discussion group of nuclear physicists in western Japan because institutions in eastern Japan were preoccupied with responses to the disaster.

11.5.1 15 March: Call for an Urgent Nuclear Physics Community by Email List (ml-np) held on 16 March at RCNP, Osaka University

In spite of the late-night call, researchers from various areas of western Japan in addition to the Kinki area filled the fourth floor lecture room of RCNP to make status reports and participate in a discussion of actions to take. Tanihara himself shortened his stay in Beijing and attended the meeting in the morning of 16 March. In the meeting they decided to undertake measurements of soil radioactivity and air dose rates in the Fukushima area, which is the most relevant and useful support for professionals of radiation measurements.

In this period, many strange and even suspicious pieces of information were distributed. Among them was a TEPCO report about detecting ^{38}Cl. A comment in the abovementioned mailing list suggested that this phenomenon indicated absorption of neutrons by chlorine in the salt in the seawater injected into the nuclear reactor, which thus indicated that nuclear reactions were still underway. If that were the case, however, ^{24}Na would have been detected in the process of neutron absorption by sodium, but the nuclide was not observed. The discussion concluded that the detection information might have been wrong and the nuclear fission had stopped. Several months later, TEPCO announced that the detection had been incorrect. This incident caused widespread concern because a serious mistake in the detection of such an important nuclide had not been corrected at any step before becoming public. This also raised questions about the skills of the people who made the measurements.

11.5.2 17 March: MEXT Contact with Osaka University and the Decision to Participate

The group made contact with MEXT on the night of the meeting to propose measurements of soil radioactivity and air dose rate regarding the FDNPS accident. In addition, on the afternoon of 17 March we had a liaison meeting with the president of Osaka University and its safety and health management office to convey our wishes and plan. The president encouraged us to do our best to contribute to resolving the situation. Fujiwara and Tanihata visited MEXT on 18 March and presented their plan. MEXT understood the importance of the radiation mapping investigation but had prioritised the screening activities of the residents to investigate the effect of radioactive materials, and asked for our cooperation with the screening. On the same day, the board of directors of Osaka University issued a resolution calling for support of all the possible actions by the university to aid the recovery effort. This was a declaration by President Kiyokazu Washida of Osaka University and other directors for emergency cooperation after this unprecedented accident.

Regarding the screening activities, it was also decided to carry out full-scale support for the safety and health management department directly under the leadership of headquarters. Responding to this decision, the RCNP began calling nuclear physics researchers across the country and asking them to participate in the screening in Fukushima. This activity spread not only to nuclear physicists but also to scientists in the fields of radiochemical and medical physics.

11.5.3 20 March: Discussion of Detailed Action Plan at the RCNP and Agreement of First Action to Participate in Screening Activities

Under such a situation, we met again on 20 March to discuss the action plan, and we decided our first action would be to participate in the screening activities and then find an appropriate time for measurements of soil radioactivity and air dose rate. On 21 March, the first team participated in the screening on site and was successful in starting thyroid measurements on 24 March. By 9 April we had screened some 6100 people and conducted thyroid screening of 890 children. This screening was directed by MEXT and the Ministry of Health, Labor and Welfare, organised around the Fukushima support headquarters, and attended by people also from the Kyoto University Nuclear Furnace Research Laboratory, Hiroshima University, Nagasaki University, local governments and the Federation of Electric Power Companies (*Denjiren*). About 100 people gathered at 07:30 every day for a briefing and then went to work in their respective municipalities. Among the screening activities, the most urgent was the monitoring of ^{131}I, which may have accumulated in the thyroid glands of children. ^{131}I has a short half-life, so that it is impossible to know the amount of the radiation dose unless it is measured promptly.

In the Chernobyl accident in 1986, soil radioactivity measurements were carried out three years after the accident. It was a well-known fact that residents in the vicinity of Chernobyl received excessive amounts of radiation that resulted in health damage over five years until the Communist Soviet Union collapsed and a new Russia was born. In particular, it was known that many children suffered from exposure to ^{131}I, of which no traces had been found and the danger of which was thus not noticed before. In the Fukushima accident there was also a strong possibility of a large amount of ^{131}I emission, and therefore urgent action was necessary.

In command of Fukushima Prefecture-based screening activities, the nuclear physics group had been requested to make urgent measurements of ^{131}I that might have accumulated in children's thyroid glands. In areas of high radioactivity, it was expected that measurements would be difficult due to the large background signal, but nonetheless measurements were necessary because of the possibility that ^{131}I had flowed to Iitate Village to the northeast and Iwaki City to the south, as suggested by the results of SPEEDI's predictions. Urgent measurements of the throats of children were started with an NaI survey meter. Initially, measurements were carried out using public halls in the Yamakiya area of Kawamata Town, but measurements were difficult under the large background of 2–3 μSv/h. In order to improve the situation, the measurement location was changed to a room with 0.1–0.4 μSv/h because γ-rays are well shielded from outside by the concrete walls of the building. No significant accumulation of radioactivity was detected in the thyroid glands of children exposed to more than 0.1 μSv/h, although the exposure below that level could not be measured. The measurement of ^{131}I exposure was finished on 30 March after two cycles of the half-life of eight days because it was expected that detecting a significant accumulation of ^{131}I in children would be difficult. Thyroid exposure examinations were completed on 890 people by 30 March. The estimated radiation exposure dose to children was small enough to confirm that the probability of thyroid abnormality was very low. In the early period, the main screening subjects were people who had evacuated from the high dose area, but with time the screening was extended to people who had temporarily returned home to retrieve household goods and cars that had been left behind. Screening was also done for people who planned a temporary home visit by bus.

Participants in the screening were registered as radiation workers as stipulated in laws and regulations because they had to be capable of performing work at locations with high levels of radiation as well as to properly answer radiation-related questions and others from residents. Many participants in the screening activities from universities and other research organisations had confidence in themselves being able to answer various questions accurately in situations in which other participants, such as those from *Denjiren* and the SDF, might not have sufficient knowledge about radiation. The MEXT staff in charge of the screening activities appreciated such cooperation by the researchers.

Participants in the screening activities were monitored by their own pocket dosimeters, and in the early period they were exposed to radiation of up to 30 μSv per day. Moreover, in this period, many participants – including Hiro-o Hasebe of RIKEN – found time to try to decontaminate structures with various solvents and detergents, yet they found it to be very difficult.

The screening activities were continued until August with support from 102 registered participants and 30 institutions, which yielded a total of 361 participants. When the screening activity become stable, we started to consider again the original plan of radiation survey and soil measurement activities.

In late March, the Nuclear Science Research Center of the University of Tokyo (CNS) announced their participation in the support activities, which expanded the research field options. Under the larger collaboration group, we submitted a proposal titled 'Radiation Dose Survey after Fukushima Nuclear Power Plant Accident' to MEXT through Osaka University on 31 March. The radiation dose at a height of 1 m and also radioactivity in soils was measured by this group of 858 scientists from 90 universities, research institutes, and three private companies for the collection of soil samples. In addition, 21 universities and research institutes collaborated on the γ-ray measurements of the samples. These data are summarised on the Osaka University website (www.rcnp.osaka-u.ac.jp/dojo). Thus, with the help of hundreds of people within the scientific community and from elsewhere, screening activities and radiation measurement activities were undertaken. It was the first time such detailed data had been taken immediately after similar accidents. The data have been helping the government to define their policies.

11.6 Large-Scale Investigation of Deposited Radioactive Materials

TOKUSHI SHIBATA, ISAO TANIHATA, MAMORU FUJIWARA, TAKAHARU OTSUKA AND SUSUMU SHIMOURA

In this section the results of the large-scale investigation of deposited radioactive materials will be reported. These results are also available at http://ramap.jaea.go .jp/map.

11.6.1 Start of the Soil Radioactivity Investigation by the Government and Academia Cooperation

Actions in March 2011

As discussed in the previous section, during the screening activity for evacuees, many researchers in nuclear physics thought that the soil radioactivity investigation had to be performed as soon as possible to determine the ^{131}I deposition distribution because of the serious effects on many children, including cancer, from the intake of short-lived ^{131}I reported after the 1986 Chernobyl accident. The half-life of ^{131}I is as short as eight days. Therefore, the radioactivity of ^{131}I was expected to decrease to 1/1000 in the two months following the FDNPS accident, resulting in difficulty in detection. The researchers continued discussion with the staff of the Emergency Response Center of MEXT and the staff of Fukushima Prefecture on the soil radioactivity investigation during their screening activities. Furthermore, they conducted necessary studies on soil sample collection and methods to measure radioactivity in soil samples. At that time, both MEXT and Fukushima Prefecture indicated that soil investigations were necessary. The researchers' plan was conveyed to Dr Tokushi Shibata, part of the JAEA at that time and also a member of the SCJ, to take actions to obtain SCJ support for their plan. Under such circumstances, Professor Takaharu Otsuka of the Graduate School of Science at the University of Tokyo noted that the research group at the University of Tokyo strongly wished to join this programme, which included scientists not only in nuclear physics but also in other relevant fields. This idea was conveyed to scientists working in radiochemistry, geoscience and atmospheric and marine sciences. These scientists in many fields gathered for an in-depth discussion before drafting a proposal titled 'A Detailed Proposal on Support Action for Radiation Measurement by Universities Volunteers' on 29 March 2011. They further discussed and completed their proposal and presented it to MEXT on 31 March 2011. The contents of their proposal were as follows:

1. The distribution map of radioactive materials should cover a wide area that includes the neighbouring prefectures far beyond 30 km from the FDNPS. The investigation of the deposition distribution of radioactive materials should be performed based on global standard methods by teams consisting of experienced researchers who know experimental methodology.
2. Because the scattered deposition of radioactive materials was observed in the case of the Chernobyl accident, radiation measurements over a wide area are necessary, and information related to the soil sampling locations should be provided using GPS measurements.
3. The investigation results should be used for determining detailed evacuation areas. The distribution map of radioactive materials should cover an area of 50 km from north to south and 60 km inland, and it should be created by collecting soil samples with a mesh of 2 km. The radiation dose rate should be measured at a height of 1 m above the ground surface where the soil samples are collected.

At the same time, a group from the University of Tokyo submitted a proposal to MEXT that included a survey of residential exposure, countermeasures after the accident and other items. The SCJ had been discussing countermeasures for the accident from many aspects and released a second emergency recommendation titled 'Regarding the Necessity of the Investigation of Radiation Levels after the Accident of the Fukushima Daiichi Nuclear Power Plant' on 4 April in the publication 'Issues of the Great East Japan Earthquake and the Fukushima Nuclear Power Plant Accident'. This recommendation included the further extension of the investigation area with a small mesh for measurements in the area where the radiation rate was high (www.scj.go.jp/en/report/houkoku-110502-2.pdf).

Actions in Early April

A discussion meeting on the investigation of radioactive material deposition on the soil and radiation dose measurements was held at the RCNP, Osaka University. Another discussion meeting that included researchers in the fields of geoscience and radiochemistry was held at the School of Science, University of Tokyo.

For the abovementioned proposals, after obtaining a positive response from MEXT, a discussion meeting was held on 4 April at the RCNP to determine a plan for conducting the soil activity investigation and a suitable method to measure the soil samples. The following actions were decided at this meeting. Osaka University would play a central role in seeking participants for collecting soil samples and would decide the protocol for measurements. Professors Isao Tanihata and Mamoru Fujiwara of the RCNP were selected to lead the gathering of participants and oversee the procurement of tools and devices necessary for the investigation. The protocols for soil sampling and radiation measurements were prepared by the core members, including Professor Masaharu Hoshi (Research Institute for Radiation Biology and Medicine, Hiroshima University), who had considerable experience at Chernobyl, and Professor Yuichi Onda (Center for Research in Isotopes and Environmental Dynamics, University of Tsukuba), who was an expert in the investigation of environmental dynamics. The meeting included several university members in fields other than nuclear physics and was organised by Professor Takaharu Otsuka, Dr Tokushi Shibata as a core member and other members who recognised the importance of investigating radioactive material deposition on the soil and the dynamics of radioactive materials in the environment.

Furthermore, Professor Teruyuki Nakajima (University of Tokyo) and Professor Mitsuru Ebihara (Tokyo Metropolitan University and the Japan Geoscience Union) both made requests for urgent surveys of the deposition of radioactive materials to academic societies, including the Japan Society of Nuclear and Radiochemical Sciences. A discussion meeting on the investigation of the deposition of radioactive materials was held on 20 April at the University of Tokyo with scientists from the fields of nuclear physics, geoscience and nuclear chemistry. Upon procurement of the necessary equipment, a trial soil sample collection was conducted at the end of April based on the results obtained from the discussion in the meeting. Furthermore, at a meeting on 2 May at the University of Tokyo, the cooperative organisation 'Joint Meeting of Environmental Radiation Nuclear Physics and Geoscience Fields' was formally established.

Around this time, Professor Onda contacted the Ministry of Land, Infrastructure, Transport and Tourism. He obtained informal consent to collect soil samples along the banks of rivers; collection along national roads could be performed without any permission from the local governments in Fukushima Prefecture.

Based on this agreement, pilot work for soil sample collection began before the Golden Week holidays in early May (a nearly week-long series of holidays). Moreover, a protocol for large-scale soil sample collection based on the results of the pilot work was also prepared. The operational costs were largely borne by Osaka University.

The Pilot Work

The collection of soil samples and radiation measurements began in early May as a pilot project and was performed with a 10 km × 10 km mesh. Although the signal of ^{131}I, which was stronger in March, was still clearly observed at this time, the intensity had decreased substantially compared to that of radioactive caesium. Thus, the importance of the early soil investigation survey was quite evident. Many other radioactive nuclides were also observed at that time, and very high concentrations of radioactive nuclides were observed in the soil. The permeation distributions of the radioactive nuclides in the soil were also measured, and they showed that most of the radioactive nuclides remained in the top 5 cm of the soil. Based on this observation, the soil sample collection procedure was determined, and samples were collected only within 5 cm of the ground surface.

Mid-May

The large-scale soil investigation project was approved by the Council for Science and Technology Policy. Thereafter budget negotiations between relevant organisations began.

The large-scale soil investigation project was announced through the mailing list of the RCNP, which supported the screening of evacuees in Fukushima Prefecture, to gather volunteers to join the project. Due to the many responses from researchers in different academic fields, the announcement was distributed to additional academic fields such as nuclear chemistry, radiochemistry, radiology, high-energy physics, environmental science and geoscience. The list of participants was prepared based on the response to this announcement. It was then decided that the Osaka University group would be responsible for the collection of the soil samples, while the University of Tokyo group took responsibility for the soil sample measurements.

The RCNP had already started to collect soil sample kits in April. All members who wished to join the large-scale investigation for radioactive deposition on the soil surface strongly hoped to measure ^{131}I in the soil samples within 10 half-lives and to finish the soil sample collection before the start of the rainy season in June. However, the work required funds from the Promotion of Science and Technology Strategy for the large-scale investigation, and obtaining these funds was very difficult. This was because so many organisations were engaged in the project, and the budget request needed to consider all necessary materials, equipment, traffic expenses and many other details. Much time was spent exchanging information between many universities and the JAEA, MEXT and Ministry of Finance. The actual work began on 6 June after hands-on training was conducted on 4–5 June, fully one month later than the planned start date. Thus, far fewer soil samples to measure the ^{131}I dose rate were obtained.

11.6.2 Collection of Soil Samples

In MEXT, a committee to oversee a project titled 'Investigation and Study of the Secondary Distribution of Radioactive Substances due to the Accident at the Fukushima Dai-ichi Nuclear Power Plant' was formed on 22 April. The committee started to draw up the protocol for the soil investigation project. The researchers started pilot work to verify the protocol for soil sampling and γ-ray measurement of soil samples. The pilot work was carried out from late April to early May. In the pilot work, meshes were set with dimensions of 10 km × 10 km in the eastern region of Fukushima, and the soil was sampled at 68 locations. The γ-ray analyses were performed at Osaka University, Tsukuba University, Tokyo Metropolitan University and the JAEA. Based on the γ-ray measurements, the original protocol draft written by Hoshi and Onda was further refined after long discussions among many scientists. Because the half-life of ^{131}I is eight days, it was necessary to begin the soil sampling project over a wide area of Fukushima as soon as possible. The final protocol was submitted to the MEXT committee. This protocol greatly contributed to obtaining highly reliable results because of a standardised process for soil sampling and γ-ray measurements with HPGe detectors, which were performed by more than 500 people who had somewhat limited experience in soil sample collection.

The budget submitted from MEXT was approved by the Ministry of Finance at the end of May 2011. Based on this information, we managed to settle on a headquarters at the Adatara Fureai Center in Nihonmatsu City on 1 June 2011; the start of the soil sampling project was planned for 4 June. The soil was sampled using U8 container samplers. In addition, we decided to collect soil core samples at approximately 300 locations using an iron pipe with a diameter of 5 cm and a length of 30 cm to study the permeation depth distribution of radioactive materials in the soil.

Radioactive materials in the environment were found to not be uniformly distributed. The radioactive intensities in soils on different ground surfaces separated by 2–3 m exhibited large fluctuations. Even in the case of soils sampled using the U8 containers, the distribution of radioactive intensities was not uniform. Based on these results, we decided to collect at least five soil samples within 2–3 m.

In the protocol, it was noted that we needed to sufficiently stir soils collected using the U8 samplers in a polyethylene zipper bag and to preserve them in U8 containers. The results of the pilot project are described in the following.

Because the measurements were performed by different groups and with various HPGe detectors at many institutes, calibrating the absolute intensity of the γ-ray measurements was essential for obtaining reliable radioactive material densities. Therefore, we prepared well-calibrated ^{134}Cs and ^{137}Cs standard sources stored in the U8 containers, and each institute measured the γ-rays using the standard source for calibration. Because several γ-rays are emitted during a short period of time following the β-decay of ^{134}Cs, the sum effect had to be taken into account when treating the full-energy peak efficiencies. This sum effect for the full-energy peak efficiencies depends on the detector size and the distance between the sample and the detector, and some corrections for coincidental sum effects are needed to obtain the absolute efficiency of γ-ray detection. However, such corrections were not necessary by using a normalised value obtained from the measurement with the standard source. In addition, to determine the reproducibility of the measurements, we required in the protocol that all reports include photos of the detector and that the γ-ray spectra should be included. Concerning the soil sampling method, the proposal written by Hoshi and Onda was discussed in the MEXT committee and was recommended to be used in the project. The budget implementation was made possible on 6 June. Osaka University supported the preparation-level activities and the procurement of the soil sampling kits.

Since the soil sampling had to begin after the test period of 4–5 June, the formal start of the project was on 6 June. Every day, the JAEA group led by Dr Kimiaki Saito selected the soil sampling points and the staff numbers necessary for the soil sampling work and arranged taxis to transport them to the soil sampling points. Each team, consisting of two or three people, collected soil samples at several locations every day. Every day, up to 30 teams worked on soil sampling. At the beginning of the project, because there were 1500 soil sampling points as described in the initial proposal, we estimated that the soil sampling could be finished within 10 days (or slightly more). However, the number of locations was increased to 2200 and samplings were added at locations in neighbouring prefectures, meaning this estimate turned out to be far too optimistic.

It was important in this project that we determine the locations for the soil sampling and get agreement from the land owners. MEXT took the role of obtaining the consent of each mayor and land owner in the cities, towns and villages of Fukushima, Miyagi and Yamagata Prefectures for the radioactive nuclide map project and its associated soil sampling work. At first, when the soil sampling began, agreements had been obtained only for the locations in the village of Kawauchi, and there were no agreements for any other locations at all. Every day, the MEXT officers made calls to mayors to obtain permission for the soil sampling work at individual locations in parks, empty spaces and rice fields. The MEXT officers obtained agreements with the respective mayors for soil sampling. It was to the credit of MEXT that this mission was completed successfully. Every night, the JAEA group led by Dr Kimiaki Saito selected the soil sampling points. The staff assignments for soil sampling could only be made thereafter. Mr Takashi Saito of the Administration of Safety and Hygiene (ASH), Osaka University, played the important role of selecting the personnel with such limited time. The JAEA staff arranged taxis to carry them to the soil sampling points.

Every day, 80–100 persons joined this soil sampling work. Although there was no person who could join the soil sampling work for the entire planned period, they returned every 3–5 days at their convenience.

The soil sampling work continued until 14 June as originally planned. However, all sampling could not be performed. The remaining soil sampling was conducted from 27 to 30 June. In this second campaign period, soil sampling was planned to include the 20 km evacuation zone centred on the FDNPS site. This soil sampling work was performed by staff members of the JAEA and the Federation of Electric Companies of Japan, and continued until 8 July.

Because the radioactive densities in the 20 km zone were higher than those in other areas, the radioactivity densities for ^{131}I were expected to be obtained even though the sampling date was late. Later, however, this was not found to be the case. Approximately half of the collected soil samples were sent to the Japan Chemical Analysis Center (JCAC), and the γ-rays emitted from the soil samples were analysed. The remaining half of the soil samples were sent to the Center for Nuclear Study (CNS), University of Tokyo. After reviewing the descriptions and labels attached to the soil samples, the collected soil samples were redistributed to the research groups working on nuclear physics and earth sciences at 20 Japanese universities and institutes. Moreover, γ-ray analyses of the soil samples were performed by the individual research groups. A total of 340 scientists and students from 21 organisations participated in obtaining these measurements. The specific γ-rays from ^{134}Cs and ^{137}Cs were observed in all soil samples. Therefore, detailed deposition density maps for ^{134}Cs and ^{137}Cs could be obtained. Unfortunately, the specific γ-rays from ^{131}I were only observed in the soil samples at approximately 400 locations. Thus, only a rough deposition density map could be obtained for ^{131}I, and the details were left for further examination. Deposition density maps for ^{110m}Ag and ^{129m}Te were also obtained.

In the soil sampling project, approximately 11 000 soil samples were collected from 2200 locations in Fukushima Prefecture and in the neighbouring prefectures. The γ-ray measurements were conducted within two months, and the radioactive materials in the soils were identified by detecting their specific γ-rays. For this project, 98 organisations, including universities and research institutes, joined in the soil sample collection, with 440 participants (Table 11.1), and 340 people from

Table 11.1 *Joined organisations and number of participants: sample collection.*

Participating Institutes	Number of participants	Participating Institutes	Number of participants	Participating Institutes	Number of participants	Participating Institutes	Number of participants	Participating Institutes	Number of participants
Aoyama Gakuin University	4	Kyoto Women's University	1	Shibukawa General Hospital	1	Tohoku University of Community Service and Science	1	Hokkaido University	9
Akita University	1	Kyoto University	6	Tokyo Metropolitan University	3	Tohoku University	15	Musashi University	1
Ibaraki Prefectural University of Health Sciences	4	Gunma Prefectural College of Health Science	1	Junshin Gakuen University	6	Dokkyo University	1	Mie University	1
Daiyukai General Hospital	2	Gunma University	2	Juntendo University	1	Nagoya City University	1	Miyagi University of Education	1
Nagoya Radiological Diagnosis Foundation	1	High Energy Accelerator Research Organization	16	Showa Pharmaceutical University	1	Nagoya University	8	University of Miyazaki	1
Meirinkai Imaichi Hospital	1	Kochi University	1	Shinshu University	3	Niigata University	14	Yamagata University	4
Utsunomiya University	1	Konan University	6	St. Marianna University of Medicine	1	Japan Atomic Energy Agency	73	Yokohama National University	1
Ehime University	1	Kobe City College of Technology	1	Chiba University	2	Nihon University	3	Real-time Earthquake Information Consortium	2
Osaka City University	1	Kobe Tokiwa University	1	Chubu University	1	Japan Chemical Analysis Center	20	RIKEN	9
Osaka University	31	International University of Health and Welfare	3	University of Tsukuba	9	National Institute of Technology, Numazu College	1	Rikkyo University	5
Okayama University	2	National Institute for Environmental Studies	1	Teikyo University	1	Hyogo Ion Beam Medical Center	1	Rissho University	3

Okayama University of Science	2	National Cancer Center Japan	1	Tokai University	2	Hiroshima International University	2	Ritsumeikan University	1
Japan Agency for Marine-Earth Science and Technology	3	National Institute of Polar Research	1	Tokyo Medical and Dental University	2	Hiroshima University	5	University of Ryukyus	3
Kanazawa Medical University	2	National Astronomical Observatory of Japan	1	Tokyo Institute of Technology	1	University of Fukui	6	Waseda University	1
Kanazawa University	11	International Christian University	1	The Jikei University School of Medicine	1	Fukushima Medical University	3	Ebara	1
Kameda Medical Center	1	National Hospital Organization	3	The University of Tokyo	14	Fukushima University	9	Japan Environment Research Co., Ltd	1
Kansei Gakuin University	3	Saitama Medical University	1	Tokyo City University	2	Fujita Health University	1	Federation of Electric Power Companies of Japan(10 electric power companies and Japan Nuclear Fuel Limited)	31
Kyushu Synchrotron Light Research Center	1	Japan Synchrotron Radiation Research Institute	1	Tokyo University of Science	3	National Research Institute for Earth Science and Disaster Resilience	1	Fujifilm RI Pharma Co., Ltd	3
Kyushu University	6	National Institute of Advanced Industrial Science and Technology	2	Toho University	3	National Institute for Radiological Science	3		
Kyoto University of Education	5	Shiga University of Medical Science	1	Touhoku Gakuin University	2	Northern District Medical Association Hospital	1		

Table 11.2 *Joined organisations and participants: γ-ray analyses.*

Participating institutes	Number of participants	Participating institutes	Number of participants
Osaka University	84	Tokyo Institute of Technology	21
Osaka Electro-Communication University	11	University of Tokyo	16
Kanazawa University	24	Tohoku University	29
Kyushu University	10	Tokushima University	7
Kyoto University	11	Niigata University	11
High Energy Accelerator Research Organization	5	Nihon University	5
Konan University	6	Japan Chemical Analysis Center	7
Saga University	1	University of Miyazaki	5
Tokyo Metropolitan University	21	Riken	30
Shinshu University	1	Rikkyo University	34
University of Tsukuba	1		

21 organisations (Table 11.2) assisted with the γ-ray analyses. In this deposited radioactive materials measurement project, which was formally called the 'Investigation and Study of the Secondary Distribution of Radioactive Substances due to the Accident at the Fukushima Dai-ichi Nuclear Power Plant', numerous scientists from different universities and research institutes comprised an excellent collaboration team. This fact itself was very surprising. In Japan, this type of large-scale collaboration with numerous scientists had never been realised in the past. It was definitely of importance that Osaka University showed a strong desire to be involved immediately after the FDNPS accident. Before the budget implementation of MEXT was usable on 6 June, Osaka University supported all necessary budget items: (1) buying soil sampling kits, GPS devices, U8 containers, iron pipes, and other goods; (2) preparing a headquarters at the Adatara Fureai Center in Nihonmatsu City; (3) arranging hotel rooms in the Dake Onsen area; and (4) managing the travel expenses for the people working on the soil sampling project. Without this great contribution from Osaka University, the pilot work and the following large-scale campaign would have never been realised.

The results of this MEXT project are described in the 'Data Base for Radiation Substance Monitoring Data' (http://emdb.jaea.go.jp/emdb/en). These findings are a great gift from the more than 700 people who contributed to understanding the effects of the most disastrous nuclear power plant accident on record in Japan.

11.6.3 Measurement of Radiation Dose Rates

The radiation dose rate distribution at a constant height above the ground surface and the distribution of radioactive materials deposited on the soil were measured as part of the soil contamination investigation. The radiation dose rates at 2200 locations where the soil samples were collected were measured at a height of 1 m above the ground surface using NaI scintillation survey meters. Car survey measurements using a system called the KURAMA that can collect radiation dose rates at many locations along roads were also obtained.

The KURAMA (Kyoto University RAdiation MApping system), which was developed at the Kyoto University Research Reactor Institute, obtained and recorded the radiation dose rate data from the analogue output of a commercially available survey meter; it simultaneously recorded the position information using GPS. Because it was known that Fukushima Prefecture began measuring radiation dose rates using several KURAMA systems, we borrowed six systems from Fukushima Prefecture to conduct the car survey. It was difficult to select the routes for the car survey in a short time; thus, the following method was adopted. Because the distance that could be driven per day was thought to be approximately 200 km, the area to be measured was divided into small areas where the main road length was approximately 200 km. Maps indicating each measurement location were prepared and were given to each measuring team on the day of the measurement. The route was chosen by each team. This method worked well because each team selected the most effective and feasible route based on their own judgement. The car survey was conducted in the period 6–13 June, and the total route length amounted to nearly 17 000 km.

11.6.4 The Establishment of a System for Measuring γ-Rays Emitted from Soil Samples

The large-scale investigation of the deposited radioactive materials was performed in a top-down manner by MEXT using an emergency budget provided by the CSTP (Council for Science and Technology Policy). However, this project was implemented at the request of researchers in the fields of nuclear physics, geochemistry and other scientific societies, encouraging action by scientific communities and the emergency recommendations of the SCJ. Soil sample collection and the measurement of γ-rays from the samples were conducted by these educators and researchers in response to the actions of the different scientific communities and the recommendations of the SCJ. Therefore, this project can be regarded as a bottom-up project. For the measurement of γ-rays from the massive amount of soil samples, it was necessary to create a systematic network. The key issues in creating such a network are discussed in this section.

The Importance of the Hub Function

The soil samples collected from 2200 points were sent to the CNS at the University of Tokyo and to the Japan Chemical Analysis Center. The CNS functioned as a hub for the 21 groups of universities and research institutes, gathering and distributing the nearly 6000 soil samples and checking and recording information and the analysed data obtained by each group of universities and research institutes. Because these soil samples were collected from a broad area during a short period of time, there was overlap in the sample locations. The collecting group name, names of collectors and date and time of collection were recorded using small labels for all soil samples. These data were compared with the data recorded on a collection procedure record written at the collection points. This process and redundant data made it possible to identify the soil sample in the case where the soil sampling did not follow the protocol or the sampling point location was difficult to determine. The important issues for identifying the enormous numbers of samples are as follows:

1. The samples should be gathered through a few hubs, and the data recorded on the label and the record written at the collection points should be carefully reviewed.
2. The written information on the label of each sample might be occasionally insufficient; it is critical that records of redundant data are included in the protocol.
3. It is important to have a full-time staff to strengthen the checking function.

Because more than 400 researchers were involved in soil sample collection and collectors were replaced during the short period of the project, item 2 from the list was essential. Because the data checking is simple but very important and requires great attention, it is difficult for researchers to satisfy both the research work and data checking needs. Fortunately, the CNS was able to employ three efficient staff members to address this issue.

Approximately 6000 collected soil samples were sent to the CNS and approximately 5000 were sent to the Japan Chemical Analysis Center. The analysed data for each soil sample at the many institutes were gathered by the CNS and checked. In this checking process, the data for the same soil sample obtained by several institutes were compared. In the summer of 2011, the activities of ^{134}Cs and ^{137}Cs were nearly the same over the large study area; thus, the intensity ratio of these nuclides was adopted as a benchmark From this comparison, a few mistakes in the data and calibrations were found, and the CNS asked for re-measurements or corrections.

Overall, 21 institutes were involved in the γ-ray measurements, yet none of them could supply full-time support for the measurements. Therefore, the checking system for the γ-ray measurements by the busy researchers was not perfect. The re-checking and necessary correction by the CNS were recognised as important jobs.

The Limit of Non-Destructive Analysis without Chemical Separation

One of the purposes of the large-scale investigation of radioactive contamination was the quantitative measurement of ^{131}I, which has a half-life of eight days; however, in only approximately 20% of the soil samples from the 2200 locations were the ^{131}I data available. Because the soil samples collected from the area within 20 km of the reactors were thought to have enough intensity, the measurement of these samples was delayed for two weeks, resulting in poor results for ^{131}I. This occurred because the γ-ray energy from ^{131}I is 364 keV, which is lower than the 662 keV γ-ray from ^{137}Cs and the 605 and 796 keV γ-rays from ^{134}Cs; thus, the γ-rays from ^{131}I were muddled in the Compton γ-ray spectrum of ^{134}Cs and ^{137}Cs. The intensity of the γ-ray from ^{131}I became less than 1/1000 of the intensity shortly following the accident, while the intensities of ^{134}Cs and ^{137}Cs were nearly the same. Therefore, the peak-to-background ratio became too small to be measured. If we used a chemical separation technique, a weak peak of ^{131}I could be seen. It was thought at that time that all soil samples should be measured using non-destructive analysis. This was because the soil samples were very valuable and any destructive treatment of the samples was thought to be unreasonable.

If the γ-rays had been measured after chemical separation for iodine, we could have measured them. A few months later, this became apparent. Thereafter, using the chemical separation method, measurement was targeted at the intensity of ^{131}I, which could be obtained from the intensity of ^{129}I, with its half-life of 1.6×10^7 years.

11.6.5 Problems Learned from the Large-Scale Investigation of the Deposited Radioactive Materials from the Perspective of the Organisations

The following issues emerged from the large-scale investigation of the deposited radioactive materials.

1. In this project, the bottom-up researchers' desire to investigate the radioactive contamination and the intentions of the administrative bodies coincided. Therefore, the investigation could begin quickly. This was a fortunate situation because it can be very challenging to motivate several hundred researchers to join a project with a top-down approach. This project began just before the

rainy season and needed to be conducted quickly because of the short half-life of ^{131}I; an early start to the project was vital.

2. The close cooperation between researchers in the fields of nuclear physics and geoscience played an important role in the investigation. People familiar with both research fields nicely introduced each other and accelerated the investigation. The same situation may not always occur in the future. It is desirable and important to have similar systems in which researchers in different fields can cooperate in the case of natural or human disasters. The required activities depend on the situation of the disaster; thus, it is difficult to specify the required activities in advance, and the scientific communities should study desirable systems for emergency situations in each scientific field.
3. In this project, budget appropriation for necessary expenses was made over a relatively short period. Because the soil investigation required the collection of soil samples, it was necessary to collect soil from both private and public properties; the action to collect soil had to be recognised administratively. This situation was different from actions taken for screening surveys of radioactive contamination by people who attended freely. Therefore, the administration's involvement was essential, and this involvement did not require much additional time.
4. If, however, the administrative actions in a normal situation and in an emergency were allowed to differ, the investigation for soil contamination could have begun earlier, resulting in many more samples for obtaining high-quality ^{131}I results. In this project, a JAEA staff member was in charge of preparing and revising the required documents between research institutes or universities and MEXT. To obtain a budget from the CSTP, the required documents were requested in the same manner as under normal conditions. Therefore, the total budget had to be constructed as a sum of the necessary expenses, such as expenses for supplies used by hundreds of researchers and taxi fares needed for all researchers. The staff members' efforts was tremendous, and yet it took about one month longer than hoped. These issues might have been avoided if different actions had been permitted in an emergency.
5. In this project, the CNS joined as an organisation instead of giving instructions to the CNS staff. Thus, support from the University of Tokyo and the Faculty of Science were given to the CNS, which operated as a hub. This support expanded beyond the CNS. Nationally, a few organisations took a similar approach; researchers mostly joined the project as individuals. It might be effective in an emergency for universities and research institutes to take prompt actions based on their specialties. These actions must be well balanced with the fundamental principles of academia.

This report shows a small part of the large-scale investigation of the deposited radioactive materials carried out by MEXT. Nevertheless, it depicts important issues related to the project. The important results obtained about the detailed radioactive contamination map owe everything to the awareness of the danger and volunteer spirit of hundreds of researchers. Of the researchers, only one had a relationship with a disaster victim; each person had their own reason for joining the cause. These people, under their different situations, devoted themselves to this project. The amalgamation of their endeavours resulted in a remarkable body of work over a short period of time, and it is our hope that this collective effort and its resultant knowledge will guide future responses in the unfortunate occurrence of a nuclear disaster.

The authors would like to express their deep sense of gratitude to these individuals for their selfless devotion to a noble cause.

11.7 Scientists' Contribution to the Study of Forests

YUICHI ONDA

The FDNPS accident released radionuclides over a wide area, and high concentrations of radiocaesium and radioiodine were found in many places. Radiocaesium derived from atmospheric nuclear tests was used for tracing soil erosion and sediment transport before the FDNPS accident. Now that we have grasped the magnitude of the deposited radionuclides, there is great concern about the subsequent transfer of radionuclides in terrestrial environments because very few researchers have studied the behaviour of radioactive materials such as radiocaesium. A preliminary study found that many radionuclides were deposited in forested areas. Those radionuclides were confirmed to be transported via natural environments, such as through soils and rivers. Thanks to the effective efforts of many people, including Dr Teruyuki Nakajima, and the great efforts rendered by MEXT and the JAEA, a survey was initiated to investigate the transfer of radionuclides in terrestrial environments (forests, soils, groundwater and river water) in the framework of the 'strategic promotion fund' (Saito and Onda, 2015) (Figure 11.5).

Such a survey requires detailed monitoring in forests and of sediment production sources together with comprehensive monitoring of the transfer process to rivers under various conditions. Fortunately, Professor Onda is the leader of a JST-CREST project titled 'Development of Innovative Technologies for Increasing Watershed Runoff and Improving River Environment by the Management Practice of Devastated Forest Plantation'. In this project, detailed nested monitoring from hillslopes and source catchments to streams was begun. Studies using atmospheric

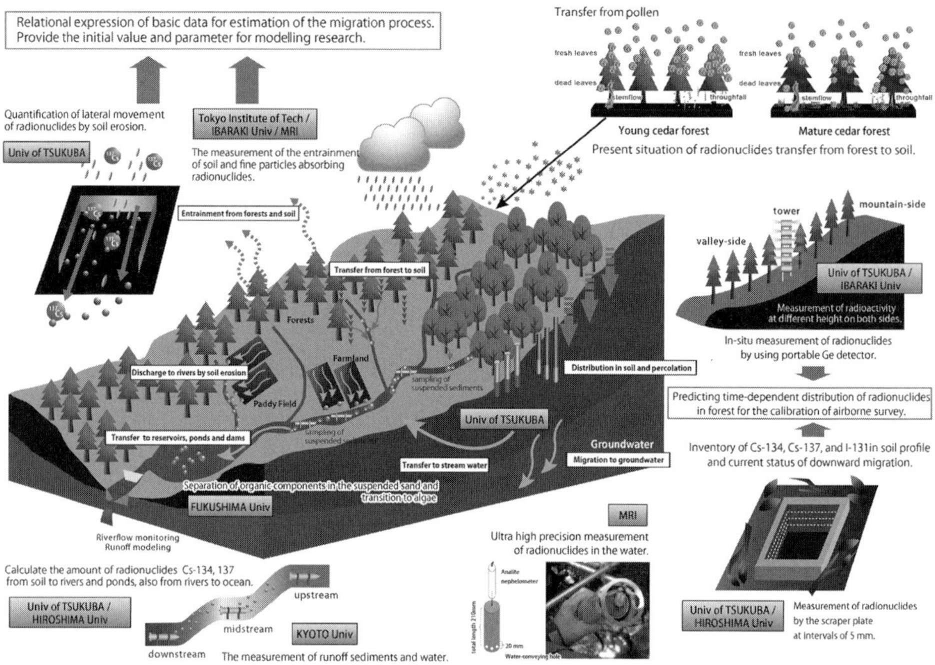

Figure 11.5 Schematic diagram of comprehensive investigation over land. (Figure reprinted from Saito and Onda, 2015).

nuclear test data and derived radiocaesium adsorbed on soil were used to determine the production source of sediments in suspended sediments with three germanium detectors installed in the laboratory. The knowledge acquired during the project was applied to the monitoring survey.

There was great support from the town of Kawamata when the abovementioned soil collection project began; the town was very cooperative with the full-scale survey. Members of the local prefectural assembly introduced us to the owners of the land used for the survey. The landowners encouraged us and gave great support, saying, ‘Since we ended up with such a tragedy, thorough investigation is highly appreciated.’ When we were looking for a monitoring site with access to power in the forests that accounted for a large portion of this area, a staff member at the town office took responsibility to obtain the landowners’ consent, which enabled us to determine the monitoring sites within a month of the survey in June 2011.

Next, measuring equipment to begin the investigation was needed. Good opportunities to purchase equipment, such as turbidity sensors and data loggers, were directly available from overseas in terms of performance and price. However, we had to wait 1–2 months for delivery. For this survey, we made contact with manufacturers in the USA and Australia, and they shortened the delivery schedule to 1–2 weeks, showing their great support for Fukushima.

To obtain comprehensive understanding of the transfer of fallout radionuclides, we had to cover a wide range of monitoring topics at the same time. Such topics included soil erosion from forest sites and farmlands, runoff from rice paddies, transfer to groundwater and runoff through rivers. Thus, we asked for cooperation from researchers who had joined the soil sampling project, including Dr Yoshio Takahashi, formerly at Hiroshima University, Dr Aya Sakaguchi and Dr Kazuya Tanaka at Hiroshima University and Dr Naohiro Yoshida at the Tokyo Institute of Technology. Although never having met us in person, Dr Kazuyuki Kita at Ibaraki University, Dr Yasuhito Igarashi at the Meteorological Research Institute and Dr Yosuke Yamashiki at Kyoto University agreed to join us with just one call. Thanks to them, a comprehensive survey of the transfer was made possible. The survey would not have been possible with the members of the universities alone. Many consulting companies rendered extraordinary support. The cooperation discussed here is only a part of the entire cooperation. Such cooperation given by many people enabled us to conduct the survey described in Chapter 6, and I would like to express my sincere gratitude to all of them.

11.8 Specific Characteristics of the Fukushima Accident

ANNE MATHIEU, DENIS QUÉLO, OLIVIER SAUNIER AND DAMIEN DIDIER

Every nuclear emergency situation has its own specific characteristics and arises in its own specific context, which includes such elements as history, sociocultural and political context, geographical area, type of facility, humans and organisations, and so forth. If, for example, the FDNPS accident had happened in a different season or at a facility away from the sea, like at Daini, it would have had very different consequences and the lessons learned for the emergency response would undoubtedly have been different.

The organisations, methods and resources used to respond to nuclear emergencies must be able to cope with all of these eventualities. The IAEA underlines the importance of having a clearly defined, well-trained organisation both aware of the risks and capable of anticipation and adaptation (IAEA, 2015). Retrospective analysis of the response to nuclear emergencies questions the efficiency of organisations. Some changes were made while emergencies were in progress. Others were subject to research before being developed into operating methods or tools. Because of its scale, the Fukushima nuclear accident constitutes a reference case.

Before the Fukushima accident, the Chernobyl accident 25 years earlier was the reference accident, especially from the point of view of radioactive releases. Aside from specific characteristics associated with their respective contexts, what was different about the Fukushima accident compared to Chernobyl was the massive number of environmental measurements taken during the course of the event.

The IAEA has analysed the accident and the way it was managed (IAEA, 2015). The simultaneous occurrence of a natural disaster caused by an earthquake followed by a tsunami and a nuclear emergency affecting several reactors is seen as one of the specific characteristics of the accident. Until Fukushima, a similar conjunction of events had not been considered in safety studies. The Fukushima accident was also the first ever nuclear disaster to cause massive releases into the sea.

Other specific characteristics of the Fukushima accident are markers of the era in which it occurred. In a globalised world where nationals of every country are spread throughout the world, a nuclear disaster leading to significant releases into the environment becomes a global emergency, as was the case at Chernobyl. The Fukushima disaster meant that countries had to inform, advise and protect their nationals in the affected country. Regardless of the location of the accident, all countries will seek to assess its consequences. Within their own territory they will use their national measurement network for environmental monitoring to ensure their land is free from contamination or to take necessary protective measures. Another consequence of globalisation is mutual international assistance; for the Fukushima accident, a number of countries provided assistance in stabilising the state of the reactors and characterising the environment in Japan. For example, around six days after the start of the accident, the USA in collaboration with MEXT ran an airborne measurement campaign to map deposits (DOE, 2011).

Consequently, the need to inform and protect the public also raises the tricky issue of access to and sharing of information required to assess the situation. In the affected country, all those involved in the emergency response can be under extreme pressure to share or not share their data and expertise at a time when their priority is still emergency response. In addition, there are now multiple sources of information and many players involved in measurement. Data can come from the public, from independent bodies or from academics with simulation and communication capabilities, particularly via social networks. For the authorities, this profoundly changes communication and emergency response. This third-party expert assessment capability, which overall is beneficial, requires a high level of transparency as regards the basic data, but also communication, explanations and justification of the decisions made.

References

DOE (2011). Radiation monitoring data from Fukushima Area 04/04/2011. www.slideshare.net/energy/ams-data-april-4v1 (accessed 19 September 2018).

Hirayama, H., H. Matsumura, Y. Namito and T. Sanami (2015). Estimation of history of I-131 concentration in air using NaI(Tl) detector pulse height distribution at monitoring posts in Fukushima prefecture. *Trans. At. Energy Soc. Jpn.*, **14**, 1–11 (in Japanese).

IAEA (2015). *The Fukushima Daiichi Accident Technical Volumes 1–5*. Vienna: International Atomic Energy Agency. http://bit.ly/2Vscu4r (accessed 19 September 2018).

Lebel, L. S., R. S. Dickson and G. A. Glowa (2016). Radioiodine in the atmosphere after the Fukushima Dai-ichi nuclear accident. *J. Environ. Radioact.*, **151**, 82–93.

Morino, Y., T. Ohara and M. Nishizawa (2011). Atmospheric behavior, deposition and budget of radioactive materials from Fukushima Daiichi nuclear power plant in March 2011. *Geophys. Res. Lett.*, **38**, L00G11, doi:10.1029/2011GL048689.

Nakajima, T., A. Watanabe, H. Tsuruta, *et al.* (2011). Nuclear power plant accident: collaboration in the crisis and the role of scientists. *Kagaku*, **81**, 934–7 (in Japanese).

NHK (2012). *A Polluted Map of Radioactive Materials No.5 Made from Networks: Search for an Early Internal Radiation Exposure Unrevealed*. NHK/ETV programme broadcast 12 March 2012.

NHK (2013). *The Great East Japan Earthquake: A Blanked Early Radiation Exposure – Search for Atmospheric Radioiodines Disappeared.* NHK special TV programme broadcast 12 January 2013.

Saito, K. and Y. Onda (2015). Outline of the mapping project. *J. Environ. Radioactiv.*, **139**, 240–9.

Science Council of Japan (2014). A review of the model comparison of transportation and deposition of radioactive materials released to the environment as a result of the Tokyo Electric Power Company's Fukushima Daiichi Nuclear Power Plant accident, Sectional Committee on Nuclear Accident. Report of Committee on Comprehensive Synthetic Engineering, Science Council of Japan. http://bit.ly/2VwjCNz (accessed 19 September 2018).

Takemura, T., H. Nakamura, M. Takigawa, *et al.* (2011). A numerical simulation of global transport of atmospheric particles emitted from the Fukushima Daiichi Nuclear Power Plant. *SOLA*. **7**, 101–4, doi:10.2151/sola.2011-026.

Terasaka, Y, H. Yamazawa, J. Hirouchi, *et al.* (2016). Air concentration estimation of radionuclides discharged from Fukushima Daiichi Nuclear Power Station using NaI (Tl) detector pulse height distribution measured in Ibaraki Prefecture. *J. Nucl. Sci. Technol.,* **53**, 1919–32.

Tsuruta, H. and T. Nakajima (2012). Radioactive materials in the atmosphere released by the accident of the Fukushima Daiichi Nuclear Power Plant. *Chikyukagaku (Geochemistry)*, **46**, 99–111 (in Japanese).

Tsuruta, H., T. Arai, K. Shiba, *et al.* (2013). A study on the polluted air masses measured at east-coast of Ibaraki prefecture in an early phase after the Fukushima Daiichi Nuclear Power Plant accident. *Hoshakagaku*, **28**, 9–16 (in Japanese).

12

Recommendations for the Fukushima Project from Foreign Scientists

12.1 Emergency Response Improvements Following the Fukushima Nuclear Accident

ANNE MATHIEU, DENIS QUÉLO, OLIVIER SAUNIER AND DAMIEN DIDIER

12.1.1 Access to Information and Databases

There has been much reflection on the issue of access to and sharing of data and information among the different agencies involved in emergency response both in the country where an accident happened and among third-party countries. A universal data exchange format has been proposed by the International Atomic Energy Agency (IAEA) as a result (IRIX Steering Committee, 2013). The creation at the start of an emergency of a database containing exhaustive entries for all measurements (e.g. location of measuring devices, measuring device types, measurement errors, producers) would facilitate the work of the different bodies while removing the burden of communication from the affected country. A public database of this kind would also improve transparency in relation to the public. In France, for example, a shared database is currently being created. The same logic could also be applied to meteorological observations, to the parameters defining the state of the facility and even to the outcomes of expert assessments.

12.1.2 Anticipation and Adaptation

Protection of the public relies on the ability to anticipate the consequences of an accident in order to define and implement appropriate protective measures. Time is of the essence. Because the aim is to limit public exposure, it is obviously better to act before the plume passes over.

There are a number of tools that can be used for anticipation. The first is the public protection plan, which defines pre-established areas where the authorities are able to take protective measures quickly. This plan is triggered by predetermined criteria linked to the state of the facility.

The second tool consists of typical releases estimated on the basis of pre-computed scenarios. If the accident is similar to one of the standard scenarios, countermeasures can quickly be proposed and implemented. The scenarios are chosen so as to encompass possible risks. At the time of the Fukushima accident, many countries based their assessments on typical releases in addition to having a public protection plan.

The Fukushima accident showed that an accident can have specific characteristics that are not covered by scenarios computed beforehand. In this case, the only way to analyse the situation, anticipate the risks, identify the actions to be taken to prevent or limit those risks, and where appropriate assess releases into the environment, is to model the state of the facility. Modelling transfers into the environment allows identification of the areas affected and assessment of the scale of the contamination and the projected doses. The predictive tools are by definition based on the modelling. They can be used in real time to adapt the measures taken to protect the population and the land, depending on the situation. Modelling offers adaptation and anticipation capabilities that make it possible to cope with all types of situations. It is therefore important to have operational models to predict releases and their consequences.

In an emergency situation, the state of the damaged facility is only partially known; the same applies to the releases and the state of the environment. The major challenge is to improve the way these uncertainties are taken into account in the response to an accident. Assessing their impact on simulations and modelling them (Haywood *et al.*, 2010; Girard *et al.*, 2014; 2016; Périllat *et al.*, 2016), learning to communicate them to the stakeholders and incorporating them within the exposure risk decision-making process constitute a significant challenge. Projects have been launched in the last few years to progress this goal; examples include the FAUNA project (Sørensen *et al.*, 2015) and the European CONFIDENCE and TERRITORIES research projects aimed at improving the management of uncertainties in emergency and existing exposure situations.

12.1.3 Assessing the Consequences of an Accident

When releases into the environment occur, they are quickly measured by monitoring networks. Regardless of the number and diversity of the measurements, they provide at best a fragmented description of the environmental contamination and of the impact of the releases on the population. For example, ground-based measuring systems document contamination at human height at specific locations but do not take account of the presence of radionuclides at altitude. Mobile monitoring systems can provide a precise description of total deposits but do not give information about the time and the importance of deposit events. Whole-body radiation count devices measure only internal exposure but do not consider other exposure pathways. No set of measurements provides a comprehensive picture of what happens to radionuclides in the environment or their impact on humans (Mathieu *et al.*, 2018). Modelling is the only tool that offers this. Consequently, modelling is essential for analysing accidents and their impact.

Before releases take place in an emergency situation, simulations are run with the aim of protecting the public, using conservative parameters. Later on, the simulations are used to validate the initial scenario and to assist with assessment of the accident's consequences. Estimation of a realistic source term is therefore necessary. Precise modelling of events at facilities and of the induced releases is sometimes difficult, especially when the data providing information about the state of the facility are insufficient, as was the case during the Fukushima accident. Tools for estimating the source term from environmental measurements are therefore used. These tools are independent of specific contexts and are thus applicable in any situation. The work done on the Fukushima accident has provided a good understanding of these approaches and allowed discerning use of environmental observations. They are now part of the range of tools that should be available in an emergency response centre, in addition to those dedicated to assessing the state of the facility.

12.1.4 Environmental Characterisation

The environmental measurement means and data processing methods used for the first time during the Fukushima accident are now included in the resources that will be used for managing any other nuclear accidents. The Fukushima accident showed the importance of being able to rely on researchers and nuclear experts working collaboratively to produce a description of the state of the environment. However, it is essential that the actions of the different players are properly coordinated and that the measurement protocols are shared to ensure the data produced are coherent and reliable (Yoshida and Takahashi, 2012; Saito *et al.*, 2014; Onda *et al.*, 2015).

In the future, there should be a particular focus on having in place measurement devices to better characterise the isotopic composition of plumes and deposits, particularly as regards the proportion of short-lived isotopes and iodine content. During the Fukushima accident, there were few measurements taken of the noble gases. There were also few measurements of ^{131}I and ^{132}I, and the measurements that were taken gave little or no information about their speciation. However, iodine content and speciation are essential for assessing doses received through exposure to plumes.

12.1.5 Practical Improvement Actions

Since 2011, efforts have focused on developing and improving the operational tools used during the response to nuclear emergencies and the associated measuring devices. Models that allow releases to be estimated on the basis of knowledge of the state of the reactors are essential. These tools are vital for providing the emergency response organisation with the anticipation and adaptation capabilities required to manage a nuclear accident properly. To complement the viewpoint provided by the analysis of the reactor state, tools to estimate the source term from environmental measurements proved crucial in reconstructing the releases in the Fukushima accident.

The benefit of modelling for emergency response is no longer in doubt because it is the main tool used for anticipating releases and their consequences. It is also an essential complement to environmental measurements for understanding the different sequences of an accident. One of the challenges is to take proper account of the inherent uncertainties of emergency situations in the response to accidents.

The measuring equipment and data processing methods used for the first time during the Fukushima accident are now included in the resources that will be used for managing any other nuclear accidents. The need for iodine measurements that distinguish between the different forms remaining in the atmosphere and for measurements of noble gases has been identified and work is being done on this.

Successful nuclear emergency response relies especially on sharing of, and having easy access to, data. The creation of a measurements database, the sharing in generic formats of input data for models or of any other data contributing to the analysis of events, and the provision of access to information that could be used to assess data reliability – all are essential to guarantee a quick and appropriate response. The tools developed since 2011 will, in the future, allow better management of accident situations at nuclear facilities and the associated nuclear emergencies.

12.2 Suggestions for Future Steps to be Taken by Japan

NICHOLAS S. FISHER

12.2.1 Cooperation of Japanese Efforts with Efforts from Scientists from Outside of Japan

It is clear that the government of Japan has supported a tremendous effort to evaluate the different aspects of the release of radionuclides from the damaged power plant at Fukushima. These aspects include environmental and public health risks, distribution and spread of these radionuclides in various environmental matrices and engineering approaches to reduce the continued release of radionuclides from the damaged site. These efforts should be lauded, and the continued support for this important work should be maintained for the foreseeable future. Given the global importance of this event, there has been, and continues to be, strong interest in this work among scientists from all countries. Unfortunately, scientists from many nations who have wanted to participate in this ongoing work have found it extremely difficult to obtain funding to support their work on scientific evaluation of the Fukushima disaster. Financial support for this work in the USA, for example, was virtually impossible to get from federal funding agencies, but modest support was available from some private funding sources (e.g. the Gordon & Betty Moore Foundation). While non-Japanese scientists enjoyed warm collaboration with Japanese scientists working on joint projects, the funding to support research efforts for the non-Japanese scientists was difficult to secure. While they did benefit from access to research cruises supported by Japan, funding to support personnel, acquisition of supplies, travel and other related items was generally not supported by Japan, even though the work that was being generated by these non-Japanese scientists was directly relevant to the goals of the Japanese government. It is suggested that a fund be established by Japan to help support these international research efforts. It is further suggested that the government of Japan should approach various other governments (e.g. the USA, Canada, Australia, EU countries, the UK, Russia, China, Korea and others) to contribute to this newly established fund to support this Fukushima work. This would be in addition to the modest Fukushima efforts supported by the IAEA.

12.2.2 Monitoring

While it is useful to monitor the concentrations of radionuclides (probably focusing on ^{137}Cs) in seawater and sediments, particular attention should be paid to monitoring biota (fish, mammals, invertebrates) for changes in tissue concentrations over time and in different regions. Some species are highly migratory, whereas others are much more residential. Monitoring efforts should be established to distinguish between these groups. Similarly, monitoring of pelagic and benthic species should be continued. It is expected that the benthic species are more likely to be contaminated for longer periods than pelagic species owing to proximity to contaminated sediments.

12.2.3 Fukushima Science Conferences

The Japanese government could consider supporting periodic (perhaps biennial) *international* conferences on all aspects of the Fukushima event. These conferences should primarily be located in Japan, but consideration could be given to one or two such conferences being held in North America and/or Europe.

12.3 Recommendations to Japanese Researchers

OLIVIER EVRARD

There has been a strong and rapid commitment within the Japanese scientific community to investigate both the spatial distribution and the fate of the radionuclides deposited on land shortly after the Fukushima Daiichi Nuclear Power Station (FDNPS) accident. Dedicated networks and appropriate organisations (e.g. the Interdisciplinary Study on Environmental Transfer of Radionuclides from the Fukushima Daiichi NPP Accident (ISET-R; www.ied.tsukuba.ac.jp/hydrogeo/isetr)) were quickly established to coordinate these research efforts at both national and international levels. These networks facilitated the investigation of radionuclide transfers in various environmental compartments (atmosphere, land and ocean) and will help build models simulating these transfers in order to better understand factors driving transport processes and to evaluate the impact of potential countermeasures implemented in the contaminated region.

Although these exceptional efforts should be commended, the scientific community now faces several challenges and will need significant and continuous support from public authorities over the medium to long term in order to meet ongoing research objectives. The first priority should be the installation of a network of permanent monitoring stations in order to measure rainfall, water, sediment and contaminant fluxes in various contexts that would be representative of the variety of land use, soil, topography and land management types observed in the region. This continuous monitoring of the hydro-sedimentary parameters is needed to calibrate, validate and improve the models used to develop a better understanding of the transfer processes and for predicting the water, sediment and contaminant fluxes from the contaminated region to the Pacific Ocean.

The second priority should be to establish a platform through which data collected by the research community can be readily shared. This aspect, which is often neglected by individual researchers, is absolutely crucial in order to avoid the loss of original data from completed research projects – for example, when PhD students and postdoctoral fellows change their institution or their field of study or when experienced researchers are retiring. A recent paper demonstrated that the availability of research data declined rapidly with article age (Vines *et al.*, 2014), but there is still time to propose a solution to avoid this loss in the case of Fukushima post-accident research. This suggestion is supported by recent results demonstrating that most researchers are open to data sharing but that they often do not make their data electronically available because of insufficient time or financial and technical support from their institution (Tenopir *et al.*, 2011).

Finally, it would be useful to organise the creation of an environmental sample repository where the invaluable material collected by the scientific community following the accident could be made available for other types of analyses or research groups. Meeting the challenge of data and sample sharing would be beneficial to the entire research community, and this could offer a unique transition for improving and disseminating scientific good practice in environmental research.

References

Girard, S., I. Korsakissok and V. Mallet (2014). Screening sensitivity analysis of a radionuclides atmospheric dispersion model applied to the Fukushima disaster. *Atmos. Environ.*, **95**, 490–500, doi:10.1016/j.atmosenv.2014.07.010.

Girard, S., V. Mallet, I. Korsakissok and A. Mathieu (2016). Emulation and Sobol' sensitivity analysis of an atmospheric dispersion model applied to the Fukushima nuclear accident. *J. Geophys. Res. Atmospheres*, **121**, 3484–96, doi:10.1002/2015JD023993.

Haywood, S. M., P. Bedwell and M. C. Hort (2010). Key factors in imprecision in radiological emergency response assessments using the NAME model. *J. Radiol. Prot.*, **30**, 23–36, doi: 10.1088/0952-4746/30/1/002.

IRIX Steering Committee (2013). International Radiological Information Exchange (IRIX) Format: Version 1.0, Reference Description.

Mathieu, A., M. Kajino, I. Korsakissok, *et al.* (2018). Fukushima Daiichi-derived radionuclides in the atmosphere, transport and deposition in Japan: a review. *Appl. Geochem.*, **91**, 122–39, doi:10.1016/j.apgeochem.2018.01.002.

Onda, Y., H. Kato, M. Hoshi, Y. Takahashi and M. -L. Nguyen (2015). Soil sampling and analytical strategies for mapping fallout in nuclear emergencies based on the Fukushima Dai-ichi Nuclear Power Plant accident. *J. Environ. Radioact.*, **139**, 300–7, doi:10.1016/j.jenvrad.2014.06.002.

Périllat, R., I. Korsakissok, V. Mallet, *et al.* (2016). Using meteorological ensembles for atmospheric dispersion modeling of the Fukushima nuclear accident. Presented at the 17th International Conference on Harmonisation within Atmospheric Dispersion Modelling for Regulatory Purposes, Budapest, Hungary.

Saito, T., H. Makino and S. Tanaka (2014). Geochemical and grain-size distribution of radioactive and stable cesium in Fukushima soils: implications for their long-term behavior. *J. Environ. Radioact.*, **138**, 11–18, doi:10.1016/j.jenvrad.2014.07.025.

Sørensen, J. H., B. Amstrup, H. Feddersen, *et al.* (2015). Fukushima Accident: UNcertainty of Atmospheric dispersion modelling (FAUNA). Nordic Nuclear Safety Research.

Tenopir, C., S. Allard, K. Douglass, *et al.* (2011). Data sharing by scientists: practices and perceptions. *PLoS One*, **6**(6): e21101.

Vines, T. H., A. Y. Albert, R. L. Andrew, *et al.* (2014). The availability of research data declines rapidly with article age. *Curr. Biol*, **24**(1), 94–7.

Yoshida, N. and Y. Takahashi (2012). Land-surface contamination by radionuclides from the Fukushima Daiichi nuclear power plant accident. *Elements*, **8**, 201–6, doi:10.2113/gselements.8.3.201.

Glossary

Aerial monitoring: The undertaking of a survey (measurement) of radiological diffusion and the composure situation from the sky using aeroplanes, helicopters or drones. On the occasion of the FDNPS accident, the monitoring was first carried out by the US DOE and by MEXT after April 2011. The Nuclear Safety Commission established a concept of 'the environmental radiation monitoring indicator' as crossing the region by aeroplane over the radioactive plume to estimate the radiological impact in order to make decisions about protection measures.

Aerosol: Dispersal system (colloid) of small liquid or solid particles floating in the atmosphere. Sulphuric acid and carbonaceous aerosols of sub-micrometre size are emitted or formed by photochemical reaction in the atmosphere, and sea-salt aerosols and mineral dust aerosols of 10 μm are emitted from the Earth's surface. A considerable part of transported radioactive materials is in a form of aerosols.

AMeDAS (Automated Meteorological Data Acquisition System): *In-situ* observation system of the Japan Meteorological Agency for meteorological parameters. In the accident, many sites were damaged and this became an obstruction to conducting accurate meteorological and transport analyses and simulations.

Atmospheric transport models: Mathematical models simulating atmospheric physical and chemical processes (emissions, transport, chemistry and removal) to reproduce behaviours of atmospheric components, including radioactive materials and air pollutants. They are also called atmospheric dispersion models. Recently, three-dimensional simulation models that calculate fundamental atmospheric processes have been commonly used, while atmospheric plume/puff models to calculate analytical solutions had been previously used as well. This book includes simulation results from various atmospheric transport models, including GEARN of the WSPEEDI system, CMAQ, CAMx, WRF-Chem, MRI-AGCM3, SPRINTARS, MASINGAR and FREXPART.

Caesium-enriched particles (Cs ball; insoluble radiocaesium-bearing solid microparticles): It has been thought that radiocaesium is released to the general environment in the form of water-soluble tiny aerosols. On the contrary, solid spherical particles 2–3 μm in size bearing highly concentrated radiocaesium, which did not dissolve in water, were first isolated from an air dust filter sample in Tsukuba, about 170 km south of the FDNPS on 15 March 2011. Thereafter the name Cs ball became popular due to the unusual nature of the particles. Major components are Si, Fe, Zn, O and Cs, and some particles were also found to contain trace uranium originating in the reactor fuel. There are still hypothetical arguments about the source and the formation mechanism of the material, but it was found that the property was quite different from the so-called Chernobyl hot particle.

Chernobyl accident: The accident occurred on 25–26 April 1986 as an explosion damaged the no. 4 light water graphite moderated reactor of the Chernobyl Nuclear Power Plant in the northern part of the Ukrainian Soviet Socialist Republic of the Soviet Union. The accident produced serious health impacts and environmental contamination in the surrounding area.

Compton scattering: The X-rays and γ-rays scattered by sharing their part of the energy with orbital electrons. Compton scattering is named after Arthur Holly Compton, who discovered the phenomenon. The total energy of scattered rays is always smaller than that of incident rays. The wavelength difference between incident rays and scattered rays is expressed by the function of the scattering angle.

CSTP (Council for Science and Technology Policy): One of the governmental committees for the top-priority science policies under the Japanese prime minister and the minister in charge of the science and technology policies.

CTBTO (Comprehensive Nuclear-Test-Ban Treaty Organization): The Comprehensive Test Ban Treaty (or Comprehensive Nuclear-Test-Ban Treaty, CTBT) is a treaty adopted in the United Nations General Assembly in September 1996 to prohibit experimental explosions of nuclear devices in outer space, the atmosphere, the ocean and underground. The CTBTO achieves the purpose of this treaty by enforcement of the treaty rules and inspections by way of an international monitoring system, on-site inspections and confidence-building measures for inspection. CTBT is not in effect, however, because ratification by all member states has not been completed as of this writing (2018). In this situation, the committee is called the Comprehensive Nuclear-Test-Ban Treaty Organization Preparatory Commission.

Deposition: The process by which gas and aerosol particles fall on the surface adjacent to the atmosphere. 'Fallout' is also used to indicate this process. In the case of the Earth's environment, aerosols are removed from the atmosphere by their deposition onto soil, vegetation, rivers/lakes and oceans through air motion and gravity falling (dry deposition) and cloud and rain (wet deposition). In the human body, atmospheric materials attach to lungs and other respiratory organs, thus causing health damage.

Detectors (Ge semiconductor, NaI scintillation, liquid scintillation, etc.): Ionising radiation is measured using a variety of methods, each with a different principle and hence different characteristics. A high-purity germanium (HPGe) detector is of a type of semiconductor detector that converts the energy of radiation directly to an electronic signal that is in proportion to the absorbed energy. Because of its fine energy resolution, this type is the first choice for applications in which a large number of nuclides are to be identified and quantified from environmental samples using γ-ray spectroscopy. This type is, however, less practical for *in-situ* measurements because it requires a cryogenic cooling system. It is therefore mainly used in laboratories. A sodium iodide (NaI) detector doped with thallium (Tl) is a type of scintillation detector in which the energy of radiation is first converted to a scintillation and then to an electronic signal by a photomultiplier. This type of detector is advantageous in field measurements for its portability and easy-to-use features. A caesium iodide (CsI) detector is of this type. Although this type has poorer energy resolution than that of HPGe detectors, it can also be used in a γ-ray spectrometer to identify a limited number of radionuclides. Liquid scintillation detectors also use scintillations from chemical compounds (scintillator) induced by radiation. An environmental sample is dissolved or dispersed in a liquid scintillator to form a cocktail, the scintillations of which are then 'counted' by this type of detector. This type is used for β-ray-emitting nuclides since they cannot be detected by the abovementioned detectors. In addition to these detectors, there are many other detectors in use. It is essential to choose a detector applicable and suited to the objective. For example, a Geiger–Müller (GM) counter is a detector that just counts the number of radiation incidents and provides us with no information for the qualification of radiation and its source in many cases.

Disaster Prevention and Nuclear Safety Network for Nuclear Environment: A nationwide data exchange system operated by MEXT at the time of the accident, through which environmental monitoring data obtained around nuclear power stations were automatically transferred to local and national governments in real time.

Dose: The fundamental concept of radiological dose refers to the amount of energy of ionising radiation received by objects or by human tissues or organs. This concept is an analogy to a dose of medicine, which is the amount of medicine administrated to a patient. For example, a human body exposed to γ-ray receives a certain amount of external dose. If radioactive materials are ingested or inhaled by a human, they will emit radiation inside the body, which constitutes an internal dose. This concept of dose is implemented with several dose-related quantities as follows:

Absorbed dose: the amount of energy of ionising radiation transferred to an object or body – see also 'Units (Gy)'.

Committed dose: Once taken into a human body, radioactivity continuously irradiates the human body, constituting an internal dose until it is excreted or decays out. A committed dose is calculated by summing up or time-integrating the dose that will be received by a person during a life-long period, usually 70 years for a child and 50 years for an adult.

Dose equivalent: Product of the absorbed dose and the quality factor of radiation, the latter quantifying the biological effects that cause damage to living tissues and organs – see also 'Units (Sv)'.

Effective dose: Sum of dose equivalent of all tissues and organs multiplied by the weighting factor respective to each tissue or organ. The effective dose is intended to quantify radiological effects to the human body – see also 'Units (Sv)'. For internal dose, a rather complex concept of committed effective dose is usually used.

Earthquake in 2011: An earthquake of magnitude 9.0 occurred at 14:46, Japanese time, on 11 March 2011, centred off the Sanriku coast in the Pacific Ocean; it was the largest magnitude earthquake in the history of Japanese modern observation. The earthquake caused a serious tsunami disaster and the FDNPS accident. There are several English translations, including the Tohoku Region Pacific Coast Earthquake, the Great East Japan Earthquake and the Tohoku District-Off the Pacific Ocean Earthquake.

Environmental Emergency Response (EER) Regional Specialized Meteorological Center (RSMC): EER is an international framework jointly established by the World Meteorological Organization (WMO) and the International Atomic Energy Agency (IAEA) with the aim of providing its member states with predictions of atmospheric transport of radioactive materials in nuclear emergency situations. The Japan Meteorological Agency (JMA) is one of the RSMCs, which are in charge of predictions based on their observed and forecast meteorological data.

EPZ (Emergency Planning Zone): The local range that should be chiefly strengthened in disaster prevention measures was established under 'Disaster Prevention Guidance' of the former Nuclear Safety Commission for a situation such as the abnormal release of radioactive material or of radiation in atomic energy facilities. Regarding measures necessary for radiation exposure reduction, a radius of 8–10 km from the nuclear power plant was considered adequate until the Fukushima nuclear plant accident. (www.rist.or.jp/atomica/data/dat_detail.php?Title_No=11-03-06-01)

ERSS (Emergency Response Support System): An information system that supports decision-making by the Japanese government to carry out nuclear emergency countermeasures on the occasion of a serious nuclear accident by providing information about the nuclear accident, which includes the reactor state and prediction of accident progression based on information sent by the power generation company. It has been operated by the former Japan Nuclear Energy Safety Organisation (presently the Nuclear Regulation Authority) since October 2003.

Evacuation zone: The Japanese government gradually expanded the 'evacuation zone' since the accident, then reorganised and renamed zones and areas. It should be noted that both of the terms 'zone' and 'area' are used synonymously for a designated geographical region, and different English translations for the same Japanese term have been used by national governmental institutions and the Fukushima prefectural government.

Evacuation zoning (reorganized as of 22 April 2011)

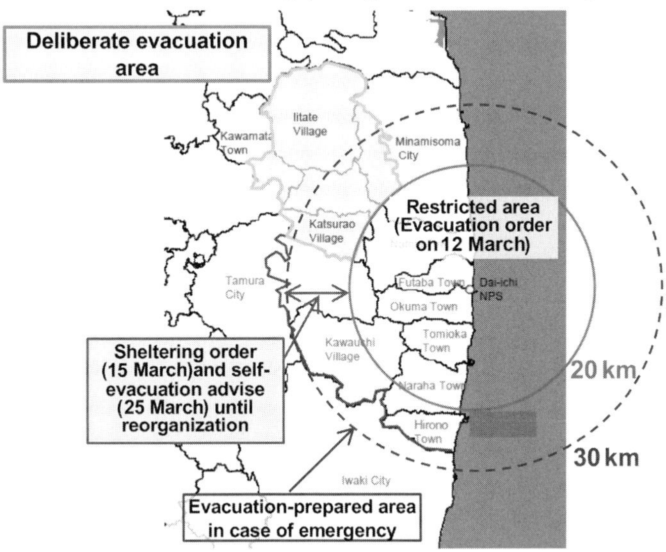

Evacuation order, sheltering order and deliberate evacuation areas: An evacuation order was issued within a 3 km radius from the FDNPS on 11 March 2011, and expanded to a 10 km radius on the morning of 12 March, then to a 20 km radius in the evening of 12 March. On 15 March, a sheltering (in-house evacuation) order was issued for areas from 20 km to 30 km of the FDNPS, and the government recommended self-evacuation for this range on 25 March. In April, the government revised its 'evacuation zone' into a 'restricted area' within 20 km and a 'deliberate evacuation area' (beyond 20 km) where

Table G1 *Timeline of evacuation orders and restructuring zones.*

Date and time	Events/criteria	Typical term used by national government and IAEA	Synonyms alternatively used by local authorities and/or other publications
Urgent protective actions			
20:50, 11 March 2011	An evacuation order was issued for an area within a 2 km radius by Fukushima Prefecture	Evacuation order	Compulsory evacuation order
21:23, 11 March 2011	An evacuation order was issued for an area within a 3 km radius by the national government	Evacuation order	Compulsory evacuation order
21:23, 11 March 2011	The order to stay indoors was issued for an area within a 3–10 km radius by the national government	Sheltering order	In-house evacuation order
05:44, 12 March 2011	The evacuation order was issued for an area within a 10 km radius by the national government	Evacuation order	Compulsory evacuation order
18:25, 12 March 2011	The evacuation order was issued for an area within a 20 km radius by the national government	Evacuation order	Compulsory evacuation order
11:00, 15 March 2011	The national government issued an order to shelter for residents within 20–30 km radius	Sheltering order	Shelter in home
25 March 2011	The national government recommended voluntary evacuation for residents within a 20–30 km radius	Recommendation of voluntary evacuation	Self-evacuation, voluntary evacuation recommendation
Restructuring of zones on 22 April 2011	The existing 20 km evacuation zone was established as a restricted area and a deliberate evacuation area was established	Restricted areas and areas to which evacuation orders have been issued	Evacuation designated zones
	20 km zone	Restricted area	Warning zone, evacuation order zone, restricted zone
	Areas beyond the 20 km evacuation zone, including Katsurao Village, Namie Town, Iitate village, part of Kawamata town and Minami-Souma city	Deliberate evacuation area	Planned evacuation zone
	A 20~30 km zone, where recommendation of sheltering and voluntary evacuation was kept, to be lifted later on 30 September Sheltering order for 30 km radius other than the designated zone was lifted	Evacuation-prepared area in case of emergency	Emergency evacuation preparation zone, evacuation prepared area
16 June 2011	National government announced a guideline which specified that these locations should be designated as specific spots recommended for evacuation	Specific spots recommended for evacuation	Specific hotspots

Rearranging the zone on 26 December 2011	Evacuation zones including restricted area and deliberate evacuation area were further reclassified to three categories according to air dose level during April 2012 to August 2013 based on this decision		
Difficult-to-return zone	Areas where it is expected that residents will face difficulties in returning for a long time	Area 3	Difficult-to-return-to zone, red zone
Restricted residence zone	Areas in which residents are not permitted to live	Area 2	Restricted residence zone, orange zone
Evacuation order cancellation preparation zone	Areas to which evacuation orders are ready to be lifted	Area 1	Evacuation order cancellation preparation zone, derestriction preparation zone, green zone
Zoning for decontamination	Areas where the national government directly implements decontamination	Special decontamination area	SDA
	Areas where local governments conduct surveys of contamination, establish decontamination plans and implement decontamination	Intensive contamination survey area	ICSA, priority contamination survey area

integral dose potentially reaches 20 mSv within a year from the accident. This is according to emergency exposure guidance levels (20~100 mSv per year) for protection from radiation set by the International Commission on Radiological Protection (ICRP) and the International Atomic Energy Agency (IAEA). A complete evacuation was required by the end of one month from 22 April for Katsurao village, Namie Town, Iitate Village and parts of Kawamata Town and Minami-Souma City. Though the term 'evacuation' is generally used, the IAEA prefers the term 'relocation' for such a case as a deliberate evacuation area. Furthermore, an area not subject to deliberate evacuation between 20 km and 30 km from the FDNPS (such as Hirono Town, Naraha Town, Kawauchi Village and parts of Tamura City and Minami-Souma City) was designated as an 'emergency evacuation preparation area' where emergency sheltering was required.

Reclassification of evacuation zones: Evacuation zones including the restricted area and deliberate evacuation area were further reclassified into three categories according to air dose levels from April 2012 to August 2013. Three classes are as follows:

- Area 3: areas where it is expected that residents will face difficulties in returning for a long time (difficult-to-return-to zones);
- Area 2: areas in which residents are not permitted to live (restricted residence zones);
- Area 1: areas to which evacuation orders are ready to be lifted (evacuation order cancellation preparation zones or zones in preparation for the lifting of the evacuation order, or shortened as derestriction preparation zones).

Exposure: 'Exposed' means to be subjected to or to come into proximity of poisonous material or radiation. There are external and internal exposures. In Japanese the expressions 'be bombed' and 'exposed' are often confused because the two words are homophones (*hibaku*). In English 'be bombed' indicates having been attacked in a bombing mission.

Fallout: Radioactive fallout is a generic name for deposition of radioactive materials from the atmosphere onto the surface of the Earth. It is caused by wind (dry deposition) and rain and snowfall (wet deposition). Deposition monitoring at appointed times is performed under the Japanese environmental radioactivity survey by MEXT to monitor the fallout every day.

FDNPS (Fukushima Daiichi Nuclear Power Station): The first nuclear power station of the Tokyo Electric Power Company (TEPCO) in Fukushima Prefecture. There are several English abbreviations, including FDNPS, FDNPP, F1NPP, FD1NPS, F1NPP and 1F NPP. The FDNPS started operations in March 1971 and units 1 through 6 were in operation in 2011. Due to inspection, nuclear fuels were not in the reactors of units 4 through 6 at the time of the earthquake on 11 March 2011. The reactors in units 5 and 6 could be cooled because these units had recovered from the power loss. All units are boiling water reactors (BWRs), especially the old type called mark I in units 1 through 5.

Fission: The process in which a nuclear fissile material such as ^{235}U or ^{239}Pu splits into two new atoms, which are called fission products. Fission yield refers to the percentage of the fission products produced per fission. For instance, the fission yield of ^{90}Sr is approximately 6%.

Hamadoori region: The coastal region of the eastern part of Fukushima Prefecture facing the Pacific Ocean to the east and Abukuma Mountains to the west.

IAEA (International Atomic Energy Agency): An international self-governing organisation affiliated with the United Nations (UN) that was established to promote the peaceful use of atomic energy. It was established when the UN Charter draft was adopted in 1957 and provides a world central forum to facilitate scientific and technical cooperation regarding nuclear energy. A total of 144 countries are currently members. It has the authority to oversee security measures, health and life and safeguards to prevent the conversion of nuclear technology to military use. The IAEA headquarters is located in Vienna, Austria, and in Japan the IAEA office is located in Tokyo.

ICRP (International Commission on Radiological Protection): An independent, international organisation with more than 200 volunteer members from approximately 30 countries across 6 continents, established in 1928. These members represent the leading scientists and policymakers in the field of radiological protection, funded through a number of ongoing contributions from organisations with an interest in radiological protection. The ICRP has developed, maintained and elaborated the International System of Radiological Protection used worldwide as the common basis for radiological protection standards, legislation, guidelines, programmes and practice, and has published more than 100 reports on all aspects of radiological protection.

Imaging plate (IP) technique: A digital version of radiographic film, autoradiography, to image radiological distributions. The X-ray photo produces an analogue image by using the medium of a photo film conventionally, while the IP applies a material (a stimulable phosphor which is activated to a metastable state by radiation energy) painted over a plastic board or coated film as an imaging medium. After the radiation exposure the luminous material (crystallite of the europium compound) emits fluorescence due to irradiation of a red laser. More information than analogue X-ray film can be obtained by digitally imaging the fluorescence. The very wide dynamic range and 1000 times higher sensitivity than normal radiographic film are advantageous, and the IP media can be used repeatedly.

International Monitoring System (CTBO-IMS): The system consists of 321 monitoring stations and 16 laboratories around the world. Established by the CTBTO (Preparatory Commission for the Comprehensive Nuclear-Test-Ban Treaty Organization), these 337 facilities monitor the planet for any sign of a nuclear explosion. The IMS uses four complementary verification methods, utilising the most modern technology available. Seismic, hydroacoustic and infrasound stations monitor the underground, the large oceans and the atmosphere, respectively. Radionuclide stations detect radioactive debris from atmospheric explosions or vented by underground or underwater nuclear explosions. Radionuclide laboratories assist radionuclide stations in identifying these radioactive substances.

International Nuclear and Radiological Event Scale (INES): The IAEA defines the International Nuclear and Radiological Event Scale (INES) as a tool for communicating the safety significance of nuclear and radiological events to the public. Member states use INES on a voluntary basis to rate and communicate events that occur within their territory. It is not a notification or reporting system to be used in emergency response. Events are rated at seven levels. The scale is logarithmic – that is, the severity of an event is about ten times greater for each increase in level of the scale. Events are considered in terms of their impact on people and the environment, impact on radiological barriers and control and impact on defence. Events without safety significance are rated as Below Scale/Level 0. Events that have no safety relevance with respect to radiation or nuclear safety are not rated on the scale. (www.iaea.org/topics/emergency-preparedness-and-response-epr/international-nuclear-radiological-event-scale-ines).

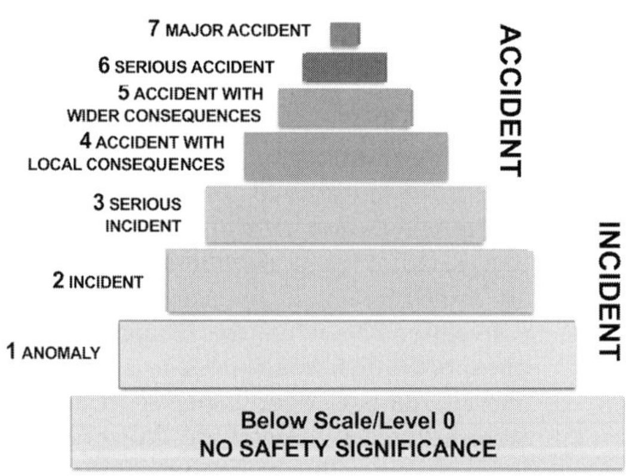

IPCC (Intergovernmental Panel on Climate Change): An intergovernmental panel organised by the United Nations Environmental Programme (UNEP) and the World Meteorological Organization (WMO) jointly in 1988 to offer scientific assessments on climate change.

JAEA (Japan Atomic Energy Agency): The largest core research and development organisation of atomic energy science and technology in Japan. It started by unification of the former Japan Atomic Energy Research Institute and the former Japan Nuclear Cycle Development Institute in October 2005. The mission is defined as 'to open up the nuclear science future, and to contribute to the welfare of humanity.' The JAEA's headquarters is located in Tokai-mura, Ibaraki Prefecture and it includes more than 10 research and technological development bases across the country, including in Tsuruga and Fukushima.

Japan Academic Network for Disaster Reduction: A network of academic associations related to disaster prevention, mitigation and recovery, established in January 2016 after the Fukushima accident, aiming to collaborate across academies and make efforts to integrate different specialties for disaster reduction. The members of the Science Council of Japan became founders and more than 50 academic societies related to disaster management gathered together and established an emergency contact network (http://janet-dr.com/020_abouteng/021_abouteng.html).

Japan Chemical Analysis Center (JCAC): A specialised analytical institution dealing with environmental radioactivity and radiation established in 1974.

J-RAPID: A programme to support urgent collaboration activities between Japanese and foreign researchers in cases of natural or anthropogenic disasters and similar unanticipated events. This was created by the Japan Science and Technology Agency and announced on 18 April 2011.

Kanto region: The region located in the eastern part of the main island (Honshu) of Japan, including the Kanto Plain and consisting of the prefectures of Gunma, Tochigi, Ibaraki, Saitama, Tokyo, Chiba and Kanagawa. There are various spellings, including Kanto, Kantou and Kantō.

KURAMA (Kyoto University RAdiation MApping system): A GPS-aided mobile radiation monitoring system that was developed by the Kyoto University Research Reactor Institute (KURRI) in response to the nuclear disaster caused by the FDNPS.

Lifetime: A half-life is the time after which the number of radioactive nuclei declines to half of the initial number. The average lifetime expected for a population of nuclei can be calculated by dividing the half-life of the nuclide by ln(2) (= 0.693).

Ministries of Japanese Government: In this book, the following acronyms are used for the following ministries:

MEXT: Ministry of Education, Culture, Sports, Science and Technology
MAFF: Ministry of Agriculture, Forestry and Fisheries
METI: Ministry of Economy, Trade and Industry
MOE: Ministry of the Environment

MOX fuel: Nuclear fuel contains blended uranium and plutonium oxides, while common nuclear fuels consists of uranium oxide only. MOX fuel was used in the FDNPS according to the so-called 'Plu-thermal plan' in Japan.

Nakadoori region: The central part of Fukushima Prefecture between the Abukuma Mountains to the east and the Ou Mountains to the west.

NISA (Nuclear and Industrial Safety Agency): The Nuclear and Industrial Safety Agency (*Genshiryoku Anzen Hoanin*) was a Japanese nuclear regulatory and oversight branch of the Agency for Natural Resources and Energy under the Ministry of Economy, Trade, and Industry (METI) established in 2001 and located in Kasumigaseki, Tokyo. NISA works with the Japanese Atomic Energy Commission as well as providing other functions. It performs oversight for the industry as requested by the Japanese government. After the FDNPS accident, NISA was criticised for its lack of independence from METI, which is also responsible for promoting nuclear power; this situation hampered a quick response to the FDNPS accident. As a result, NISA was replaced by a new agency, the Nuclear Regulation Authority (NRA), on 19 September 2012.

NRA (Nuclear Regulation Authority): The Nuclear Regulation Authority (*Genshiryoku Kisei Iinkai*, NRA) is an administrative body of the Cabinet of Japan established to oversee nuclear safety in Japan as part of the Ministry of the Environment. It was established on 19 September 2012. The NRA was formed from the Nuclear Safety Commission, which came under the authority of the Cabinet, and the Nuclear and Industrial Safety Agency (NISA). The NRA mandate included discarding the previous ineffective approach to regulatory work and stressing the importance of a field-oriented approach to achieve genuinely effective regulations and to ensure transparency and appropriate information disclosure on regulations, including the decision-making process.

The mission of the NRA is to protect the general public and the environment through rigorous and reliable regulation of nuclear activities. To do so, it strives to make decisions independently, based on the latest scientific and technological information, and to be ready to swiftly respond to all emergency situations while ensuring that in 'normal' times a fully effective response system is always in place.

NSC (Nuclear Safety Commission): Japan's Nuclear Safety Commission (*Genshiryoku Anzen Iinkai*) is a commission established within the Cabinet of Japan as an independent agency to play the main role in nuclear safety administration. Commissioners are appointed by the prime minister of Japan with approval of the Diet. The commission has greater authority than any other ordinary advisory committees, in that it can make recommendations to relevant agencies in the name of the prime minister if necessary. The NSC reviews safety inspections conducted by regulatory agencies such as the Nuclear and Industrial Safety Agency (NISA). It was abolished on 19 September 2012 and its authority was transferred to the Nuclear Regulation Authority.

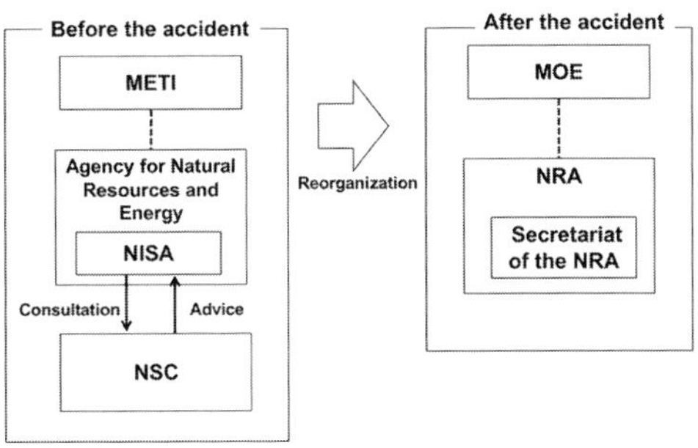

Permeation depth: The general definition of the permeation depth specifies the degree of infiltration to the ground. In the case of depth distribution of ^{137}Cs concentration in undisturbed soil, it is expected to exhibit an exponential decline with depth. Several definitions are available for permeation depths. In Chapter 6, the definition of the permeation depth is one-tenth of the attenuation depth to the surface concentration level.

Radioactive material: Material that contains radioactive atoms and exhibits radioactivity.

Radioactivity: Property of a substance in which ionising radiation such as α-, β- and γ-rays is emitted. This term refers also to the rate of disintegration, or decay, of radioactive atoms. Since the disintegration rate is in proportion to the number of the radioactive atoms, it often refers to the amount of the radioactive atoms or the substance itself that exhibits radioactivity.

Radionuclides (radioactive nuclides): The term 'nuclide' refers to a species of an atomic nucleus, which is specified by the numbers of protons and neutrons constituting it. A radionuclide or a radioactive nuclide is a nuclide that is unstable and emits ionising radiations to attain a stable state, hence exhibiting radioactivity. More than 3500 radionuclides and some 270 stable, or non-radioactive, nuclides have been found so far. Among the radionuclides, ^{134}Cs, ^{137}Cs, ^{131}I, ^{129}I, ^{133}Xe, ^{132}Te and ^{90}Sr are the typical ones released during a nuclear reactor accident. The number in the superscript corresponds to the mass of the nuclide, which is the sum of the numbers of protons and neutrons in its nucleus.

RadNet: US nationwide environmental radiation monitoring network consisting of some 140 stationary monitors covering 50 states. It is operated by the US Environmental Protection Agency and provides data on environmental radiation levels in real time.

SCJ (Science Council of Japan): A Japanese representative organisation of scientists, belonging to the Cabinet Office of Japan, from all fields including the natural sciences, the humanities and the social sciences. The SCJ's purpose is to express the opinion as a summary inside and outside the country, and it was established in 1949 as a 'special organisation' to perform duties independently under the jurisdiction of the prime minister. At the time of the Fukushima accident, the following committees were in existence:

- Sectional Committee on Nuclear Accident Committee on Comprehensive Synthetic Engineering
- Nuclear Accident Response Subcommittee, General Engineering Committee
  - Sub-Subcommittee for Environmental Contamination by Nuclear Plant Accidents
- Information Dissemination Subcommittee of Computational Science Simulation

SEM-EDX (scanning electron microscope energy dispersive X-ray spectrometry): A type of electron microscope that scans a sample with a focused electron beam, thus obtaining an enlarged image. The image is obtained by detecting and processing reflected and induced electrons from the sample surface. The technique also serves for elemental analysis, by which the composition of the sample can be known through EDX by detecting characteristic X-rays from the sample induced by electron beam irradiation. Because the EDX instrument is connected with the SEM main body, it is abbreviated as SEM-EDX.

Source term: The state of atmospheric release of radioactive materials from a nuclear facility due to an accident; the source term includes the released radionuclide composition, release rate and release duration.

SPEEDI (System for Prediction of Environmental Emergency Dose Information): A computer system to predict the atmospheric dispersion of radioactive materials and radiological doses in the event of nuclear accidents in Japan; this was developed by the former Japan Atomic Energy Research Institute (presently the JAEA). It had been operated by the Nuclear Safety Technology Center on consignment from MEXT and local governments since 1986. With the improvement of prediction models, the advanced SPEEDI had been operating since 2005. Based on SPEEDI, JAEA has developed WSPEEDI for large-scale atmospheric transport and its second version, WSPEEDI-II, which consists of the regional meteorological model MM5 and the Lagrangian particle dispersion model GEARN.

SPM (suspended particulate matter): Atmospheric aerosols less than 10 μm in diameter, slightly different from PM10 which is used in air pollution monitoring in other countries. In Japan, the mass concentrations (μg/m^3) of SPM collected on filter-tapes are hourly measured by the β-ray attenuation method in the air quality monitoring network by local governments under the Air Pollution Control Act. According to the Japanese Ministry of the Environment, 1684 stations for SPM measurement were in operation in Japan in 2016.

Tohoku region: The northeastern part of Japan consisting of the prefectures of Aomori, Iwate, Miyagi, Akita, Yamagata and Fukushima.

U8 container: A 100 ml polypropylene container often used in the field of environmental radioactivity in Japan. Often considered to be the standard of sampling or measurement because it is widely and frequently used.

Units for radioactivity and dose: Various units are used to quantify the radioactivity and dose as follows:

Bq (Becquerel): Bq is a unit of radioactivity (disintegration rate) in the SI unit system. It is named after Henri Becquerel, a French researcher who discovered radioactivity. 1 Bq is defined as the activity of a quantity of radioactive material in which one nucleus decays per second (1 disintegration per second (1 dps) = 1 Bq = 1/s). TBq is 10^{12} Bq and PBq is 10^{15} Bq.

Ci (Curie): an old unit for radioactivity (disintegration rate), which is named after Dr Marie Curie. It is a radiological quantity equal to exactly 3.7×10^{10} disintegration for one second. Originally it corresponded to radioactivity (an amount or a strength) of 1 g of ^{226}Ra.

Sv (Sievert): the unit used for dose equivalents. Most survey meters are adjusted to give readings of the ambient dose equivalent rate in this unit as Sv/h. The unit Sievert is also used for effective dose, which is an aggregative measure of radiation effects on a human body. A handy conversion of 1 Sv (effective dose) = 1 Sv (ambient dose equivalent) = 1 Gy (air absorbed dose) was often used in reporting and evaluating monitoring data, although these quantities are in principle different from each other and this conversion may result in an error of several tens of per cent.

Gy (Gray): the SI unit of absorbed energy of radiation. One joule (J) of energy absorbed by 1 kg of matter makes 1 Gy of absorbed dose. Most of the monitoring posts report physical quantities such as air-absorbed dose rate or air kerma using this unit as Gy/h.

UNSCEAR (United Nations Scientific Committee on the Effects of Atomic Radiation): The Committee was established by the General Assembly of the United Nations in 1955. Its mandate in the United Nations system is to assess and report levels and effects of exposure to ionising radiation. Governments and organisations throughout the world rely on the committee's estimates as the scientific basis for evaluating radiation risk and for establishing protective measures (www.unscear.org/unscear/en/about_us.html).

Names of Locations

Names of places that appear in this book are listed as follows and shown in the figures.

Prefectures: Hokkaido (1), Aomori (2), Akita (3), Iwate (4), Yamagata (5), Miyagi (6), Fukushima (7), Tochigi (A), Ibaraki (B), Gunma (C), Saitama (D), Tokyo (E), Chiba (F), Kanagawa (G)

Cities, towns, villages: Fukushima (FK), Fukui (FU), Futaba (FB), Hasaki (HS), Hitachi (HT), Iitate (IT), Iwaki (IW), Kohriyama (KY), Katsurao (KR), Kawasaki (KW), Kawamata (KM), Yamakiya (YK), Kashiwa (KS), Namie (NM), Niigata (NI), Nihonmatsu (NH), Naraha (NR), Nasushiobara (NS), Onagawa (OG), Ohkuma (OK), Onahama (OH), Sendai (SD), Sapporo (SP), Tokai (TO), Tsukuba (TK), Tomioka (TM), Yokohama (YM)

Bays, ports, peninsula: Hitachinaka Port (HP), Ishikari Bay (IB), Miyako Bay (MY), Tokyo Bay (TB), Tsugaru Peninsula (TP)

Rivers: Abukuma (ABR), Arakawa (AWR), Kuchibuto (KBR), Natsui (NTR), Ohta (OTR), Same (SMR)

Nuclear power stations: Fukushima Daiichi Nuclear Power Station (NP1), Daini Nuclear Power Station (NP2), Onagawa Nuclear Power Station (ONP)

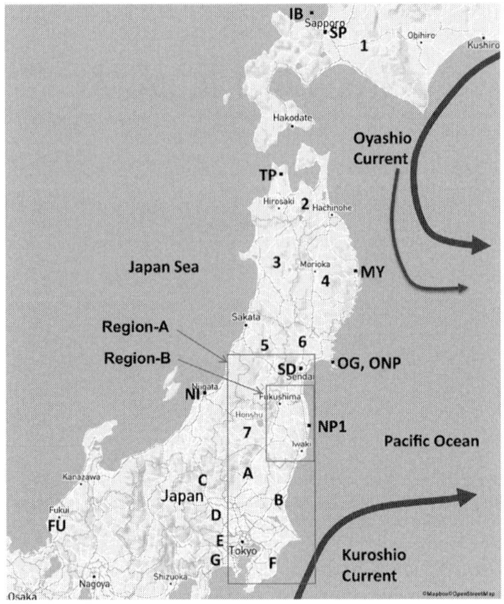
IB
SP
1
Sapporo
Obihiro
Kushiro
Hakodate
Oyashio Current
TP
Hirosaki
2
Hachinohe
Japan Sea
3
Morioka
4
MY
Sakata
Region-A
Region-B
5
6
SD
Sendai
OG, ONP
NI
Niigata
Fukushima
NP1
Honshu
7
Iwaki
Pacific Ocean
Kanazawa
C
A
Japan
B
Fukui
FU
D
E
Tokyo
G
F
Kuroshio Current
Nagoya
Shizuoka
Osaka
©Mapbox©OpenStreetMap

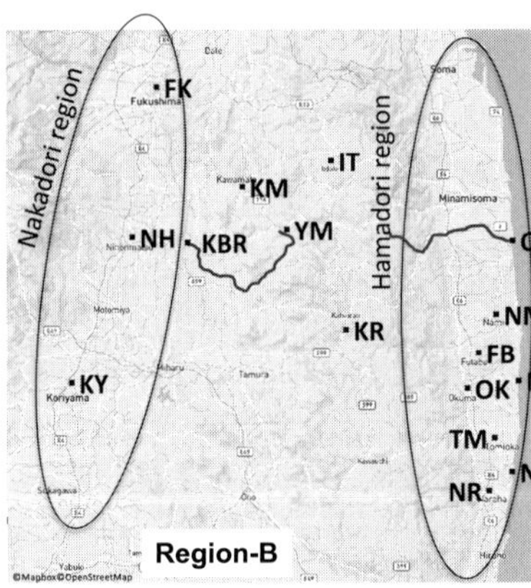
Nakadori region
FK
Fukushima
IT
KM
NH
KBR
YM
KY
Koriyama
KR
Hamadori region
Minamisoma
OTR
NM
FB
OK
NP1
TM
NP2
NR
Region-B
©Mapbox©OpenStreetMap

Region-A
SD
Sendai
ABR
Yamagata
Yonezawa
Soma
FK
KM
Minamisoma
NH
Aizu-Wakamatsu
NM
NP1
KY
NR
TM
Abkuma Mt.
IW
NTR
NS
SMR
Kitaibaraki
Nikko
Nasukarasuyama
HT
Utsunomiya
TO
Mito
HP
Oyama
Ishioka
AWR
TK
Tsukuba
KS
HS
Choshi
Tokyo
Chiba
KW
TB
Yokohama
YK
Kamogawa
Tateyama
©Mapbox©OpenStreetMap

Index